MECHANICAL E

MECHANICAL ENGINEERING SCIENCE

for Part 1 of Technicians' Course

A. OXLEY
C.Eng., M.I.Mech.E., M.I.Prod.E., Full Tech. Certs. C.G.L.I.

Senior Lecturer in Engineering at West Cheshire Central College of Further Education, Eastham, Wirral

SI UNITS

LONDON
EDWARD ARNOLD (PUBLISHERS) LTD

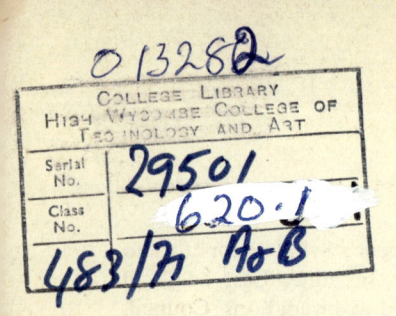

© A. OXLEY, 1970

First published 1963

Reprinted 1964, 1966, 1968

Second edition 1970

Reprinted 1971

SBN: 7131 3240 X

Reproduced photo-litho in Great Britain by
J. W. Arrowsmith Ltd., Bristol 3.

PREFACE TO SECOND EDITION

This book is designed to cover the Engineering Science Syllabus for the two years course of the City and Guilds of London Institute, East Midland Educational Union, Northern Counties Technical Examinations Council, Union of Educational Institutions, Yorkshire Council of Further Education and Union of Lancashire and Cheshire Institutes, for Mechanical Engineering Technicians. Since technicians are concerned with the practical applications of science, I have concerned myself with helping them to appreciate the fact that such applications are all around them in the factory and workshop.

I have attempted to inform by example. To this end, more space is given to worked examples than to the explanatory text. Included too are details of experiments which might be used for the purpose of 'finding out' on the student's part rather than as a means of proving a formula that has already been accepted. The experiments can be set as class exercises in which the student reports on his findings without necessarily going through the full ritual of a 'lab. write up'. This is especially so with the experiments which are qualitative rather than quantitative.

Where desirable the exercises are split into two sections, so that each section can be used separately in the appropriate year of the course. The answers to the problems and to the worked examples are generally given in decimal form, to the third decimal place, where such accuracy is required. Some are given to the nearest whole number where this is more practical.

Units from the Système International, i.e. SI units, have been used throughout the book.

I hope that the book will prove to be useful in helping the Apprentice Technician to master his chosen vocation.

A. Oxley
West Cheshire Central College of Further Education,
Eastham,
Wirral

CONTENTS

Chemistry in the Workshop

INTERNATIONAL SYSTEM OF UNITS (SI UNITS) - - vii

CHAPTER 1. MATTER - - - - - - - - - 1
Solids, liquids and gases. Atoms and molecules. Elements, compounds and mixtures. Ores. Alloys. Exercise 1.

CHAPTER 2. COMPOSITION OF THE ATMOSPHERE - - 13
Dust and moisture in the atmosphere. Exercise 2.

CHAPTER 3. CHEMICAL REACTIONS - - - - - - 20
Oxidation, reduction, carburisation, combustion and corrosion. Heat-treatment process. Exercise 3.

Electricity in the Workshop

CHAPTER 4. ELECTRICITY - - - - - - - - 32
Current, voltage and resistance. D.C. and A.C. electricity. Ohm's law. Energy and power. Calculations on series circuits. Calculations on parallel circuits. Exercise 4.

CHAPTER 5. MAGNETISM - - - - - - - - 51
Magnetism, magnetic materials, permanent magnets, electromagnets. The electric motor, The dynamo, The ammeter, The voltmeter.

Materials and Machines in the Workshop

CHAPTER 6. FORCE AND STRESS - - - - - - 70
Force, measurement and application of force. Tensile, compressive and shear force. Stress. Exercise 5. Strain. Elasticity, Hooke's law, modulus of elasticity, percentage elongation and reduction in area, Safe working stress, factor of safety. Exercise 6.

CHAPTER 7. MOMENT OF A FORCE - - - - - - 97
Calculation of the moment of a force about a point, types of moments, reaction, couple, moment of a couple. Centre of area, position of centroid by calculation. Centre of gravity. Exercise 7(a). Torque, constant and variable torque. Exercise 7(b).

CHAPTER 8. WORK AND POWER - - - - - - 135
Work done by a constant force. Work done by a variable force. Energy. Power, power transmitted by a belt drive, efficiency. Exercise 8.

CHAPTER 9. BELT AND PULLEY SPEEDS - - - - - 152
Calculations. Exercise 9.

CHAPTER 10. GEAR WHEELS - - - - - - - 164
Gear wheel speeds and gear trains, velocity ratios. Exercise 10.

CHAPTER 11. FRICTION - - - - - - - - 179
Uses of friction, disadvantages and advantages of friction, calculation of friction force between two surfaces in contact. Lubricated surfaces and friction. Exercise 11.

CHAPTER 12. TORQUE AND POWER - - - - - - 195
Relationship between torque and power. Relationship between torque and gear wheel speed. Exercise 12.

CHAPTER 13. MECHANICAL ADVANTAGE AND VELOCITY
RATIO - - - - - - - - 208
Lifting machines, mechanical advantage, velocity ratio, law of a machine. Exercise 13.

CHAPTER 14. PARALLELOGRAM AND TRIANGLE OF FORCES 220
Parallelogram of forces. Triangle of forces. Exercise 14.

CHAPTER 15. RESOLUTION OF FORCES - - - - - 231
Horizontal and vertical component forces. Exercise 15.

CHAPTER 16. TEMPERATURE AND HEAT - - - - - 239
Temperature, temperature measurement and scale, °C. Heat, sensible heat, latent heat, measurement of heat, heat quantity, specific heat capacity. Exercise 16.

CHAPTER 17. ENERGY CONVERSION - - - - - 252
Conversion of energy from one form to another, heat–mechanical–electrical.

CHAPTER 18. EXPANSION OF SOLIDS, LIQUIDS AND GASES 261
Coefficient of linear expansion. Exercise 18.

REVISION EXERCISES - - - - - - - - 270
(Grades 1 and 2).

ANSWERS TO EXERCISES - - - - - - - - 280

INDEX - - - - - - - - - - - 285

LOGARITHMS AND ANTILOGARITHMS - - - - 289

INTERNATIONAL SYSTEM OF UNITS (SI UNITS)

In Britain we are about to discard our system of units and adopt a system which has been recommended by the International Organization for Standardization. The system is based on metric units. The following table shows those SI Units, with their English counterparts, with which the Engineering Technician will need, eventually, to become familiar.

Quantity	Unit and Abbreviation	Multiple	Sub-multiple	Existing Unit
Time	second, s	kilosecond, ks	millisecond, ms	second
Length	metre, m	kilometre, km	millimetre, mm	foot
Velocity	metre per second, m/s	kilometre per second, km/s	—	ft/s
Rotational Frequency (Speed)	reciprocal second, s^{-1} (rev/s)	—	—	rev/min rev/s
Area	square metre, m^2	square kilometre, km^2 $km^2 = 10^6 m^2$	square millimetre, mm^2 $mm^2 = 10^{-6} m^2$	ft^2 or in^2
Volume	cubic metre, m^3	—	cubic millimetre, mm^3	ft^3 or in^3
Mass	kilogramme, kg	megagramme, Mg or tonne, t	milligramme, mg	lb or ton
Density	kilogramme per cubic metre, kg/m^3	—	—	lb/in^3
Force	newton, N	meganewton, MN kilonewton, kN	millinewton, mN	lbf or tonf
Pressure	newton per square metre, N/m^2 (10^5 N/m^2 = 1 bar)	meganewton per square metre, MN/m^2 kilonewton per square metre, kN/m^2	millinewton per square metre, mN/m^2	lbf/in^2
Stress	N/m^2	MN/m^2 or kN/m^2	—	lbf/in^2 or $tonf/in^2$

Quantity	Unit and Abbreviation	Multiple	Sub-multiple	Existing Unit
Moment of force (torque)	newton metre, N m	kilonewton metre, kN m	micro-newton metre, μN m	lbf ft
Energy, Work	joule, J	kilojoule, kJ	millijoule, mJ	ft lbf
Power	watt, W	kilowatt, kW	milliwatt, mW	horsepower
Temperature	degree Celsius, °C	—	—	°C or °F
Heat quantity	joule, J	kilojoule, kJ	millijoule, mJ	Btu or Chu
Specific heat Capacity	joule per kilogramme deg C, J/kg deg C	kilojoule per kg deg C, kJ/kg deg C	—	Chu/lb deg C Btu/lb deg F
Linear expansion co-efficient	per degree C, /°C	—	—	/deg C /deg F
Electric Current	ampere, A	kilo-ampere, kA	milli-ampere, mA	A
Electromotive force, Potential difference	volt, V	kilovolt, kV	millivolt, mV	V
Electrical Resistance	ohm, Ω	kilo-ohm, kΩ	micro-ohm, $\mu\Omega$	Ω
Angle	radian, rad	—	milliradian, mrad	Degree, Radian

1 joule = 1 watt second
Mega = 10^6 = 1 000 000, one million
kilo = 10^3 = 1 000, one thousand
milli = 10^{-3} = 0·001, one thousandth part
micro = 10^{-6} = 0·000 001, one millionth part.

CHEMISTRY IN THE WORKSHOP

Chapter 1. Matter

ALL around us we see objects shaped by man—motor vehicles, houses, ships, aircraft, and so on. In the workshop we see benches, machine tools, marking blue, drills, hammers, and the materials that we shape. All these objects and the many others not mentioned are made from substances which come from the earth. The substances appear in mineral, vegetable or animal form. If we think of materials in their original state we may list them under the general heading of *matter*. For example, the mild steel pressing of a motor-car body or the shell of an ocean liner were once lumps of matter known as iron ore. Matter may appear in any one of three forms under normal conditions of temperature and pressure.

1. *Solid*. A solid is firm and fixed in shape and volume; e.g. iron ore, coal, granite.
2. *Liquid*. A liquid is of fixed volume, and is shaped by the container which holds it; e.g. water, mineral oil, sap from the rubber tree, whale oil. Liquid will flow under suitable conditions.
3. *Gas*. A gas is not fixed in volume or shape, for it fills any shaped container into which it is put; e.g. coal gas, fire damp from coal workings, marsh gas, natural gas from the bed of the North Sea.

Solids, liquids and gases

The examples given are solids, liquids or gases in their natural state. However, most substances can be made to exist in any one of the three states depending on the surrounding temperature and pressure conditions. The ice–water–steam conditions are the commonest, though steam is not strictly a gas; it is a vapour or gaseous fluid. Most of us have seen or heard of liquid oxygen. As likely as not we have seen liquid steel, iron or brass. If not, we have surely seen molten lead and very likely a vapour rising from it if its temperature were sufficiently high. Some substances, such as fat, do not follow the solid–liquid–gas pattern. Fat can be liquified but, when attempts are made to gasify it, it just carbonises; it changes to a black solid substance which is basically carbon, like burnt oil in motor vehicle engines. Some substances when heated change from solid to gas, missing out the liquid stage. This process is known as *sublimation*. Such a substance is naphthalene; another is camphor, from which camphor blocks are made.

Atoms and Molecules

All matter is made up of atoms. To obtain some idea of the size of an atom, imagine that we take a single filing from a block of pure aluminium and divide this into millions of pieces, i.e. split it up so that any one of the extremely small pieces is incapable of further division. We know that this operation is not practically possible, but it should help us to understand just how small an atom is. Briefly, an atom is the smallest part of an element which can take part in a chemical change. The atom is so minute that it has never been seen. Electrically the atom can be broken down further, for it is known that the atom is made up of *electrons, protons* and *neutrons*. These can be considered as electrical particles smaller than the atom itself. Electrons carry a negative electrical charge, protons carry a positive charge, whilst neutrons carry no charge. The result of the atom-splitting process is to divide the atom into these electrical particles. It must be understood that splitting of a single atom has never been observed; as we have already noted, the atom is too small to be seen even with an electron microscope.

A *molecule* of an element or compound is the smallest part which can exist independently. The molecule of an *element* is made up of one or more atoms of the element. For instance, the molecule of helium contains one atom; the molecules of hydrogen, oxygen and chlorine each contain two atoms joined together. The molecule of a *compound* contains one or more atoms from at least two elements, e.g. *one* atom of zinc combines with *two* atoms of chlorine to give *one* molecule of zinc chloride (killed spirits).

Elements, compounds and mixtures

In the previous paragraph reference was made to an element and a compound. We will now try to define these.

An *element* is a pure piece of matter of a particular type, which cannot be chemically broken down into any simpler substance. There are 103 known elements, and these are the basis of all other substances. Pure aluminium (Al) is an element, so is pure copper (Cu). Neither can be broken down chemically into a simpler form than pure aluminium and copper. Other examples of elements, with their standard abbreviations, which are present in common workshop materials and processes, are as follows:—

Hydrogen	H	Chromium	Cr	Indium	In
Carbon	C	Manganese	Mn	Tin	Sn
Nitrogen	N	Iron	Fe	Antimony	Sb
Oxygen	O	Cobalt	Co	Tungsten	W
Magnesium	Mg	Nickel	Ni	Platinum	Pt
Silicon	Si	Zinc	Zn	Gold	Au
Phosphorus	P	Germanium	Ge	Mercury	Hg
Sulphur	S	Molybdenum	Mo	Lead	Pb
Chlorine	Cl	Silver	Ag	Bismuth	Bi
Vanadium	V	Cadmium	Cd	Uranium	U

We can add to an *element* but we cannot decompose it.

MATTER 3

Experiment. Weigh a small crucible, put some copper filings into it and re-weigh. Obtain the 'weight' of the filings. Heat the crucible and contents over a bunsen flame in the open atmosphere for a period. Remove crucible from flame and re-weigh.

If the copper has decomposed, i.e. some part of it has been driven off into the atmosphere, it will weigh less than before heating. On the other hand, if it has become more complex it will have taken something out of the atmosphere and will therefore weigh more. It should weigh more, oxygen from the atmosphere having combined with the heated copper to form copper oxide, CuO. Here we have a substance which is more complex than an element as it is made up of atoms from two elements. It is known as a compound.

A *compound* is a chemical combination of two or more elements. In order to form a compound, a chemical reaction has to take place. A chemical reaction is a process in which atoms of one element become attached to atoms from another element, or other elements. Zinc chloride (killed spirits), a flux used in soft soldering, is a compound of the elements zinc and chlorine.

Experiment. Pour a small quantity of hydrochloric acid into a beaker and add to it a small quantity of zinc filings. Watch the reaction take place, bubbles of hydrogen escaping into the atmosphere.

Hydrochloric acid, HCl, is a compound of *hydrogen* and *chlorine*. When the element zinc is added, each atom of chlorine releases the hydrogen atom with which it is chemically combined and becomes combined with a zinc atom. This happens because the chlorine atom has a greater affinity (liking) for the zinc atom than it has for the hydrogen atom; a kind of natural selection. In the chemist's language this reaction would be stated as follows:

$$2HCl + Zn \rightarrow ZnCl_2 + H_2$$

$$\begin{pmatrix}\text{2 molecules}\\\text{hydrochloric acid}\end{pmatrix} \begin{pmatrix}\text{1 atom}\\\text{zinc}\end{pmatrix} \text{turns to} \begin{pmatrix}\text{1 molecule}\\\text{zinc chloride}\end{pmatrix} \begin{pmatrix}\text{1 molecule}\\\text{hydrogen}\end{pmatrix}$$

N.B.—One molecule of hydrogen contains two hydrogen atoms and is written as H_2. Similarly with chlorine Cl_2 and so on.

This is a way of expressing what happens to all the atoms and molecules present.

Here we have the elements zinc and chlorine combining to form an entirely new substance, the compound zinc chloride.

Other examples of compounds are:

$$\underset{\text{(1 molecule oxygen)}}{\underset{O_2}{\text{Oxygen}}} + \underset{\text{(2 molecules hydrogen)}}{\underset{2H_2}{\text{Hydrogen}}} \underset{\text{to give}}{\rightarrow} \underset{\text{(2 molecules water)}}{\underset{2H_2O}{\text{Water}}}$$

In this case one molecule of oxygen combines with two molecules of hydrogen to produce two molecules of water.

$$\underset{\begin{pmatrix}\text{1 atom}\\\text{copper}\end{pmatrix}}{\underset{Cu}{\text{Copper}}} + \underset{\begin{pmatrix}\text{1 atom}\\\text{sulphur}\end{pmatrix}}{\underset{S}{\text{Sulphur}}} + \underset{\begin{pmatrix}\text{2 molecules}\\\text{oxygen}\end{pmatrix}}{\underset{2O_2}{\text{Oxygen}}} \underset{\text{to give}}{\rightarrow} \underset{\begin{pmatrix}\text{1 molecule}\\\text{copper sulphate}\end{pmatrix}}{\underset{CuSO_4}{\text{Copper sulphate}}}$$

Copper sulphate is used to give a ' marking out ' coat to a steel surface. It is also used to provide a protective coat against carburisation on certain surfaces of components which are to be case-hardened by the ' salt bath process '. Incidentally, the sodium cyanide salt used in this process is a compound of sodium, carbon and nitrogen:

$$\underset{(1\ \text{atom sodium})}{\text{Na}} + \underset{(1\ \text{atom carbon})}{\text{C}} + \underset{(1\ \text{atom nitrogen})}{\text{N}} \underset{\text{to give}}{\rightarrow} \underset{\binom{1\ \text{molecule}}{\text{sodium cyanide}}}{\text{NaCN}}$$

The preparation of these compounds is a chemist's job and it is not just a question of mixing the ingredients together. Certain elements will not combine, whilst some have a greater preference for one element over another. In the case of the ones that do combine, it may be one atom of one element with one, two, three or even more atoms, i.e. one molecule of another. They combine in definite weight ratios and orders which are laid down by nature and cannot be changed; e.g. the chemical composition of water is always H_2O.

A *mixture* is an association of two or more substances, there being no chemical reaction between them. It is a mechanical combination, an intermingling of grains of one substance with grains of another substance, or other substances. A mixture may be of solids, e.g. salt with sugar, copper filings with zinc filings; of liquids, e.g. cream with milk; of gases, e.g. oxygen, nitrogen and carbon dioxide as in the atmosphere; of solid and liquid, e.g. sugar dissolved in tea; of gas and liquid, e.g. carbon dioxide in water to give soda water, or mineral water.

A pure substance cannot be a mixture. Pure water consists of water only, whereas tap water consists of water and impurities; it is therefore a *mixture*. It is extremely difficult to obtain a pure substance and rarely achieved though it is possible to get very close to perfection. *Note:* distilled water is very nearly pure water.

A mixture is obtained when salt, or sugar, is dissolved in water and on dissolution no trace of the salt, or sugar, can be seen. A perfect mixture has been made and the name given to it is a *solution*. The substance being dissolved is known as the *solute*, whilst the substance used to dissolve it is the *solvent*. When we use paraffin to clean dirt and grease from machine parts, the paraffin is a solvent because it dissolves the grease. We are left with a mixture of paraffin, grease and dirt. All degreasing substances are solvents.

$$\text{SOLUTE} + \text{SOLVENT} \rightarrow \text{SOLUTION}$$

Cutting solution used in machining operations may be a mixture of oil (soluble) in water, i.e. an *emulsion*. The mixture of salt in water produces a quenching solution used in the heat-treatment of metals. Pigment powder dissolved in methylated spirit produces a mixture which is used as a marking out paint. The workshop atmosphere is a mixture of gases, often containing dust.

Pigment powder + Methylated spirit → Marking out paint
(SOLUTE)   (SOLVENT)   (SOLUTION)

A solvent is usually in liquid form.

If we add some pieces of tin to molten lead, the lead will dissolve the tin. We have produced a mixture of lead and tin, i.e. a solution of tin in lead. On being allowed to solidify the mixture would become a *solid solution* of tin and lead, i.e. a perfect mixture. We have produced an *alloy* of lead and tin known as white metal.

An *alloy*, in its simplest form, is really a mixture of liquid metals which has been permitted to solidify.

If pieces of zinc were added to molten copper, a solution of zinc and copper would be produced. Upon cooling the mixture would become a solid solution, i.e. an alloy known as brass. Lead paint is a mixture of white lead powder and linseed oil. The mixture of red lead powder in linseed oil makes a good jointing solution for joints in water pipes.

Experiment. Put some iron filings and some silver sand into a jar and shake the jar vigorously so as to mix the contents thoroughly. Now lower a bar magnet into the mixture.

Has there been any chemical reaction between the sand and the filings? Put some of the mixture into a test tube and heat over a bunsen flame for a little while, holding the tube with tongs. Allow to cool before checking again with bar magnet to see if filings can be separated from the sand. Add water to the jar and stir vigorously then allow contents to settle. Has any chemical reaction taken place?

Experiment. Put 7 g of sulphur powder and 4 g of iron filings into a test tube and mix thoroughly. Lower a bar magnet into the tube and note what happens. Mix the contents of the tube again and heat over a bunsen flame holding the tube with tongs. Watch for the contents of the tube to glow and on doing so remove it from the flame. Note whether the glowing continues when the tube is removed from the heat source and make appropriate deductions.

Allow both tube and contents to cool and then examine them closely. Has a new substance been formed? Try to separate the iron filings from the sulphur with the bar magnet.

In fact a chemical reaction should have taken place in which the elements iron and sulphur combined to form iron sulphide.

Fe + S → FeS
(1 atom iron) (1 atom sulphur) to give (1 molecule iron sulphide)

Incidentally the masses 7 g and 4 g are important. It so happens that 'nature' has determined that sulphur and iron combine in the ratio of

7 to 4 by mass. If more than 7 g of sulphur are used with 4 g of iron the surplus will burn off as gas, or remain in the test tube to form a mixture with the iron sulphide. Excess iron filings could be removed with a magnet, whilst remaining excess sulphur could be dissolved in carbon disulphide and then poured off.

These two experiments produce different results and show clearly the difference between a compound and a mixture, i.e. between a *chemical change* and a *physical change*.

Remember that compounds are formed to a definite chemical law, the elements always combining in definite constant proportions and having specific preferences for one another. The reason why a copper film is deposited on a steel surface from the compound copper sulphate is because the sulphate molecule has a greater preference for the iron atom than it has for the copper atom. Hence it exchanges copper for iron by depositing copper atoms on the steel and in return takes up iron atoms. It is natural for it to do this. The process stated chemically is:—

$$\underset{\substack{1 \text{ molecule} \\ \text{copper sulphate}}}{CuSO_4} + \underset{(1 \text{ atom iron})}{Fe} \underset{\text{to give}}{\rightarrow} \underset{\substack{1 \text{ molecule} \\ \text{iron sulphate}}}{FeSO_4} + \underset{\substack{1 \text{ atom} \\ \text{copper}}}{Cu}$$

A common name for copper sulphate is *blue vitriol* and for iron sulphate *green vitriol*. Watch for the colour change next time you use the substance.

Ores

Most of the metals that we use in engineering practice occur in their natural ores as compounds. It is from the smelting, or refining, process that the element required, e.g. the 'pure metal', is obtained. Examples of this are:

Iron ore
This occurs in the forms indicated below.
Hematite is impure ferric oxide (iron oxide) with the chemical formula Fe_2O_3.
Magnetite is impure ferric oxide chemical formula, Fe_3O_4.
Spathic is impure ferrous carbonate (iron carbonate) chemical formula, $FeCO_3$.

We know from our workshop technology notes that these ores are smelted in the *blast furnace*. The blast furnace process is a chemical process in which the element iron is separated from its earthy compound. The process is essentially one of reducing the oxygen in the ore by adding carbon, in the form of coke, to the furnace.

Zinc ore
This ore is found as a carbonate, or sulphide, compound.
Calamine is a zinc carbonate, $ZnCO_3$.
Blende is a zinc sulphide, ZnS.

In the refining of each of these ores the process is begun by roasting the ore in air. This results in both carbonate and sulphide becoming oxide compounds. The carbonate gives off a carbon dioxide molecule whilst the sulphide takes oxygen from the atmosphere and gives off sulphur dioxide.

Stated in the chemist's language:

$$\underset{\substack{\text{1 molecule} \\ \text{zinc carbonate}}}{ZnCO_3} \underset{\text{to give}}{\rightarrow} \underset{\substack{\text{1 molecule} \\ \text{zinc oxide}}}{ZnO} + \underset{\substack{\text{1 molecule} \\ \text{carbon dioxide}}}{CO_2}$$

$$\underset{\substack{\text{2 molecules} \\ \text{zinc sulphide}}}{2ZnS} + \underset{\substack{\text{3 molecules} \\ \text{oxygen}}}{3O_2} \underset{\text{to give}}{\rightarrow} \underset{\substack{\text{2 molecules} \\ \text{zinc oxide}}}{2ZnO} + \underset{\substack{\text{2 molecules} \\ \text{sulphur dioxide}}}{2SO_2}$$

Note: In statements of chemical reactions the number of atoms on the right of the arrow is equivalent to the number on the left, i.e. atoms of matter are neither created nor destroyed by man.

From the reaction above, the zinc oxide is then put into a fireclay retort furnace along with carbon in the form of coke. The whole is then heated and the oxygen from the zinc oxide combines with the carbon, forming carbon monoxide gas. In the process the zinc becomes a vapour and this is piped away from the retort to a condenser. Here it becomes a liquid metal which is poured into moulds to form ingots of pure zinc. The chemical statement for the reaction is as follows:

$$\underset{\substack{\text{1 molecule} \\ \text{zinc oxide}}}{ZnO} + \underset{\substack{\text{1 atom} \\ \text{carbon}}}{C} \underset{\text{to give}}{\rightarrow} \underset{\text{(1 atom zinc)}}{Zn} + \underset{\substack{\text{1 molecule} \\ \text{carbon monoxide}}}{CO}$$

Lead ore

This is found as a compound, lead sulphide, PbS, known as *galena*. The galena is roasted in air and during the process a lead oxide compound known as litharge is produced.

$$\underset{\substack{\text{2 molecules} \\ \text{lead sulphide}}}{2PbS} + \underset{\substack{\text{3 molecules} \\ \text{oxygen}}}{3O_2} \underset{\text{to give}}{\rightarrow} \underset{\substack{\text{2 molecules} \\ \text{lead oxide}}}{2PbO} + \underset{\substack{\text{2 molecules} \\ \text{sulphur dioxide}}}{2SO_2}$$

The refining process is continued by heating the lead oxide compound in a small blast furnace with coke. Oxygen from the oxide combines with the carbon from the coke forming carbon monoxide gas.

$$\underset{\substack{\text{1 molecule} \\ \text{lead oxide}}}{PbO} + \underset{\substack{\text{1 atom} \\ \text{carbon}}}{C} \underset{\text{to give}}{\rightarrow} \underset{\substack{\text{1 atom} \\ \text{lead}}}{Pb} + \underset{\substack{\text{1 molecule} \\ \text{carbon monoxide}}}{CO}$$

Actually lime is added to combine with earthy impurities that may be present. Also some iron is added to neutralise any lead sulphide that may be charged into the furnace. Both are floated off with the slag, leaving pure lead which is cast into ingots.

CHEMISTRY IN THE WORKSHOP

Copper ore

This appears mainly in the form of a compound known as *pyrites* which contains copper, iron and sulphur ($CuFeS_2$). The chemistry involved in breaking down this compound is very complex and well beyond the scope of this book. Some copper ore is found in a purer state and this is known as *boulder copper*. This ore is refined by an electro-chemical process, see Fig. 1.

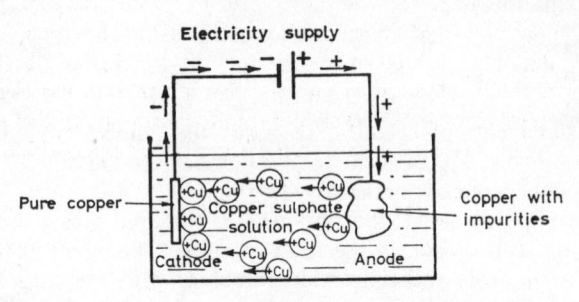

Fig. 1

The flow of an electric current, see page 31, causes electrically charged atoms to pass into the copper sulphate solution from the lump of impure copper. The impurities are left behind. This enriches the copper sulphate solution with copper atoms which carry a positive electric charge. These atoms proceed to the piece of pure copper which is negatively charged where they give up their charge and are deposited. This flowing process continues, so long as the current flows, until all the copper atoms have left the impure lump. *Anode* is the name given to the member that carries the positive electrical charge and *cathode* is the name of the negatively charged member.

Note: Electrically charged atoms are known as *ions*.

Experiment. *To demonstrate the flow of copper atoms.*

Partly fill a glass trough with copper sulphate solution. Cut two tee-shaped pieces to suitable dimensions from a sheet of copper, say 1 mm thick. Carefully determine their masses on a chemical balance. Now arrange them in the copper sulphate solution as shown in Fig. 2. Connect them to a 2-volt battery, and let the current flow for approximately 60 minutes. Now disconnect the plates from the circuit, dry them thoroughly in warm air and re-measure their masses.

Was there a change in the masses of the plates?

If flow of copper atoms did take place, was it from anode to cathode?

Incidentally, this experiment illustrates the process involved in the electro-plating of components, e.g. some types of cutlery, automobile fittings, etc.

The process is known as *electrolysis*. The solution in the trough is called an *electrolyte*.

Aluminium ore

There are a number of ores containing aluminium, e.g. *bauxite*, which is the commonest, is an oxide compound containing water, $Al_2O_3 2H_2O$.

Corundum used in the manufacture of grinding wheels is another. This is an oxide compound, Al_2O_3.

Kaolin (china clay) is an aluminium–silicon–oxide containing water, $Al_2Si_2O_7 2H_2O$. This is used in the production of ceramic tips for cutting tools. Also in the production of porcelain which is used for electrical and heat insulators, e.g. the protective sheath on the thermo-couple of an immersion type pyrometer. It is also used in the seger cone temperature measuring process.

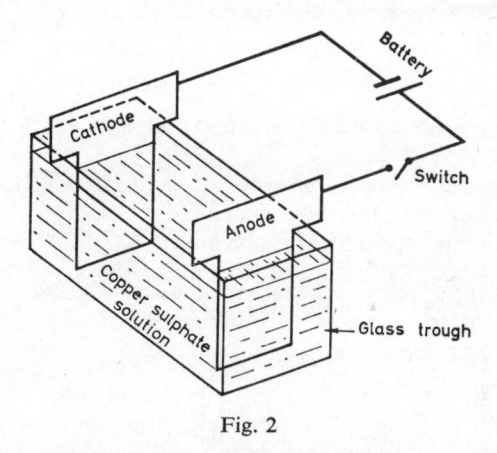

Fig. 2

Mica is yet another compound containing aluminium. It is in fact a potassium–aluminium–silicon–oxide, $K_2Al_2Si_6O_{15}$, sometimes referred to as *felspar*. This too is used in seger cones, also as an electrical insulator.

Cryolite is a sodium–aluminium–fluorine compound, Na_3AlF_6. This is actually used in the production of pure aluminium from the basic ore.

Bauxite is the chief ore used and the process for obtaining the pure aluminium from the ore is part chemical and part electrical. The ore is partly purified by a chemical process, then it is refined by the electrolysis process, the electrolyte being cryolite.

Tin ore

This is found chiefly as a tin–oxide, SnO_2, known as *tinstone*. After crushing and washing, the ore is roasted in a reverbatory furnace along with some material rich in carbon. Anthracite coal is often used. The tin compound is split up during the process, the oxygen leaving the tin to combine with carbon from the coal to form carbon dioxide gas.

$$\underset{\substack{\text{1 molecule} \\ \text{tin oxide}}}{SnO_2} + \underset{\substack{\text{2 atoms} \\ \text{carbon}}}{2C} \to \text{to give} \underset{\substack{\text{1 atom} \\ \text{tin}}}{Sn} + \underset{\substack{\text{2 molecules} \\ \text{carbon monoxide}}}{2CO}$$

The molten tin is run off from the furnace leaving the slag behind. Otherwise, the slag may appear on the surface of the tin when it is poured into the moulds.

Antimony ore

In the natural state, antimony appears as a sulphide compound called *stibnite*, Sb_2S_3. The pure metal is obtained from the ore by means of heating the stibnite in a reverbatory furnace with scrap iron. The sulphur leaves the antimony because of its greater preference for iron.

$$\underset{\substack{\text{1 molecule} \\ \text{antimony sulphide}}}{Sb_2S_3} + \underset{\text{(3 atoms iron)}}{3Fe} \to \text{to give} \underset{\substack{\text{2 atoms} \\ \text{antimony}}}{2Sb} + \underset{\substack{\text{3 molecules} \\ \text{iron sulphide}}}{3FeS}$$

The antimony floats on the ferrous (iron) sulphide and is tapped off into moulds.

This concludes the simple chemistry of the extraction of the common engineering metals from their ore compounds. Some of them are used in their single state, e.g. copper for wire and boiler plate, lead for cable sheathing and plumbing work. Others are used to make alloys, e.g. tin and lead to give soft solder, copper and zinc to give brass.

Alloys

We saw earlier that a solution was a perfect mixture, and we used the example of salt in water to illustrate this, page 4, e.g. a solid dissolved in a liquid. If we mix a liquid in a liquid, e.g. alcohol in water as in wines and spirits we should again have made a solution. If we place some pieces of copper and tin in a crucible and melt them down, they too will form a solution. If we permit the solution to cool, then it will solidify with one metal perfectly dissolved in the other. We shall in fact have produced a *solid solution* (this is the metallurgist's name for it). We in the workshop would call it an *alloy*; this alloy being known as bronze. An alloy is a solid solution of one or more metals in another metal. The metal which is the solvent is the one which is present in the largest quantity. The white metals are alloys of lead, tin, copper and antimony. Steel can be loosely said to be an alloy of iron and carbon though the carbon is chemically combined with some of the iron to form the compound iron carbide, Fe_3C. It is this compound which is in solid solution in the iron. Hence it would be more correct to say that steel was an alloy of iron and iron carbide. See Workshop Technology notes for further details of this.

MATTER

Exercise 1

Section A

1. What are the *three* states in which matter can exist? Give one example, from the workshop, of each state.
2. Give two examples from the workshop of solutions which embrace a solid and a liquid.
3. Give two examples of a gaseous solution in the workshop.
4. What is a mixture? Give two examples from the workshop which illustrate this.
5. Complete the following table by filling in the spaces correctly.

Substance	Element	Compound	Mixture
Coal gas			
Iron			
Steel			
Tap water			
' Cutting compound '			
Grinding paste			
Grinding wheel			
Lubricating oil			
Brass			

6. If you were given a mixture of iron filings and borax powder, explain how you would separate the two.
7. How does distilled water differ from ' tap water '?
8. What is a solvent? Give two examples of solvents used in the workshop.
9. Is soft solder an element, compound or alloy? Give reasons for your answer.
10. A paint brush with lead paint on it may be cleansed in turpentine, whereas one with cellulose will not be affected by the turpentine. Why is this?

Section B

1. Make an alloy of 60% lead and 40% tin, by 'weight', in a small crucible over a bunsen flame and determine the melting point of the alloy. Compare it with the melting points of the lead and tin separately.
2. Explain the term *solid solution*.
3. Complete the following table by filling in the spaces correctly.

Substance	Element	Compound	Alloy or Mixture
Carbon			
Iron carbide			
Copper sulphate			
Belt paste			
Micrometer blue			
Cast iron			
Tin			
Zinc			
Wrought iron			
Antimony			

4. Explain the difference between a compound and a mixture and give workshop examples to illustrate the difference.
5. To remove oil, or grease, from your hands you may wash them in paraffin. Explain the chemistry of this operation.
6. You are given two beakers, one containing silver sand in water, the other common salt dissolved in water. Explain in chemical terms the difference between the contents of each beaker. How would you recover the salt and sand?
7. Explain the chemistry of (i) a degreasing, (ii) a de-scaling process.
8. What is pure water, an element, compound or mixture? Give reasons for your answer.
9. What is the difference between a chemical change and a physical change? Explain how this could be demonstrated experimentally.
10. After making a soft solder joint with 'killed spirits' as the flux, it is advisable to wash the joint in a soap solution. Explain why this is desirable.
11. Explain the function of the flux in the soft soldering process, stating the chemistry involved.
12. State any changes which will take place in the borax powder flux during a brazing operation.
13. Explain why on completion a brazed joint is immersed in sulphuric acid.
14. What are the differences involved in the chemistry of a soft soldering process when white vaseline, instead of 'killed spirts', is used as the flux?

Chapter 2. Composition of the Atmosphere

WE ALL KNOW that air surrounds us in the workshop and factory. In fact, it surrounds the earth and everything on it. If we were asked to show that this was a fact, could we do it? Maybe we could demonstrate its presence by holding a piece of newspaper by the gap at the foot of a closed door. The newspaper would very likely flutter, the movement being caused by the air current coming under the door. We could demonstrate the fact more scientifically in a Technical College laboratory.

Experiment. Arrange an empty flask, i.e. one containing only air, upside down on a retort stand. The mouth of the flask should be fitted with a rubber bung through which a piece of glass tube passes. The lower end of the tube should be immersed in water contained in a beaker as shown in Fig. 3. Having fixed up the apparatus apply a bunsen flame gently to the flask.

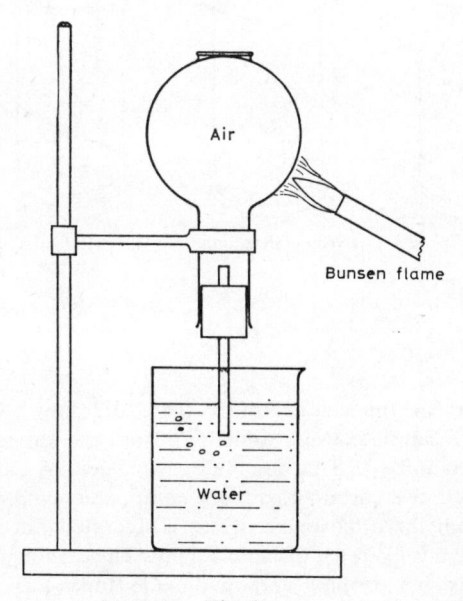

Fig. 3

This heating causes the contents of the flask to expand, see page 261, and overflow. Since we believe that it is air which is in the flask we should expect to see air bubbles coming from the glass tube. Does this in fact happen?

For further evidence of the presence of air we could look to the pump that we use for inflating tyres on our cycle wheels.

To demonstrate the presence of air around us we have had to turn to physical means because the air is colourless and odourless. We cannot see or smell it.

Having established that a gas known as air is around us, we might now seek to find what it contains. Previously, page 4, we have noted that it is a mixture of gases. The task at this stage is to find out what these gases are.

Experiment. Place a lighted candle in a glass jar, Fig. 4(a), and observe its behaviour. It will burn satisfactorily for a while, then it will flicker and may go out. Next, place a lighted candle in the jar and cover mouth of the jar with a piece of flat cardboard, see Fig. 4(b). The candle will burn for a while and then become extinguished. Re-light the candle and place a tee-shaped piece of cardboard in the mouth of the jar as shown in Fig. 4(c). The candle should continue to burn brightly.

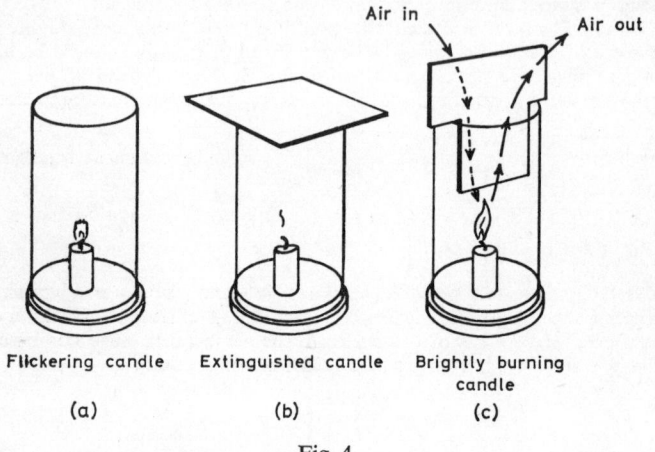

Flickering candle (a) Extinguished candle (b) Brightly burning candle (c)

Fig. 4

The behaviour of the candle under these differing conditions clearly indicates that the candle is using something from the atmosphere around it in order to maintain itself. That something is oxygen. A flame is a chemical reaction in which the carbon–hydrogen compound candle wax combines with oxygen from the atmosphere. Heat is liberated during the reaction, in fact enough heat is liberated to cause a light. The carbon and the hydrogen combine with oxygen forming carbon dioxide and water vapour. In the chemist's language,

$$C + O_2 \rightarrow CO_2$$

$$\underset{\text{candle wax}}{2H_2} + \underset{\text{oxygen}}{O_2} \underset{\text{to give}}{\rightarrow} \underset{\text{products of combustion}}{2H_2O}$$

If the oxygen supply to the flame is reduced, or cut off, as it is in the first two parts of the experiment, the chemical reaction is slowed down, or cut

COMPOSITION OF THE ATMOSPHERE

off completely. In the third part of the experiment, by fitting the cardboard partition, a chimney is created. This directs the products of combustion away from the flame and channels the air containing oxygen to it.

We could further demonstrate the presence of oxygen in the atmosphere in the following way:

Experiment. Cut a strip of copper from a brightly polished sheet; if necessary, polish the strip with emery cloth. Hold the strip by means of a pair of tongs in a bunsen flame for a while. On withdrawing the copper strip from the flame it will be seen to have a black film on it. This film is copper oxide, copper atoms having combined with oxygen atoms from the atmosphere to form it.

Two elements have combined to form a compound:

$$2Cu + O_2 \rightarrow 2CuO$$
$$\begin{pmatrix} 2 \text{ atoms} \\ \text{copper} \end{pmatrix} \quad \begin{pmatrix} 1 \text{ molecule} \\ \text{oxygen} \end{pmatrix} \quad \text{to give} \quad \begin{pmatrix} 2 \text{ molecules} \\ \text{copper oxide} \end{pmatrix}$$

Note: Oxygen atoms generally work in pairs, i.e. as a molecule and each molecule of oxygen combines with two atoms of copper.

The black film of copper oxide can easily be scraped off showing unaffected copper beneath it.

The scaling of steel when heated for forging purposes involves a reaction similar to the one in this experiment.

Experiment. Put some lead foil into a small crucible and place over a bunsen flame. The lead melts. Stir the molten lead whilst continuing to heat it. Watch it change to a yellow powder. Atoms of oxygen from the atmosphere have combined with atoms of lead to form the compound lead oxide known as litharge.

$$2Pb + O_2 \rightarrow 2PbO$$
$$\begin{pmatrix} 1 \text{ molecule} \\ \text{lead} \end{pmatrix} \quad \begin{pmatrix} 1 \text{ molecule} \\ \text{oxygen} \end{pmatrix} \quad \text{to give} \quad \begin{pmatrix} 2 \text{ molecules} \\ \text{lead oxide} \end{pmatrix}$$

Not all the atmosphere is oxygen. There are other gases. The bulkiest of these is nitrogen. As far as the technician is concerned, oxygen is the most important, for besides breathing it, he needs it for many industrial processes. Chief amongst these may well be combustion, see page 22.

The composition of the atmosphere by volume is 78% nitrogen, 20% oxygen and 2% other gases. This 2% includes carbon dioxide, which we breathe out, and plant life breathes in, and inert gases. Nature apparently ensures that a steady balance is maintained. There are various other constituents, solid, liquid and gaseous, which may be generally referred to as impurities.

Experiment. *To find the proportion of oxygen in a quantity of air.*

Add to a 0·5 litre glass flask a known volume of water, the volume being such that the flask will be approximately one-third full. Put some copper turnings into a hard glass tube which is sealed at both ends with rubber bungs through which glass tubes pass. Arrange a graduated cylinder on a beehive shelf so that it can be used as a collecting jar for nitrogen. Connect these with glass and rubber tubes into the arrangement as shown in Fig. 5. Arrange a bunsen with a fish-tail jet under the hard glass tube. Fill the collecting jar and the pneumatic trough about two-thirds full with water. Invert jar with its mouth placed under the water in the trough. Next light the bunsen flame and heat the copper turnings to red heat. Now cause the air sample in the flask to flow over the hot turnings by adding a known volume of water to the flask.

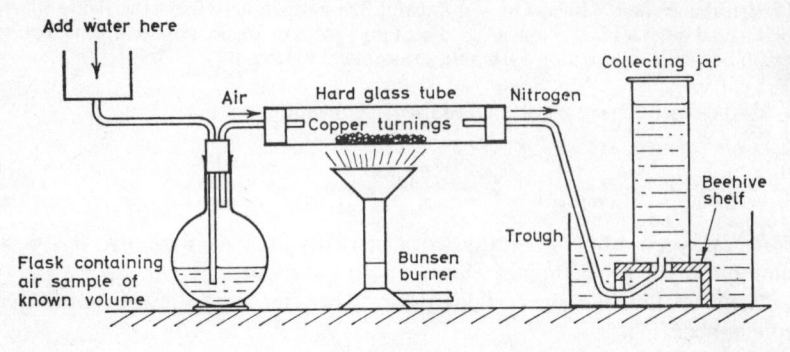

Fig. 5

The water level in the flask rises, pushing an air sample out over the hot copper turnings which extract oxygen from the air stream, leaving the nitrogen to pass into the collecting jar. It is really a burning of copper, see page 25. When the air sample is all used, shut off the bunsen flame. Note the volume of nitrogen in the collecting jar and subtract this volume from the initial volume of the air sample. The result is the volume of oxygen used in burning the copper. Express this volume as a percentage of the initial volume of the air sample; it should be 20%. Greatest accuracy in measuring the volume of nitrogen will be achieved if the water levels in the trough and jar are the same, e.g. pressure in the jar equal to that outside.

An alternative method for measuring the oxygen–nitrogen ratio in the atmosphere is as follows:

Experiment. Put some iron filings into a ' hollowed ' piece of timber which will float on water and is small enough to be enclosed by a test tube. Float the dish in a beaker which is half full with water, and enclose it in a test tube held inverted in a retort stand as shown in Fig. 6.

Record the height of the air column in the test tube. Leave for about a week, then record the new water level in the tube.

The level should have risen to reduce the air volume in the tube by one-fifth of the initial volume since the oxygen in the tube atmosphere should have reacted with the iron. The reaction in this experiment will have happened at a slower rate than that between the copper and oxygen. Nevertheless, it will have taken place.

$$4Fe + 3O_2 \rightarrow 2Fe_2O_3$$
$$\begin{pmatrix}\text{4 atoms}\\ \text{iron}\end{pmatrix} \begin{pmatrix}\text{3 molecules}\\ \text{oxygen}\end{pmatrix} \text{to give} \begin{pmatrix}\text{2 molecules}\\ \text{iron oxide}\end{pmatrix}$$

Repeat this experiment using magnesium filings.

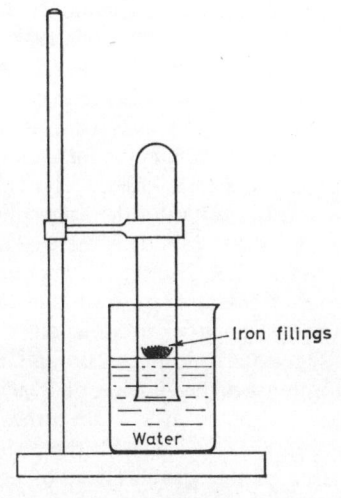

Fig. 6

Dust and Moisture in the atmosphere

During a summer day, when the sun is shining brilliantly, have you ever noticed a shaft of sunlight breaking through a chink in the roof into some dark corner of the workshop? If so, you will first have noticed the dust shown up by the beam, floating in the workshop atmosphere. Alternatively you may have wiped your brow with a clean white handkerchief after walking through a factory, or workshop, and been amazed to find the dirt on it, when you had believed your brow to be clean. Dust from the atmosphere had settled on your forehead. The atmosphere is laden with minute particles of floating dust put there by smoking chimneys, quarrying and mining operations and some by grinding operations. The wind, whipping dust from the earth is also responsible for putting some of it there, whilst some is the residue of exhaust fumes from motor vehicle engines. In the best regulated workshops efforts are made to lessen the pollution of the atmosphere by fitting dust extractors to grinding machines, woodworking machines and their like. Similar precautions are taken in coal mines and cement and grain mills. This dust, which is more prevalent in the atmosphere around industrial areas than it is in non-industrial areas, is injurious to the human body, causing chest and lung illnesses. The dust does tend to settle on objects and if these have a film of oil on them, as in the case of machine parts, serious trouble might arise, for it can cause stoppage of an oil-way resulting in bearing seizure. Dust is a menace in paint spraying and polishing shops. Gritty dust is especially disadvantageous in the latter case for it may cause the scoring of a surface. A layer of dust on a surface may be the cause of a false measurement being taken and this could be critical in cases where extremely accurate sizes are required. For this reason efforts are made to eliminate dust from the atmosphere of metrology laboratories, gauge inspection rooms and similar places. Also it may impede the working of sensitive mechanisms fitted to precision measuring instruments.

Atmospheric dust plays a major part in the formation of fog. This explains why fog is more menacing in industrial areas. The mention of fog brings us to another substance which is floating in the atmosphere, water vapour, i.e. fine particles of water. This water vapour is there as a result of evaporation from oceans, seas and rivers. The level of the vapour content in the atmosphere is dependent upon the temperature of the atmosphere. Warm air has a higher vapour-carrying capacity than cooler air.

It is easy to show that the atmosphere around us contains moisture. Watch for the mist forming on the outside of a glass which contains an iced drink, the glass being situated in a warm atmosphere. The warm air in contact with the cold glass is cooled, with the result that water particles are formed on the surface of the glass. Those of us who wear spectacles will have had the experience of a mist forming on the lenses on moving from a cold atmosphere into a warm room. Again warm air on coming into contact with the colder lenses sheds some of the moisture it contains, condensation.

Dew, fog and frost are formed in a similar way. At least the reasoning behind their formation is exactly the same as in the two previous examples. During autumn nights, plants, grass, bushes, trees, fences, etc. suffer a drop in temperature, so does the air in contact with them. This causes water vapour to be condensed from the air, and so a dew is formed. As winter approaches the late autumn nights become colder and the temperature fall is so great as to cause a drop in temperature of the whole atmosphere and not just plant life, pavements and other surfaces. The result is that water particles are precipitated over a wide space and a fog is formed. The water particles usually condense on dust particles floating in the atmosphere, hence the reason for fog being denser in industrial areas. In winter, the fall in the night temperature is to such a level as to cause the precipitated vapour to appear as a hoar frost, i.e. the dew freezes the instant it forms.

We have then around us ample proof of the atmosphere containing moisture. There is proof too in the reverse direction, drying clothes, evaporation of perspiration from the brow. The moisture goes into the atmosphere.

COMPOSITION OF THE ATMOSPHERE

The moisture content (humidity) of a factory, or dance hall atmosphere is important, for the comfort of the occupants is as much dependent upon it as it is upon the temperature. In the case of factories, such as cotton mills, the moisture of the atmosphere is vital to the successful spinning of the cotton. The atmosphere of a coal mine may become unbearable due to high temperature which in turn may mean high moisture content. This could happen if the ventilation system was not satisfactory. It would be equally unbearable if an atmosphere lacked moisture content, as it may do in a warm building during winter-time if the temperature becomes too high and there is no water about to be evaporated. A damp atmosphere may be injurious to machines and components and for this reason alone we are concerned with the moisture content of the atmosphere, because it may cause corrosion. This will be dealt with in the next chapter.

Exercise 2

Section A
1. Give a list of the gases and other substances contained in the atmosphere.
2. Is the moisture content of the air always the same? Explain your answer.
3. Why is the moisture content of the atmosphere in a paper mill important?
4. The evaporation of perspiration is nature's way of keeping the human body cool. Explain why it is difficult to be cool and comfortable in a humid atmosphere.
5. Explain why it is desirable to fit a filter to the air intake on the carburetter of an internal combustion engine.
6. Certain areas in many industrial cities and towns have been declared ' smokeless zones '. What general effect has this ruling had on the atmosphere of such places?
7. A good ventilating system for a factory would be fitted with an intake filter and a humidity control. Explain what these are and why it is necessary to have them.
8. Why is it that on a warm humid day one's clothes tend to be damp and sticky even though the temperature of the atmosphere is high?
9. Why are filters fitted to some power station chimneys?
10. Why is it important that a metrology laboratory should have a ' manufactured ' atmosphere?
11. Explain the formation of mist on the interior surfaces of motor car windows during cold weather.
12. Why might it be a disadvantage to oil, or grease, the spindle of a dial indicator?

Chapter 3. Chemical Reactions

Oxidation, reduction, carburisation, combustion and *corrosion* are processes in which a chemical reaction takes place. Corrosion apart, we are concerned with the others because they are present in the processes of heat-treatment, forging, casting and refining of metals.

Oxidation

This is the name given to a chemical reaction in which:

(*a*) oxygen is added to an element, or compound, e.g.
when oxygen combines with copper to form copper oxide,

$$2Cu + O_2 \rightarrow 2CuO,$$

or combines with zinc to form zinc oxide,

$$2Zn + O_2 \rightarrow 2ZnO$$

or combines with carbon to form carbon dioxide,

$$C + O_2 \rightarrow CO_2$$

(*b*) hydrogen is removed from a compound, e.g.
when zinc is added to hydrochloric acid to make the flux zinc chloride (killed spirits),

$$2HCl + Zn \rightarrow ZnCl_2 + H_2$$

(*c*) when a compound ending in *-ous* is changed to one ending in *-ic*, e.g. when *ferrous* chloride is changed to *ferric* chloride by adding more chlorine.

$$\underset{\substack{\text{2 molecules}\\ \text{ferrous chloride}}}{2FeCl_2} + \underset{\substack{\text{1 molecule}\\ \text{chlorine}}}{Cl_2} \xrightarrow{\text{to give}} \underset{\substack{\text{2 molecules}\\ \text{ferric chloride}}}{2FeCl_3}$$

This full definition would be required by a chemist. However, since most of our work is covered by part (*a*) we may state that: oxidation is a chemical reaction in which oxygen is added to an element, or compound.

The following are *oxidising* agents:

oxygen, chlorine, nitric acid, hydrogen peroxide.

Reduction

Similarly we might define a reduction as a chemical reaction in which oyygen is removed from a compound.

A chemist would require a fuller definition, namely, a reduction is a chemical reaction in which:

(a) Oxygen is removed from a compound, e.g. oxygen is removed from zinc oxide by adding carbon, producing zinc and carbon dioxide.

$$2ZnO + C \rightarrow 2Zn + CO_2$$

(b) Hydrogen is added to an element, or compound, e.g. the addition of hydrogen to chlorine producing hydrochloric acid.

$$H_2 + Cl_2 \rightarrow 2HCl$$

(c) A compound ending in -*ic* is changed to one ending -*ous*, e.g. the *reduction* of ferric sulphate to ferrous sulphate by adding hydrogen sulphide.

$$Fe_2(SO_4)_3 + H_2S \rightarrow S + H_2SO_4 + 2FeSO_4$$

Reduction is the *reverse* of *oxidation*.

The following are reducing agents:

hydrogen, carbon, carbon monoxide

Carburisation

We engineering technicians know what the case-hardening process is, or at least we have heard of it. Most likely we have a note in our Workshop Technology notebook which informs us that it is a process in which low-carbon steel components are ' baked ' with some material rich in carbon, e.g. charcoal. The ' baking ' taking place at a particular temperature for a specific length of time. During this time carbon from the charcoal becomes chemically combined with the iron in the surface layers of the component to form the compound iron carbide (cementite). This process in the chemist's language is known as *carburisation*. The carburised component is then heat-treated in order to give it the desired physical properties. The carbon used in the carburising process may be in solid, liquid, or gaseous form.

For the carburising process in which the components are packed into boxes, solid carbon is used. The process is known as *pack carburising* and the solid carbon may be used in the form of coal, coke, wood or animal charcoal, e.g. bone dust. The wood and animal charcoals and the coke are produced by roasting out of contact with the air (oxygen from this) wood, animal bones and coal respectively. The charcoal from wood would be carbon plus a small amount of potassium carbonate. Coke would be almost pure carbon. Animal charcoal contains carbon, calcium carbonate (CaO_3), and calcium phosphate (Ca_3)(PO_4)$_2$.

Note: Chalk is calcium carbonate (CaO_3).

Barium carbonate ($BaCO_3$) is usually added to carburising compounds because it has the effect of speeding the actual carburising.

The *salt bath process* uses a liquid carburising agent, sodium cyanide (NaCN). The compound breaks down at the bath temperature of 900°C to 950°C, the carbon combining with the iron in the surface layers of the component. The nitrogen too has a hardening effect.

'Town gas' properly treated is used where a gas carburising agent is required.

Combustion

Combustion is another word for burning. We can often detect the process by sight insofar as we can see a flame, or red glow. However, there are cases of burning which are not accompanied by a flame or red glow, e.g. the action of 'battery' acid on our overalls, bleach (nitrogen peroxide, NO_2) on a piece of rag, oil in a quenching bath. We might ask the question, what is meant by burning? The answer, it is a chemical reaction in which heat is produced. It is also a process in which compounds are broken up and the elements so liberated form new, different, compounds.

We read on page 14, that if a candle were to burn successfully it required oxygen. It could be shown that to burn coal, or gas, or timber, oxygen is required. Thus oxygen is required to support combustion. When we see coke burning we are in fact witnessing a chemical reaction between carbon and oxygen. In the process they combine to form carbon dioxide and heat is liberated; so much heat in fact that there is a red glow accompanying the reaction.

$$\underset{\substack{\text{carbon} \\ \text{(1 atom carbon)} \\ \text{fuel}}}{C} + \underset{\substack{\text{oxygen} \\ \text{(1 molecule oxygen)} \\ \text{oxygen}}}{O_2} \xrightarrow{\text{to give}} \underset{\substack{\text{carbon dioxide} \\ \text{(1 molecule carbon dioxide)} \\ \text{products of combustion}}}{CO_2}$$

The complete reaction, of course, is dependent upon there being the correct amounts of carbon and oxygen. We read on page 4 that elements combine with other elements on a preferential basis and in definite quantities determined by nature. If there is a surplus of oxygen then the reaction shows up as follows:

$$\underset{\binom{\text{1 atom}}{\text{carbon}}}{C} + \underset{\binom{\text{2 molecules}}{\text{oxygen}}}{2O_2} \xrightarrow{\text{to give}} \underset{\binom{\text{1 molecule}}{\text{carbon dioxide}}}{CO_2} + \underset{\binom{\text{1 molecule}}{\text{oxygen}}}{O_2}$$

If there is not enough oxygen, the combustion of the carbon will not be complete. The products of combustion will then be carbon monoxide instead of carbon dioxide, or a combination of both.

$$\underset{\binom{\text{2 atoms}}{\text{carbon}}}{2C} + \underset{\binom{\text{1 molecule}}{\text{oxygen}}}{O_2} \xrightarrow{\text{to give}} \underset{\binom{\text{2 molecules}}{\text{carbon monoxide}}}{2CO}$$

or,

$$\underset{\binom{\text{5 atoms}}{\text{carbon}}}{5C} + \underset{\binom{\text{4 molecules}}{\text{oxygen}}}{4O_2} \xrightarrow{\text{to give}} \underset{\binom{\text{3 molecules}}{\text{carbon dioxide}}}{3CO_2} + \underset{\binom{\text{2 molecules}}{\text{carbon monoxide}}}{2CO}$$

CHEMICAL REACTIONS

The oxygen supply for combustion usually comes from the atmosphere, though for the oxy-acetylene flame this is not true. The coal fire is a process of the combustion of carbon in air, and a chemical formula similar to that used for coke would result from such a combustion. The combustion at the flame of a candle is the burning of candle wax, it first being liquified then vapourised. Candle wax is a compound of the elements hydrogen and carbon; it is often referred to as a *hydro-carbon*. Most fuels are hydro-carbons

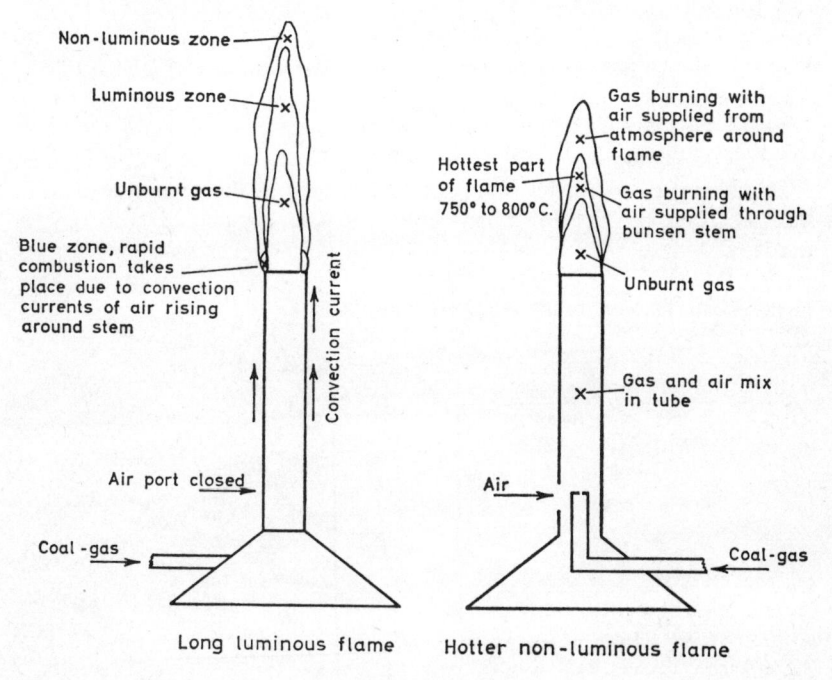

Fig. 7

including oil and natural gas. Both the hydrogen and the carbon are combustible. The combustion of acetylene gas in oxygen provides a very hot flame as we may observe in the welding shop. Acetylene gas is a hydro-carbon with the formula C_2H_2. The carbon is oxidised to carbon dioxide and the hydrogen to water from condensing steam.

$$2C_2H_2 \text{ (2 molecules acetylene)} + 5O_2 \text{ (5 molecules oxygen)} \to 4CO_2 \text{ (4 molecules carbon dioxide)} + 2H_2O \text{ (2 molecules water)}$$

Pure oxygen is supplied to the flame usually from a steel bottle, and this mixes with the acetylene at the burner nozzle. The supplying of pure oxygen to the flame ensures a better combustion, resulting in a much hotter flame, than if the flame had to get its own oxygen from the atmosphere. The mixing of the oxygen and acetylene just before combustion helps considerably towards producing a hotter flame.

A study of a bunsen flame will help us to appreciate the significance of mixing the gas to be burnt with the oxygen before combustion takes place, see Fig. 7. The non-luminous flame is much hotter because there is a better combustion of the coal gas. If a cold article such as a crucible is held in the luminous part of the luminous flame, particles of soot (unburnt carbon) will be deposited on it.

Experiment. *To determine the temperature of the two types of bunsen flames.*

Insert a thermo-couple, which is connected to a pyrometer, into different parts of each flame so as to determine the hottest part.

In the foregoing examples of combustion, we have used coal, coke and coal gas as the combustible material, and oxygen as the supporter of combustion. The two are interchangeable, i.e. it is possible to burn oxygen in coal gas.

Experiment. *To burn air in a coal gas atmosphere.*

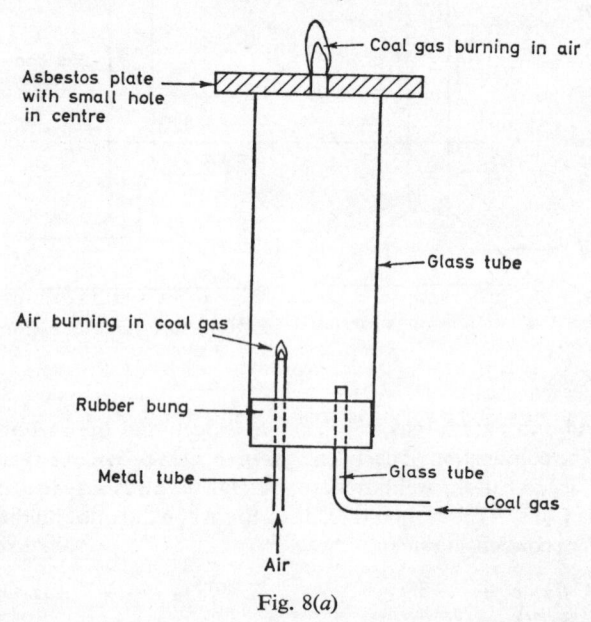

Fig. 8(a)

Arrange a glass tube with air and coal gas supplies as shown in Fig. 8(a). Turn the gas on and momentarily seal the hole in the asbestos cover whilst igniting the air issuing from the metal tube by pushing a lighted taper up the metal tube. Unseal the hole in the asbestos cover and ignite the gas issuing from it. Note now the flame from the metal tube. What are your conclusions?

CHEMICAL REACTIONS

We may be inclined to believe that only fuels and certain other materials will burn. This, of course, is not true. Metals will burn. I am sure we have all seen the scale skin on a forging. This scale is iron oxide. It is the result of a reaction in which the elements iron and oxygen have combined to form iron oxide. We might say that the iron in the outer layers of the forging has been burnt. It is worth noting that, in this particular reaction, no heat has been given off, therefore iron could not be classified as a fuel.

$$4Fe + 3O_2 \rightarrow 2Fe_2O_3$$
$$\binom{4 \text{ atoms}}{\text{iron}} \quad \binom{3 \text{ molecules}}{\text{oxygen}} \quad \text{to give} \quad \binom{2 \text{ molecules}}{\text{iron oxide}}$$

Of course this scaling can affect the dimensions of a forging considerably. The engineering technician must watch this factor carefully.

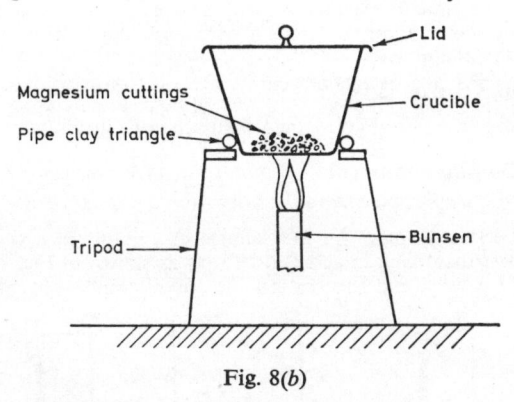

Fig. 8(b)

Experiment. *To burn the metal magnesium.*

Put about ½ g of magnesium in a crucible the mass of which has previously been determined. Set up the crucible on a tripod as shown in Fig. 8(b) and light the bunsen burner. Raise the lid, slightly, occasionally to admit oxygen to the crucible, but try to avoid smoke (fine particles of magnesium oxide) escaping from the crucible. When all the magnesium has burnt, extinguish the bunsen flame and allow the crucible, with lid in position, and contents to cool. Determine the mass of the crucible and contents.

$$2Mg + O_2 \rightarrow 2MgO$$
$$\binom{2 \text{ atoms}}{\text{magnesium}} \quad \binom{1 \text{ molecule}}{\text{oxygen}} \quad \text{to give} \quad \binom{2 \text{ molecules}}{\text{magnesium oxide}}$$

Has the mass of the crucible contents increased or decreased? Give reasons for your answer.

Corrosion—(Rust)

We know only too well, often to our dismay, what the effects of rust are, but do we know what is actually going on during the formation of the rust. Surely it would be of some benefit to us if we did, for we might be able to deter, or slow down, its formation. Corrosion is a chemical, or electro-chemical process, and most of it that we see around us is the result of oxygen in the atmosphere combining with iron, e.g. in steel, to form iron oxide. The reaction is similar to combustion since there is no difference, only in speed of action, between the process which produces scale on a hot piece of steel and that under the wing of a motor vehicle. Both scales are iron oxide and are produced by the *oxidation* of the iron in the steel. The action when zinc filings are dissolved in hydrochloric acid, as referred to on page 20, is a corrosion process. So is the chemical reaction involving the zinc case of a flash-lamp battery, when the electric current flows.

Experiment. Place a small quantity, say 1 g, of iron filings in a clock glass of known mass. Then determine the combined mass of both glass and filings. Place a small quantity of water on the filings and leave them for a week in a damp atmosphere if possible. After this time has elapsed, dry off any surplus moisture from the filings by heating slightly, over a bunsen flame, if necessary. Again determine the mass of the clock glass and its contents.

Have the filings corroded?

Has the mass of the filings changed? Explain your findings.

Experiment. *To find out whether or not iron, under atmospheric conditions, does combine with a substance from the atmosphere when it corrodes.*

Put some mild steel turnings (or iron filings) in a muslin bag, soak the bag and contents in water, then hang in a sealed bell jar as shown in Fig. 9(a). Leave the whole set-up for a week and then return for observations to be carried out. The

(a) At beginning of experiment

(b) At end of experiment

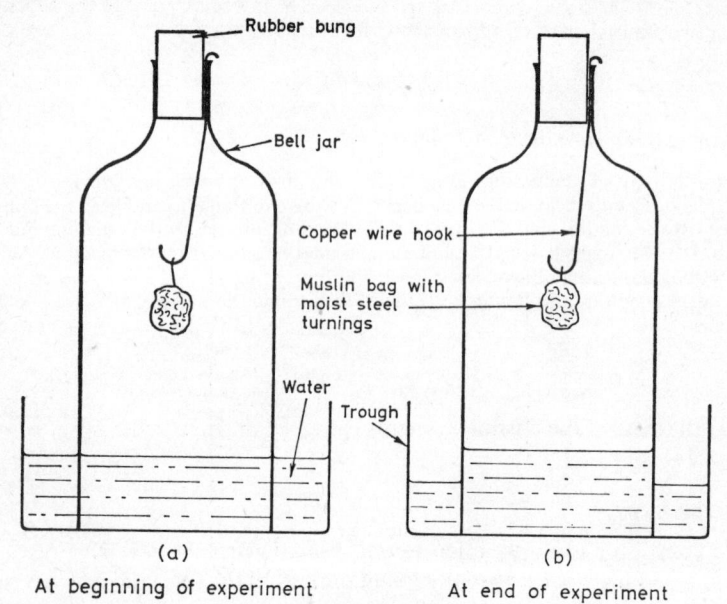

Fig. 9

CHEMICAL REACTIONS

condition after a week is shown in Fig. 9(b). On returning, add water to the trough until the level is the same as that in the jar. Next light a taper, remove the stopper from the bell jar, and gently place a lighted taper in the jar.

The taper should be extinguished.

Why did the level of water rise in the bell jar during the experiment?

Why was it advisable to adjust the level of the water in the trough to that of the level in the jar before removing the stopper?

Why was the lighted taper extinguished?

Rust is iron oxide which is formed during the process of corrosion and the last experiment indicates that some constituent of the atmosphere assists in the forming of rust.

Experiment. *To show that iron (or steel) will not rust in dry air, or water free from air.*

Boil some tap water in a beaker for about half an hour to liberate the dissolved air. Let the boiling be such that the water is well agitated so as to assist the liberation. Allow to cool. Place some mild steel turnings at the bottom of two test tubes. Fill one test tube with boiled water and seal it with a rubber bung through which passes a glass tube fitted with a rubber tube and sealing clip. The glass tube should be flush with the underside of the rubber bung. In the other tube put some cotton wool followed by calcium chloride to dry the air in the tube. Next fit a rubber bung to the tube, see Figs. 10(a) and (b) respectively. Boiled water will rise in the glass and rubber tubes, when the bung is fitted to the test tube. Arrange the clip below the water level so as to ensure exclusion of air. Leave the tubes for one or two weeks then inspect the turnings for rust.

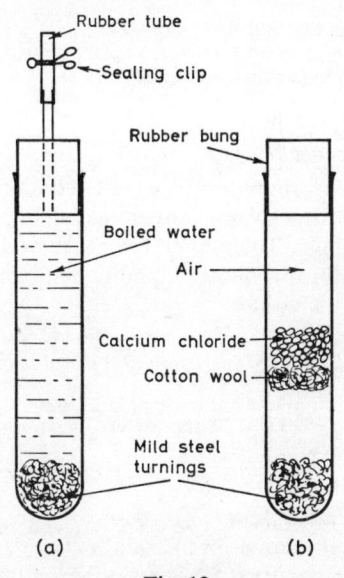

Fig. 10

Neither set of turnings should have rusted. This shows that corrosion will not take place in air, or water, alone.

To make sure that the turnings can corrode, empty both test tubes, except for the turnings, add a small quantity of tap water to the tubes and leave for a week. Note the condition of the turnings after subjection to air and water combined. What are your conclusions?

So far we have used only iron, or iron-bearing steel, to illustrate corrosion. However, other metals do corrode. For instance aluminium, copper and zinc corrode in a normal atmosphere but it is at a much slower rate than that of iron. The rate can be speeded up either in a laboratory test, or under certain atmospheric conditions. Zinc car fittings, even though coated with chromium, will corrode at a relatively rapid rate if subjected to sea-spray (brine solution). The corrosion of aluminium will be speeded up if the metal is subjected to strong alkaline (caustic soda) or certain strong acid solutions. Tin corrodes relatively quickly when exposed to acid solutions in the atmosphere, e.g. a tinplate vessel containing fruit juice. Lead is corrosion resistant to the atmosphere and certain acids. It is, however, attacked by certain other acids, e.g. hydrochloric and nitric acids. Under normal conditions copper becomes coated with green rust known as *verdigris*. This corrosive condition is accelerated by dampness, e.g. copper pipes carrying cold tap water when water from the atmosphere condenses on them.

Experiment. *To observe the rate of corrosion when a metal is subjected to different conditions.*

Take three beakers; in one put some tap water, in another some brine solution, and leave the third beaker dry. In each beaker place a strip of zinc, the strips being partly immersed in the liquid in the two wet beakers.

Which strip corroded the most?

What does this experiment show?

A metal which will corrode in atmospheric conditions is more likely to do so in an industrial atmosphere containing sulphur fumes than it would do in cleaner surroundings. The sulphur fumes are likely to become dissolved in water and thereby form a dilute sulphuric acid which would attack the metal more vigorously than water.

Experiment. *To show the rate of corrosion of a metal under different conditions*

Using three beakers, add dilute sulphuric acid to one, dilute hydrochloric acid to another and to the third beaker add tap water. Place a strip of copper in each beaker and leave for a week.

Which strip corroded most?
What does the experiment show?
Repeat the experiment using strips of lead.
How do the results compare with the previous experiment?

Experiment. *To show the corrosion rates of aluminium under different conditions.*

Make a solution by dissolving some common soda in a beaker of hot water. In another beaker put some tap water and a third beaker leave dry. Place in each a strip of aluminium and leave for a week. At the end of this period, examine the specimens and state your conclusions.

Experiment. *To observe the corrosion rates of mild steel, lead, and copper under the same conditions.*

Add to three beakers some battery acid (sulphuric acid); in one place a strip of mild steel, in the second a strip of copper and in the third a strip of lead. Leave for a week and then examine the metals.

What are your conclusions?

Repeat this experiment using the same metals together with an aluminium strip in a fourth beaker, the solution in each beaker being a common soda solution. Repeat using hydrochloric acid.

Introduce stainless steel and magnesium strips into some of the foregoing experiments and observe the effects.

We should, after these experiments, appreciate what corrosion is, how it takes place, and how different metals are affected under varying conditions. Clearly we must ensure that cutting compounds and quenching solutions do not corrode components, or the machines on which they are used, or the containers in which they are stored. The acidity of the solution is an important factor. Slight corrosion whilst not affecting the dimensions of a component can mar its appearance. Severe corrosion could reduce the size of a component and render it scrap.

Great efforts are made to prevent, or at least reduce, the rate at which corrosion takes place. Lead sheet linings are used in tanks which carry highly corrosive substances. Stainless steel and monel metal castings are used for centrifugal pump parts. Motor vehicle fittings are chromium plated. The shell of a motor car is *bonderised* (the forming of a phosphate film on the mild steel by dipping shell in a hot bath of acid phosphates of zinc and manganese). To add further to the corrosion-resisting coat, the shell is then painted. Mild steel parts for sink units, refrigerators and cookers are enamelled. Other components may be just painted. Chromium-plated components in transit are often lacquered or taped to screen them from the atmosphere, since the plate is porous. Aluminium parts are anodised (the production of an aluminium oxide skin electrolytically).

Aluminium readily oxidises, and fittings made in this metal for outdoor use, e.g. on modern buildings, are protected against corrosion by a coat of aluminium oxide. This skin is about 0·008 mm thick if produced by electrolytic process, or 0·000 016 mm if produced atmospherically.

Oxidation, reduction, carburisation processes in heat-treatment

We have learned what is meant by oxidation, reduction and carburisation. We shall now consider their relevance to heat-treatment of metals. A heat-treatment furnace may have an oxidising, reducing or neutral atmosphere; which of these it has is very important to the success or failure of a heat-treatment. The fuel in these furnaces is often coal gas but may be natural gas or oil. Whatever the fuel, the burning process is a chemical one in which hydrogen and carbon combine with oxygen to provide the heat. The oxygen supply should be adequate to burn the fuel completely. If, however, too much oxygen is supplied there will be surplus oxygen molecules floating around in the furnace—an oxidising atmosphere. This oxygen will tend to oxidise the iron and cargon in steel, forming scale on the steel and reducing the carbon content of the surface layers. Clearly this is unsatisfactory, especially if the process is annealing bright drawn bars, or similar components or material. On the other hand, if the oxygen supply is inadequate the fuel will not be fully burnt and the combustion products will contain hydrogen and carbon atoms in search of oxygen atoms. We then have a furnace atmosphere the opposite of oxidising, i.e. a *reducing*, or carburising, atmosphere. Such an atmosphere extracts oxygen from the oxides in the steel and deposits carbon, increasing the carbon content of the surface layers of steel. Scaling will be lessened, though not completely eliminated because some of the oxygen, even though in short supply, will combine with iron to form iron oxide.

Furnace atmospheres are usually made neutral by using an inert gas, i.e. one that will not react with the constituents in the steel. More will be said about this in Workshop Technology lectures; the purpose here is to point out the importance of understanding clearly the chemistry of these processes.

Exercise 3

Section A

1. Describe the chemical process known as oxidation, and give two examples of its occurrence in the workshop.
2. Water or oil could be used for the hydraulic feed mechanism on a milling machine. Why is oil preferred, and what particular property must it possess?
3. An oxy-acetylene flame can be adjusted to have an oxidising, or reducing, effect on the component it is heating. Explain the difference between the two types of flame. What disadvantages might ensue from the use of either?
4. An oil used to lubricate the bearings of a machine tool contained a small quantity of sulphuric acid. Explain what serious effect this could have on the bearing components.
5. A blow pipe flame used for brazing is long and luminous before the air is turned on. When the air is turned on the flame loses its luminosity and is apparently much hotter. Explain why this is so.
6. State three important properties that a cutting solution must have.
7. A forge fire appears to give out more heat when the air blast is applied than it does without the air blast. Is this true? Give reasons to support your answer.

CHEMICAL REACTIONS

8. In the annealing of bright drawn bars, the bars are heated in a furnace atmosphere which contains no oxygen. Explain why this is done.
9. A process known as carburising is used in the heat-treatment of some components. What is carburising?
10. The metal magnesium is alloyed with aluminium in some instances. What difficulties are likely to be encountered in the alloying process?
11. Etching is a process used for marking size details on some drills and milling cutters. Explain the chemistry of the process.
12. Salted grit is used on ice-bound roads. What serious consequences could this have on the underside of a motor vehicle?

ELECTRICITY IN THE WORKSHOP AND FACTORY

Chapter 4. Electricity

Current, voltage and resistance

Engineering technicians know only too well the value of *electricity* in helping to carry out various processes. It provides energy, through electric motors, for driving the various machine tools, e.g. lathes, milling and drilling machines, punching and shearing machines, special purpose machines, and the like. It also provides illumination to enable us to see more readily the fine graduations on the jig boring machine traverse vernier gauge. We use it to heat and ventilate the shops and offices in which we work.

What is electricity? This is a question which is not easily answered. Nevertheless, if we are to be good technicians, it is our duty to try and find out.

We have learned from page 1 that matter is made up from atoms. The chemist tells us that the atom is the smallest particle of matter that can exist chemically. The electrical engineer informs us that atoms are made up of still smaller particles known as *protons, electrons* and *neutrons*. A proton is approximately 1,850 times heavier than an electron and it carries a positive electrical charge. An electron carries a negative charge. A neutron is a combination of one proton plus one electron and is electrically neutral; it has *no electrical charge*. The positive charge of the proton is cancelled by the negative charge of the electron. Electrons orbit protons in the same way as the earth orbits the sun. The hydrogen atom has one electron and one proton, i.e. one neutron. The atoms of all other elements have more than one electron and one proton.

To help us grasp the idea of the structure of matter, let us think of it as being like the universe, e.g. of planets moving around suns in vast space, the whole being known as a *galaxy*. There are billions of these galaxies. There are billions of protons and electrons which are as infinitely small as the planets are large. The space in which these protons and electrons move is as infinite, relative to their size, as that in which the planets move.

An electric current flowing along a wire is, in effect, a stream of electrons flowing along the wire. There are billions of them taking part in the flow. This flow takes place more readily in some materials than others. Some elements, e.g. silver, copper and aluminium readily lose electrons from their atoms. Once electrons have broken away from their atoms they wander aimlessly about from one atom to another until an electric current is made to flow through the metal by connecting it to a cell or some other source of electricity supply. When the current is flowing a great number of the electrons move together in a direction opposite to that of the current. By convention (rule) we think of the current flowing from positive to negative. The electrons are flowing in the reverse direction. It is as though the convention of + and − had been wrongly interpreted originally, may be due to a misunderstanding. The larger the number of free electrons in a material, the greater the ability of that metal to conduct electricity. Here we have the reason why silver, copper and aluminium, in that order, are good *conductors* of electricity. Materials such as ebonite, rubber, glass and mica are known as non-conductors of electricity, or *insulators*, because they are short of free electrons; i.e. electrons do not break away from their atoms so readily. You see electrons break away from atoms with differing degrees of ease in different materials. Hence these materials have differing conductive capacities, so we have good and bad conductors of electricity.

Simple hydraulic circuit   Simple electric circuit

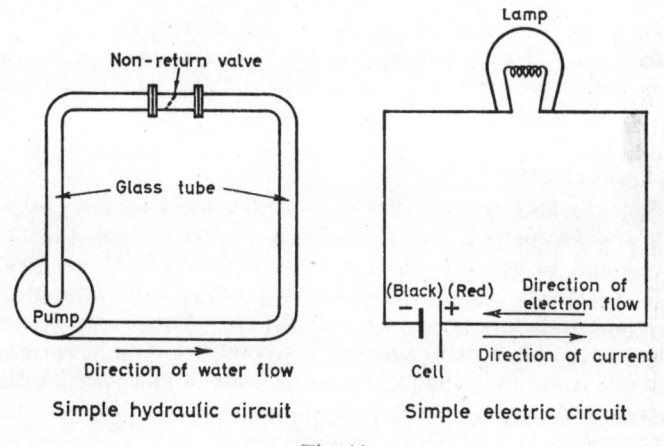

Fig. 11

In order that an electric current will flow, two conditions must be satisfied, these are:

1. There must be a complete circuit of conducting materials.
2. There must be a driving unit which will force the electrons round the circuit, see Fig. 11.

The driving unit is an electric cell (a battery is a collection of cells), or dynamo. The electric circuit can be likened to a hydraulic circuit, the driving unit in the latter instance being a pump. The water, like the electricity flows because of the driving force caused by the pump. In the case of the cell, the driving force is known as an *electro-motive force* (e.m.f.). Just as the flow of water, i.e. the water current, is resisted by the pipe, so the flow of the electricity, i.e. the electric current, is resisted by the wire. For a given speed of flow there will be more water flowing through a large diameter pipe than will flow through a smaller diameter pipe. Similarly there will be more electricity flowing through a larger diameter wire. This resistance to flow of the electric current is known as the *resistance* of the wire, or conductor. We must remember that a small diameter wire of a given material will have a greater *resistance* than a large diameter wire of the same material.

In overcoming the resistance of the pipes, the water loses energy. This results in a pressure drop between the pump outlet and a point further along the circuit. Similarly, the electricity loses energy in overcoming the resistance of the conductor. This results in a drop in the pressure behind the electric current, though in the case of electricity it is spoken of as being a drop in the *potential* of the e.m.f. Its potential for doing work has been reduced in forcing current from one point in the circuit to a point further along the circuit. This drop in potential of the e.m.f. is referred to as the *potential difference* (p.d.) In the circuit shown in Fig. 11, there will be a p.d. across the lamp because electrical energy has been used in forcing the current through the lamp. Similarly, there will be a p.d. across the circuit, i.e. a drop in electrical pressure between the positive outlet and the negative inlet to the cell, since energy has been used in overcoming the circuit resistance.

D.C. and A.C. electricity

The electricity that we have referred to so far, flows continuously in one direction and because of this is known as a *direct current* (d.c.). It is, however, found to be advantageous in some instances to use electricity which flows one way and then reverses its direction. This type of electric current is known as an *alternating current* (a.c.). The change in direction of flow may take place many times in a second, the flow being for equal time intervals in each direction. The normal rate of alternation is 50 hertz (cycle per second). One cycle is made up of a flow in each direction, i.e. in 1/50 of a second the current flows to and fro once.

Measurement of an electric current—Volt, ampere and ohm

The *electro-motive force* (e.m.f.) pushing electricity round a circuit is measured in *volt*, just as the pressure pushing water round a circuit is measured in newton per metre^2. The e.m.f. in an electric circuit, or some part of it, is measured by means of a voltmeter. The *potential difference* (p.d.) ' across ' two points (i.e. between two points) in a circuit is also measured in *volt* since the p.d. is a measurement of the e.m.f. used to push the electricity between the two points.

Instead of talking about the electron flow, when electricity flows in a circuit, we speak of the current flowing. The *current* flowing is measured in *coulomb per second*, just as the flow of water is measured in cubic metre per second. To make this easier, one coulomb per second is known as an *ampere*. We measure the *current flow* in *ampere* by means of an ammeter.

As we have seen, an electrical conductor resists the flow of electricity through it. This *resistance* is measured in *ohm*. Resistance to flow in the

ELECTRICITY

pipe is measured in newton. There is a relationship between the p.d. and the current flowing between two points in a circuit. This relationship is stated in a law known as Ohm's law.

Ohm's law

The p.d. between two points in a circuit is directly proportional to the current flowing between the two points providing the temperature of the circuit remains constant.

Stated mathematically, we have:

$$\text{p.d.} \propto \text{Current.}$$
$$\text{(varies as)}$$

Writing this as an equation by introducing a constant,

$$\text{p.d.} = \text{Current} \times \text{Constant}$$

$$\frac{\text{p.d.}}{\text{Current}} = \text{Constant.}$$

The constant is representative of the *resistance* of the circuit between the two points, so we have the Ohm's law formula,

$$\text{Resistance} = \frac{\text{p.d.}}{\text{Current}}$$

or,

$$\text{Current} = \frac{\text{p.d.}}{\text{Resistance}}, \quad \left[\text{Amp} = \frac{\text{Volt}}{\text{Ohm}}\right]$$

Experiment. *To investigate the truth of the Ohm's law equation.*

Arrange a resistor of known value, say 2 ohm, in a circuit with a switch, ammeter and a cell of say 2 volt. Connect a voltmeter across the resistor, see Fig. 12.

Make sure the switch is open during the connecting of the wires. Also make sure the positive terminal of the ammeter is connected to the positive terminal of the cell. Close the switch and note the readings of the ammeter and the voltmeter. Using these readings calculate the resistance of the resistor by means of the Ohm's law formula and check the result with the known value. Remember the ammeter gives the current flowing through the resistor, whilst the voltmeter records the p.d. (drop in e.m.f.) across the resistor. Repeat the experiment using two cells, then three cells, also repeat using different resistors.

Example. A current of 12 amp flows through an electrical conductor and the p.d. across the two ends of the conductor is 3·5 volt. What is the resistance of the conductor?

Using the Ohm's law formula,

$$\text{Resistance} = \frac{\text{p.d.}}{\text{Current}}$$

$$\text{Resistance} = \frac{3 \cdot 5}{12} = \underline{0 \cdot 292 \text{ ohm.}}$$

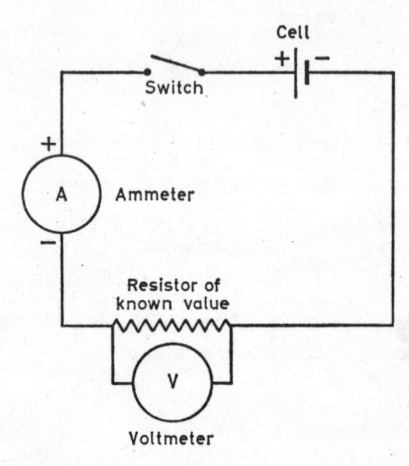

Fig. 12

Energy in electrical form: power

An electric cell is a source of energy, and as such it is capable of doing work. The cell in the circuit shown in Fig. 11 will expend energy driving the current round the circuit and lighting the lamp. Workshop examples of the use of energy in electrical form are the electric motor, the electric soldering iron and heater. The method used for calculating energy in mechanical form, or work, see page 135, can be applied to the calculation of work done by energy in electrical form. There will be no work done if there is no current flowing. The work done by an electric current is given by,

$$\underset{\text{current}}{\text{Work done by}} \overset{\text{(joule)}}{=} \text{e.m.f.} \overset{\text{(volt)}}{\times} \underset{\text{flowing per second}}{\text{Quantity of current}} \overset{\text{(ampere)}}{\times} \text{Time} \overset{\text{(second)}}{}$$

or,

$$\underset{\text{expended}}{\text{'Electrical' energy}} = \text{e.m.f.} \times \underset{\text{flowing per second}}{\text{Quantity of current}} \times \text{Time}$$

If this is related to the work done by a current flowing between two points in a circuit, then the 'electrical' energy is given by:

$$\underset{\text{expended}}{\text{Energy}}\text{ (joule)} = \text{p.d. (volt)} \times \underset{\text{per second}}{\text{Current flowing}}\text{ (amp)} \times \text{Time (second)}$$

The power used to do this work (remember that power is the rate of doing work, see page 141), is the 'electrical' energy expended per unit of time.

$$\text{Power (joule/s)} = \text{e.m.f. (volt)} \times \text{Current flowing per second (amp)}$$

To obtain this expression we have divided the 'electrical' energy equation by time in order to obtain the energy used per second. One joule per second is called a watt.

Hence:

$$\text{Power (watt)} = \text{e.m.f. (volt)} \times \text{Current flowing (amp)}$$

If the problem is to determine the power used between two points in a circuit, the equation becomes:

$$\text{Power (watt)} = \text{p.d. (volt)} \times \text{Current flowing (amp)}$$

i.e. $\qquad\text{Watt} = \text{Volt} \times \text{Amp}$

Power may be measured in kilowatt,

$$1 \text{ kilowatt} = 1\,000 \text{ watt.}$$

Example. It is found that in a particular turning operation the electric motor driving the lathe is using 6·5 amp of current. If the motor is connected to a 250 volt d.c. supply, determine the 'electrical' power being used.

$$\text{Power} = \text{e.m.f.} \times \text{Current flowing per second}$$

$$\underset{\text{used by motor}}{\text{Power}} = \frac{250 \times 6\cdot5}{1\,000}$$

$$= \underline{1\cdot625 \text{ kW.}}$$

Note: The amount of current used by a motor will be governed by the demand on the motor by whatever it is driving. There is, of course, a limit governed by the maximum power available at the particular voltage. Had a deeper cut been used, at the same feed, in the above example, more energy would have been required at the tool point. Consequently a greater amount of energy would have been needed at the motor, with the result that it would have taken more current.

38 ELECTRICITY IN THE WORKSHOP

Experiment. *To measure the current flowing in a circuit when a number of lamps are connected,* (i) *in series,* (ii) *in parallel.*

Connect one 12-volt car lamp in series with an ammeter, switch and 12-volt battery. Make sure the switch is open whilst connecting the wires. Also ensure that the positive terminal on the ammeter is connected to the positive terminal on the battery. Close the switch and note the reading on the ammeter. Open the switch and add a second lamp to the circuit, it being connected in series to the first. Close the switch and note the ammeter reading. Repeat the procedure with three, four etc. lamps connected in series, see Fig. 13(a).

A series circuit
(a)

A parallel circuit
(b)

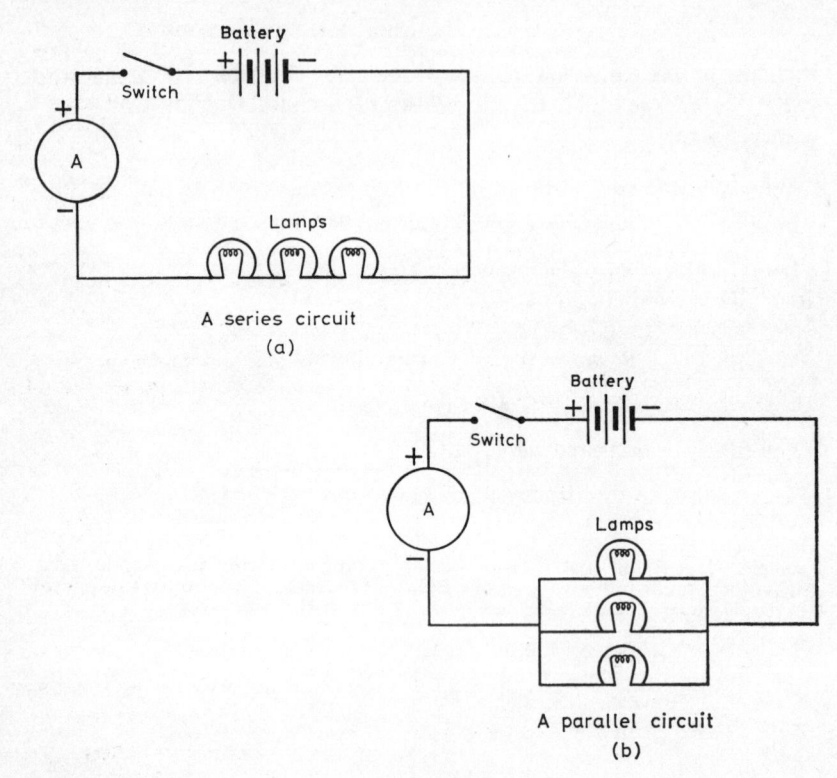

Fig. 13

Note: A *series circuit* is a circuit in which the components are connected one following the other.

It will be observed that as the number of lamps in the circuit is increased, the current flowing in the circuit will decrease, so will the brightness of the bulbs. This is due to the fact that when resistances (each lamp has a specific resistance) are connected in series, the total resistance of the circuit is increased.

Connect the lamps in parallel, beginning with two, then increase the

ELECTRICITY

number to three, four, etc., see Fig. 13(*b*). Close the switch and note the ammeter reading after each lamp is connected into the circuit.

With this type of circuit, it will be observed that the current taken by the lamps increases as the number of lamps connected in the circuit is increased and the light output of the lamps remains constant.

Note: A *parallel circuit* is one in which the components are connected in the circuit in such a way that they are parallel to each other.

Experiment. *To measure the current flowing in a circuit under differing conditions.*

Connect an ammeter, switch and 12-volt battery in series with a 12-volt car headlight lamp. Close the switch and observe the ammeter reading. Open the switch, replace the lamp by a trafficator coil and observe the quantity of current taken by this component. Repeat the procedure using a windscreen wiper motor followed by a starter motor in place of the lamp.

What do you notice about the ammeter readings; were they all alike?

Incidentally, with a suitable rig you could load the starter motor to varying degrees and observe the ammeter reading at each load.

Repeat the experiment using coils of differing resistance values, in turn.

Experiment. *To measure the p.d. across a battery, a coil and a circuit.*

Connect a battery, switch and coil in series, then connect a voltmeter across the battery terminals, one across the coil, and another across the circuit, see Fig. 14. Close the switch and observe the voltmeter readings.

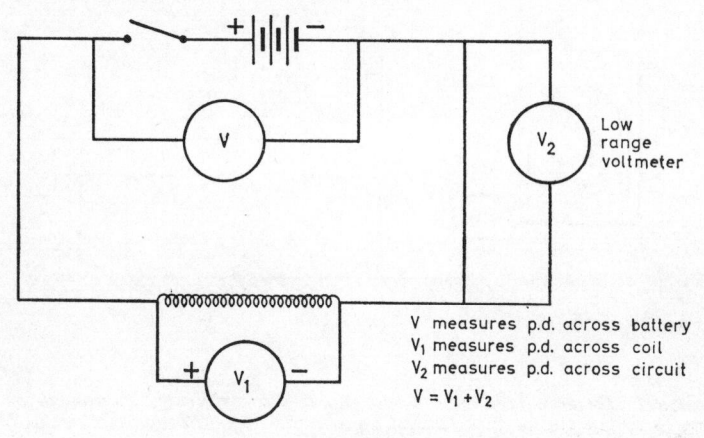

V measures p.d. across battery
V_1 measures p.d. across coil
V_2 measures p.d. across circuit
$V = V_1 + V_2$

Fig. 14

The voltmeter connected across the battery terminals records the e.m.f. of the battery, i.e. the voltage available to the circuit. Its reading will be less than 12 volt because e.m.f. is used to overcome the internal resistance of the battery. The reading of the voltmeter connected across the coil gives the voltage (e.m.f.) used to push the current through the coil, i.e. to overcome the coil resistance. The e.m.f. used in overcoming the circuit resistance is recorded by the voltmeter V_2.

Example. An electric motor driving a small shaping machine is wired to a 230-volt d.c. supply. During a certain cutting operation an ammeter shows that it is using 13·5 amp. How many kilowatts of power are being supplied to the motor?

$$\underset{\text{(watt)}}{\text{Power}} = \underset{\text{(volt)}}{\text{e.m.f.}} \times \underset{\text{(amp)}}{\text{Current flowing}}$$

Power used by shaping machine $= (230 \times 13\cdot5)$ watt

but 1 000 watt $= 1$ kilowatt

$$\text{Power used by shaping machine} = \frac{230 \times 13\cdot5}{1\ 000} = \underline{3\cdot1 \text{ kW.}}$$

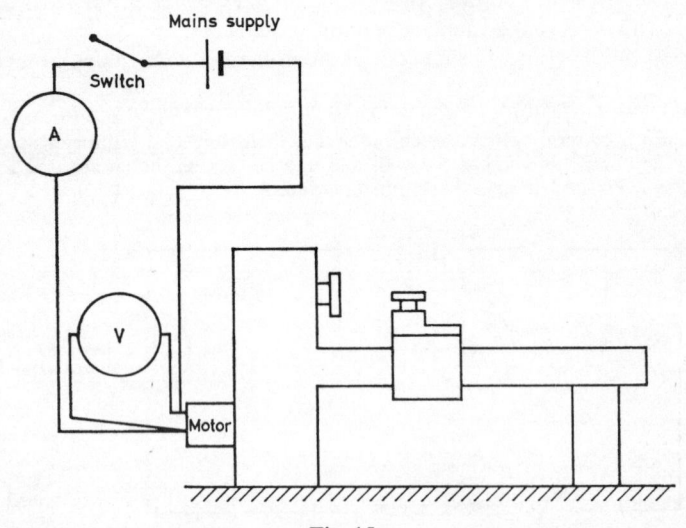

Fig. 15

Experiment. *To show how the current supplied to an electric motor varies with the variation in depth of cut on a centre lathe.*

Arrange an ammeter in series with a switch and the electric motor driving a centre lathe. Then connect a voltmeter across the terminals of the motor, see Fig. 15. Set a job in the lathe, e.g. a piece of round mild steel bar in 3-jaw chuck. Close the switch and note the voltmeter and ammeter readings whilst the lathe spindle is stationary. Open the switch. Apply a 0·50 mm cut and a 0·30 mm per rev. feed, close the switch and begin cutting the bar. Note the meter readings.

ELECTRICITY

Repeat this procedure using cuts of 1·00, 1·50, 2·00, 2·50 and 3·00 mm, leaving the feed rate constant at 0.30 mm. Tabulate the results.

Depth of Cut (mm)	Voltmeter Reading (volt)	Ammeter Reading (amp)

Plot a graph of *ammeter reading* (vertical axis) against *depth of cut* (horizontal axis).

How does the current taken by the motor vary with the depth of the cut?

Does the power taken by the motor increase or decrease as the depth of the cut increases? Explain your answer.

What practical value is a test of this type; i.e. what knowledge does the technician gain?

Repeat the experiment keeping the depth of cut constant, and varying the feed rate.

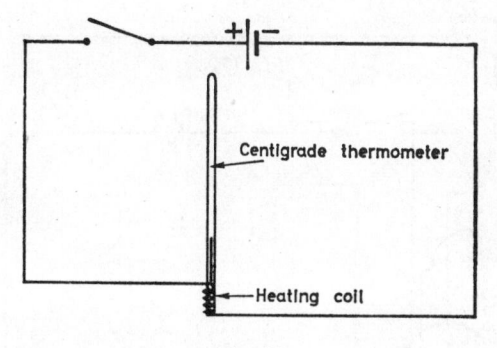

Fig. 16

Experiment. *To show how the heating effect of electricity through a coil varies with voltage supplied to the coil.*

Arrange a 1-volt cell, switch and coil in a series circuit. The coil should be made from light gauge copper wire wound round the bulb of a 0°–400° Celsius thermometer, then bound with asbestos tape, see Fig. 16. Close the switch and watch the mercury in the thermometer rise; record its highest value. Open the switch and add a further 1-volt cell to the circuit. Close the switch and again record the highest temperature reached by the mercury column. Repeat the procedure for 3, 4, 5 and 6-volt supplies. Plot a graph of voltage supply (horizontal axis) against coil temperature (vertical axis).

Can you state one example in which the heating coil principle is used in industry?

Experiment. *To show how the current flowing in a circuit varies with the resistance of the circuit.*

Arrange a 12-volt battery, switch, ammeter and a 2 ohm resistor in series as shown in Fig. 17.

Close the switch and note the ammeter reading. Open the switch and replace the 2 ohm resistor by one of 3 ohm. Again close the switch and note the ammeter reading. Repeat the procedure using 4, 5, 6 and 7 ohm resistors. Tabulate the results and plot a graph of resistor resistance (horizontal axis) against ammeter reading (vertical axis).

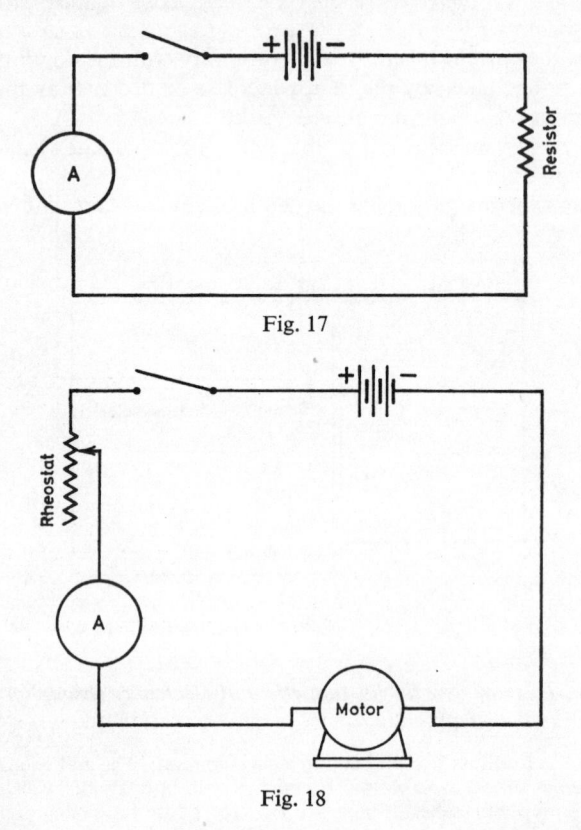

Fig. 17

Fig. 18

Experiment. *To find out how the current flowing in a circuit affects the speed of an electric motor.*

Arrange a 12-volt battery, switch, rheostat, ammeter and a windscreen wiper motor in series as shown in Fig. 18 (A *rheostat* is a variable resistor, i.e. by adjusting the rheostat the current supply to the motor can be reduced, or increased).

Adjust the rheostat gradually, stopping at about six selected intervals to measure motor speed and record ammeter reading. Tabulate the results, and plot a graph of ammeter reading (horizontal axis) against motor speed (vertical axis).

The principle involved in this experiment has industrial applications. Can you name two of these?

Experiment. *To find out how the heat in a heating coil varies when the amount of electrical current flowing is varied.*

Arrange a 12-volt battery, switch, rheostat, ammeter and heating coil in a series circuit as shown in Fig. 19. Use a 0°–400° Celsius thermometer and wind the

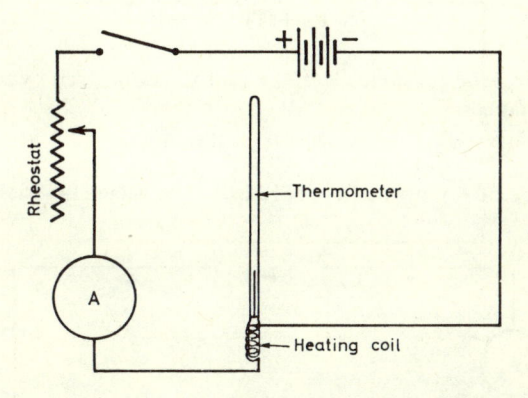

Fig. 19

coil round its bulb, then wrap with asbestos tape. Commence with the rheostat adjusted so as to give maximum resistance. Close the switch and observe the mercury column in the thermometer rise; record its highest level, also the ammeter reading. Adjust the rheostat gradually, stopping at about six selected points to record the temperature of the coil and the ammeter reading at that temperature. Tabulate the results and plot a graph of current (horizontal axis) and coil temperature (vertical axis).

We have seen from the previous experimental work that there are two types of circuits. Now we will look at these more closely and do some calculations on them.

Calculations on series circuits

A resistor resists the flow of electricity; all electrical equipment and wiring are resistors to some degree. In this sense we will consider four resistors (it could be any number) *connected in series*, see Fig. 20. The p.d. across each resistor is given by V_1 V_2 V_3 and V_4, whilst the resistance of each resistor is R_1 R_2 R_3 and R_4.

The p.d. across the four resistors $= V_1 + V_2 + V_3 + V_4 = V$

The current flowing through each resistor is I ampere, i.e. reading of each ammeter would be the same.

We already know that voltage, current and resistance are connected by the Ohm's law equation; hence we have:

$$\text{Voltage} = \text{Current} \times \text{Resistance}$$

therefore $V_1 = I \times R_1$, $V_2 = I \times R_2$, $V_3 = I \times R_3$ and $V_4 = I \times R_4$

p.d. across the four resistors $= I.R_1 + I.R_2 + I.R_3 + I.R_4$

or
$$V = I(R_1 + R_2 + R_3 + R_4)$$

$$\frac{V}{I} = R_1 + R_2 + R_3 + R_4$$

So we have the total resistance R of the four resistors given by the sum of the separate resistances.

$$R = R_1 + R_2 + R_3 + R_4$$

This **holds** good for any number of resistors connected in series.

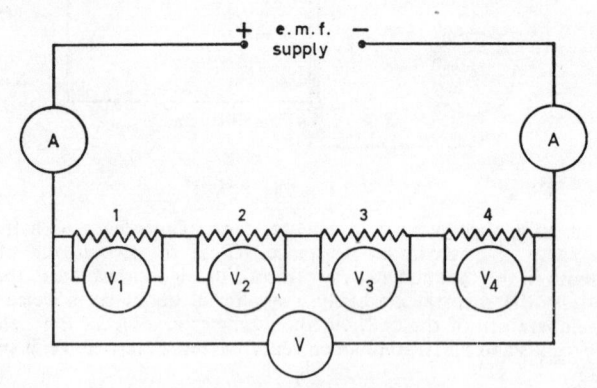

Resistors arranged in series

Fig. 20

Examples

1. Three resistors connected in series have resistance values of 6, 5 and 2 ohm respectively. If the current flowing in the circuit is 21 ampere, find the voltage drop across each resistor and the voltage drop across the three resistors.

From Ohm's law we have:

$$\text{p.d.} = \text{Current} \times \text{Resistance}$$

ELECTRICITY

Hence, since voltage drop is the same as p.d. we have,

   p.d. across 6 ohm resistor = 21 × 6   = 126 volt

   p.d. across 5 ohm resistor = 21 × 5   = 105 volt

   p.d. across 2 ohm resistor = 21 × 2   = 42 volt

   p.d. across three resistors = 126 + 105 + 42 = 273 volt

or p.d. across three resistors = 21(6 + 5 + 2) = 273 volt

2. Two resistors connected in series have a p.d. across them of 11·5 volt and 6 volt respectively when a current of 3 ampere flows in the circuit. Calculate the resistance values of the resistors and their combined resistance.

From Ohm's law we have:

$$\text{Resistance} = \frac{\text{p.d.}}{\text{Current}}$$

Resistance of first resistor $= \dfrac{11 \cdot 5}{3} = 3 \cdot 8$ ohm

Resistance of second resistor $= \dfrac{6}{3} = 2$ ohm

Combined resistance $= 3 \cdot 8 + 2 = 5 \cdot 8$ ohm

e.g. voltage drop across combined resistance $= 3 \times 5 \cdot 8 = \underline{17 \cdot 4 \text{ volt}}$

Calculations on parallel circuits

We will now consider four (it could be any number) resistors *connected in parallel*, see Fig. 21. In the parallel circuit, the current does not pass through all four resistors in turn but divides and passes through the four at the same time, the quantity going through each being dependent on the resistance values of the separate resistors. It is similar to arranging four pipes in parallel, the pipes being of different bore diameters and therefore capable of passing different quantities of water; the total quantity entering being equivalent to the total quantity leaving. The total current flowing in the circuit equals the sum of the currents flowing through each resistor. The p.d. across the resistors is the same across each individual resistor.

Therefore, using the Ohm's law formula to express the current flowing through each resistor, we have:

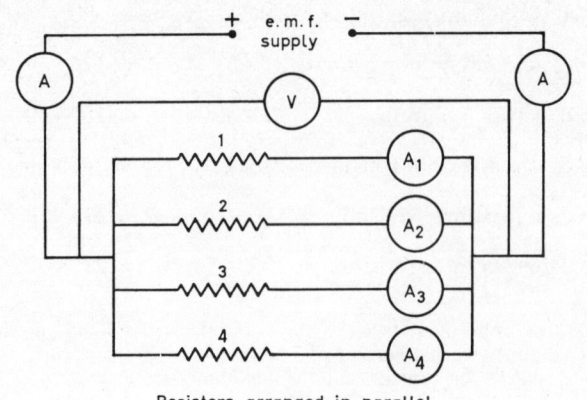

Resistors arranged in parallel

Fig. 21

Current flowing through Resistor 1 = $\dfrac{\text{Potential Difference (Pressure Drop)}}{\text{Resistance of Resistor 1}}$

Current flowing through Resistor 2 = $\dfrac{\text{p.d.}}{\text{Resistance 2}}$

Current flowing through Resistor 3 = $\dfrac{\text{p.d.}}{\text{Resistance 3}}$

Current flowing through Resistor 4 = $\dfrac{\text{p.d.}}{\text{Resistance 4}}$

Total current flowing through all resistors:

$$I = I_1 + I_2 + I_3 + I_4$$

Where I_1 I_2 I_3 and I_4 are the readings of the ammeters A_1 A_2 A_3 and A_4.

Suppose the total resistance of the four parallel resistors be R, and the p.d. across them V, then for the four parallel resistors we have:

$$I = \dfrac{V}{R}$$

or

$$\dfrac{V}{R} = I_1 + I_2 + I_3 + I_4$$

Expressing the current flowing through each resistor in terms of voltage and resistance,

$$\dfrac{V}{R} = \dfrac{V}{R_1} + \dfrac{V}{R_2} + \dfrac{V}{R_3} + \dfrac{V}{R_4}$$

ELECTRICITY

Dividing both sides of this equation by V we have,

$$\frac{1}{R} = \frac{1}{R_1} + \frac{1}{R_2} + \frac{1}{R_3} + \frac{1}{R_4}$$

To show the significance of this relationship and to emphasise the difference between the arrangement of resistors in *parallel* and those in *series* we will assume that

$$R_1 = 2, \ R_2 = 3, \ R_3 = 1 \text{ and } R_4 = 5 \text{ ohm}$$

The total resistance, R, of the arrangement is as follows:

$$\frac{1}{R} = \frac{1}{2} + \frac{1}{3} + \frac{1}{1} + \frac{1}{5}$$

$$\frac{1}{R} = \frac{15 + 10 + 30 + 6}{30}$$

$$\frac{1}{R} = \frac{61}{30}$$

$$R = 0{\cdot}5 \text{ ohm}$$

From this we see that the total resistance is approximately 0·5 ohm, which is much less than the resistance of any one of the individual resistors. If connected in series the total resistance would be:

$$\text{Total resistance} = 2 + 3 + 1 + 5 = 11 \text{ ohm}$$

Carry out experiments on series and parallel circuits to check the foregoing theory.

Examples

1. Three resistors having resistances of 10, 20 and 50 ohm respectively are arranged in parallel and connected to a 250-volt supply. Find the current flowing through each resistor and the total current flowing in the circuit.

Using a sketch similar to the one in Fig. 21, except that we have three resistors instead of four, and assuming the total resistance in the circuit to be R:

$$\text{Current flowing through 1st resistor} = \frac{250}{10} = 25 \text{ amp}$$

$$\text{Current flowing through 2nd resistor} = \frac{250}{20} = 12{\cdot}5 \text{ amp}$$

$$\text{Current flowing through 3rd resistor} = \frac{250}{50} = 5 \text{ amp}$$

Total current flowing through resistors = $25 + 12{\cdot}5 + 5 = 42{\cdot}5$ amp

Alternatively we may find the total current in the circuit as follows, beginning by finding the total resistance R of the resistors.

$$\frac{1}{R} = \frac{1}{10} + \frac{1}{20} + \frac{1}{50}$$

$$\frac{1}{R} = \frac{10+5+2}{100}$$

$$\frac{1}{R} = \frac{17}{100}$$

$$R = \frac{100}{17} \text{ ohm}$$

Total current flowing through resistors $= \frac{V}{R} = 250 \div \frac{100}{17} = 250 \times \frac{17}{100}$

Total current flowing through resistors = <u>42·5 amp</u>

2. Four coils having resistances of 5, 7, 12 and 17 ohm, are first connected in series, then in parallel. The supply voltage to the circuit in each case is 230 volt. Calculate the total current flowing in each case, to three significant figures.

Series connection

Total resistance of coils, R $= 5+7+12+17 = 41$ ohm

Current flowing through coils $= \dfrac{230}{41}$ $= \underline{5\cdot 6 \text{ amp.}}$

Parallel connection

Total resistance of coils is given by:

$$\frac{1}{R} = \frac{1}{5} + \frac{1}{7} + \frac{1}{12} + \frac{1}{17}$$

$$\frac{1}{R} = \frac{1428 + 1020 + 595 + 420}{7140}$$

$$\frac{1}{R} = \frac{3463}{7140}$$

$$R = \frac{7140}{3463} \text{ ohm (2 ohm approx.)}$$

Current flow through coils $= 230 \div \dfrac{7140}{3463} = 230 \times \dfrac{3463}{7140}$

$= \underline{112 \text{ amp.}}$

ELECTRICITY

Clearly there is quite a difference in total resistance of the two circuits, 41 ohm against 2 ohm, and consequently in the current flowing, 5·6 amp against 112 amp.

Exercise 4

Section A
1. Give four examples from the workshop of processes, or tools in which electricity is now used but which originally were manual operations.
2. What is the advantage of a non-conductor of electricity? Give examples of the use of such a component.
3. Give a list of four conductors of electricity putting them in order of the best to the worst.
4. Assuming that the supply voltage to an electric motor driving a lathe is constant, state which of the operations will use most current: a finishing operation, or a roughing operation. Give reasons for your answer.
5. It is required to replace a hand driven fan on a forge fire by a motorised fan which must be capable of giving a variable air supply. Explain how this could be done and make a sketch of the motor circuit.
6. An ammeter was put into the circuit of the motor driving a radial arm drilling machine. It was noticed that the meter gave a higher reading when a 25 mm diameter hole was being drilled than it did for a 10 mm diameter hole. Explain why this should be so if the feed rate and speed were the same in each case.
7. A milling machine motor was wired to a 250-volt d.c. supply. During a certain operation the motor was using 17·25 ampere. How many kilowatt of power were being used?
8. It is required to dim the light in an inspection room so as to facilitate the reading of shadow graphs of thread profiles. Using a sketch of a suitable circuit, explain how this could be done.
9. A two-heat electric heater uses ¾ kilowatt of power when the switch is in the 'low' position and 2 kilowatt when in the 'high' position. How much current is used in each case when the heater is connected to a 230-volt d.c. supply?
10. Calculate how many ampere of electric current are used when the following lamps are connected separately into a 250-volt circuit; 60 watt, 150 watt, 200 watt.

Section B
1. The voltage drop across a 2·5 ohm resistor is found to be 3·25 volt. What is the value of the current flowing?
2. A fully charged 6-volt battery was estimated to have an internal resistance of 0·075 ohm. What max current is it capable of giving?
3. An infra-red lamp was connected in series with a 12-volt battery and an ammeter. The ammeter reading was 2·72 ampere when the current was flowing. Neglect the resistance of the battery and find the resistance of the lamp.
4. Two coils having resistances of 3 ohm and 1 ohm respectively are connected in series to a d.c. supply. The current flowing in the circuit is 20 ampere. What is the p.d. across each coil?
5. A lamp and heating coil of resistance 6 ohm and 15 ohm respectively are connected in parallel to a 50-volt d.c. supply. Calculate the total resistance of the two resistors and the total current flowing in the circuit.
6. A heating coil, resistance 25 ohm and a lamp, resistance 85 ohm, are connected in series to a 250-volt d.c. supply. How many kilowatt of power, to three significant figures, do they use?

7. If a 25-ohm heating coil and a 85-ohm resistor were connected in parallel to a 250-volt d.c. supply, what power in kilowatt would be used? Give your answer to three significant figures.
8. Four resistors of 2, 7, 9 and 4 ohm are connected first in series, then in parallel to a 24-volt d.c. supply. How much electrical power in watt would be used in each case? Give your answer to three significant figures.
9. Explain why the lamps in a 6-volt car circuit were burnt out when a 12-volt battery was connected into the circuit in error. What would happen if two 6-volt batteries were connected in series in the same circuit? If the same two batteries were connected in parallel in the same circuit, what would happen?
10. Two resistors of 10 ohm and 15 ohm respectively are connected in parallel and then connected in series with a motor of resistance 4 ohm. The d.c. supply voltage to the circuit is 220 volt. Calculate the current taken by the resistors and that taken by the motor.

CHAPTER 5. MAGNETISM

In this chapter we shall discuss magnetism, magnetic materials, magnetic effect of an electric current, permanent and electro-magnets and magnetic workshop tools.

Magnetism

We all know what a magnet is. We will also have done the experiment in which either a bar or horse-shoe magnet is lowered into a heap of iron filings, some of which are seen to have become attached to the magnet when it is withdrawn from the filings. We may then have tried to do likewise with the magnet placed into brass or copper, or indeed any non-ferrous metal filings. The non-ferrous filings would not become attached to the magnet. From such an experiment we conclude that there is some attraction between the bar of iron or steel known as a magnet and the iron or steel filings. This attraction is the result of a magnetic force, commonly referred to as magnetism. The bar has to be given this magnetism either electrically or by stroking it (in one direction only) on a bar which is already magnetised. In earlier days some of the first magnets to be produced were made by the blacksmith. The smith would hammer a red-hot bar of iron on the anvil whilst it cooled, the bar lying in a north-south direction in the earth's magnetic field. The magnetism was induced into the bar from the earth's field.

It is a fact that when heated beyond a certain temperature iron and steel cease to be attracted by a magnet, i.e. they become non-magnetic. This occurs at the upper critical point in steel above 0·55% carbon content (see Workshop Technology notes) at a temperature varying according to the composition of the steel, from about 770°C upwards. Nickel ceases to be magnetic at 320°C, whilst cobalt is non-magnetic above 1,120°C and pure iron becomes non-magnetic above 770°C. The magnetism is regained in all cases on the cooling of the metal. It is interesting to note this loss of magnetism is made use of in the hardening of steel tool bits. These are attached to a soft iron bar which is magnetised electrically and used to hold the bits whilst they are heated. Upon reaching the appropriate temperature, they fall off the magnet into a quenching bath below.

Experiment. Suspend vertically, some short pieces of high carbon steel wire from a bar magnet, then apply a bunsen flame to the wire. Observe the temperature, by colour unless a thermo-couple is available at which wire falls from magnet. Try pieces of fallen wire for magnetism with a second magnet. Repeat the experiment using nickel wire. Repeat the experiment using wrought iron then mild steel wire.

Did the various wires all fall off the magnet at the same temperature?

Magnetic materials

Not all materials are magnetic. In fact the only materials that are magnetic are metals and not all of these react positively to a magnet. It appears to be generally understood that ferrous metals are magnetic, whilst non-ferrous metals are non-magnetic. This is not strictly true, though apparently it is a useful guide where the common engineering metals are concerned. We already know that iron and steel are strongly magnetic. In fact there are no other metals so readily attracted by a magnet. Nickel and cobalt, two metals used in tool steel alloys, are slightly attracted. Bismuth and antimony, two metals used in bearing alloys, are repelled, whilst the metals copper, tin, lead, magnesium and aluminium do not appear to react either way. However, it is found that an alloy of copper, manganese and aluminium is strongly magnetic. Since there is more iron and steel used in the engineering industry than any other metals, and these are most responsive to magnetism, we shall concern ourselves chiefly with these metals.

Magnetism is useful in the workshop and in the engineering industry generally. Apart from its uses in electric motors and generators, which we shall deal with later, see page 67, it has uses in magnetic chucks for holding work on grinding machines, magnetic bases of some scribing blocks (surface gauges), solenoids for brake mechanisms, electric magnets on cranes in scrap yards and steel works for loading and unloading scrap metal, billets, sheets and so on.

All magnets can be divided into two classes:

1. Permanent magnets.
2. Electro-magnets.

We will deal with these in turn.

Permanent magnets

A compass needle is a permanent magnet, pivoted so that it will rotate; so is a bar magnet. The earth is a permanent magnet, and like all other magnets it has what are known as *magnetic poles*. These are diametrically opposite and referred to as the magnetic north pole and the magnetic south pole. The straight line joining them is known as the *magnetic axis*. A compass needle gives us the direction of the earth's magnetic axis; it is slightly inclined to the geographic axis. In the case of the bar magnet the poles are situated near the ends of the bar. They are situated near the ends of the limbs of a horse-shoe magnet. A permanent magnet is one which retains its magnetism over a long period of time without having external connections. We will attempt to find out more about these magnets by experimenting.

MAGNETISM

Experiment. *To observe the behaviour of a magnet under certain conditions.*

Support a bar magnet in a balanced position, by a piece of fine thread, from the arm of a retort stand as shown in Fig. 22. Take hold of another bar magnet and place the N-pole end of it near to the N-pole end of the suspended one. Repeat this procedure using the S-pole ends of the magnets. Next bring the S pole of the magnet in your hand near to the N pole of the suspended magnet.

What have you learned about the behaviour of the like poles?
What have you learned about the behaviour of the unlike poles?

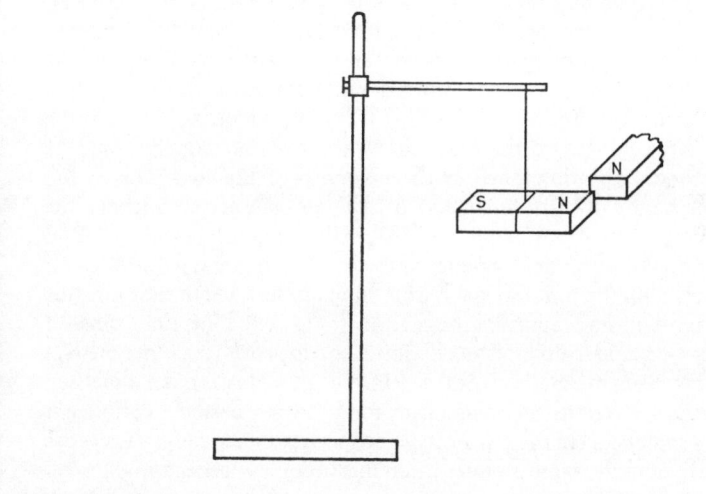

Fig. 22

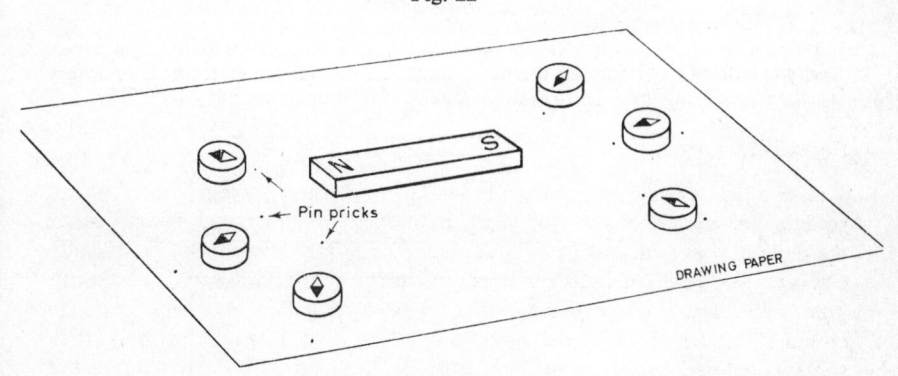

Fig. 23

Experiment. *To locate the poles of a bar magnet.*

Place a bar magnet on a piece of drawing paper and carefully draw round it with a well-sharpened pencil. Take a small compass and place it near one of the poles of the magnet as shown in Fig. 23. With a pin mark the direction in which the compass needle is pointing by pricking the drawing paper at both ends of the needle. Repeat the procedure with the compass in at least three different positions around each pole.

Sketch lines through each pair of pin-pricks for every compass position. These lines will intersect at a common point at each end of the magnet profile. The intersections are the locations of the two poles. A straight line drawn through the pole points gives the magnetic axis of the magnet.

Stroke the pin used in this experiment a number of times along the bar magnet, in line with its magnetic axis, in a N to S direction, or from S to N, whichever direction you choose must be kept to. Now try to pick up the pin first from one end, then from the opposite end using the same end of the magnet each time. Obtain a pin which has not been stroked on a magnet and try to pick this up with the magnet, first from one end and then from the other, using only one end of the magnet. What are your conclusions? Is there, for instance, any difference in the response of the two pins to the magnet? You might use the other end of the bar magnet and repeat the procedure taking careful notice on the way in which the pins react. You have in fact created a permanent magnet in the pin which you have stroked on the bar magnet. Dip the pin into some iron filings to test whether or not this is true. The pin will not retain its magnetism for a very long time, since it will very likely be made from mild steel. Soft steel does not retain magnetism long though it is easily magnetised. Hard steel, e.g. a tool bit, is not magnetised as readily but it will retain the magnetism for a longer period. Permanent magnets are usually given their magnetic properties electrically. They are usually neutralised in the same manner when this becomes necessary.

Experiment. *To investigate the condition of the space surrounding a magnet.*

Place a bar magnet on a bench and place over it a piece of cellophane paper. Sprinkle iron filings thinly on the paper and note the pattern formed. If necessary tap the paper gently in order to even the spread of the filings, see Fig. 24.

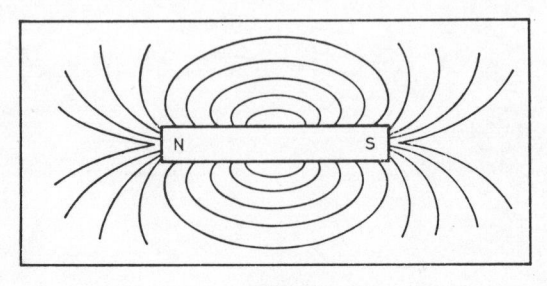

Fig. 24

The iron filings indicate that there is an unseen area of magnetic influence around the magnet. This is known as a *magnetic field*. The filings can be seen

MAGNETISM

to be formed into lines linking the pole areas. These are known as *magnetic lines of force*. There is a magnetic field around the earth.

Care of Permanent Magnets. You will have noticed that the bar magnets you were given to use in these experiments were brought from the store in pairs and contained in a wooden box. If you were sufficiently observant to inspect them whilst they were in the box, you would see that the N pole of one magnet and the S pole of the other were at the same end of the box, with a pole piece connecting them at each end. The pole pieces are made of soft iron. Sometimes the pole piece is referred to as a *keeper*. Horse-shoe magnets are stored singly since they have a pair of unlike poles side by side. These again are connected by a keeper during storage, see Fig. 25. The

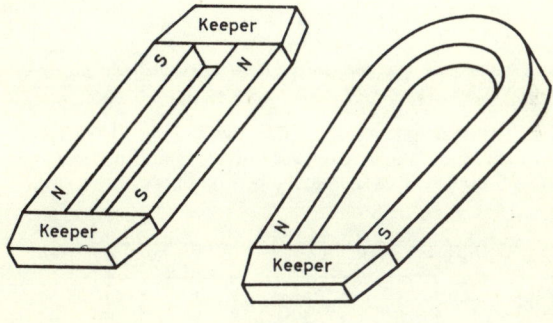

Fig. 25

keepers enable the *lines of force* to be kept in a closed circuit. This preserves the magnetism and ensures a longer life for the magnet. If magnets are not properly stored, they soon lose their magnetism.

Experiment. *To investigate the magnetic fields around two pairs of bar magnets when they are arranged in two different ways.*

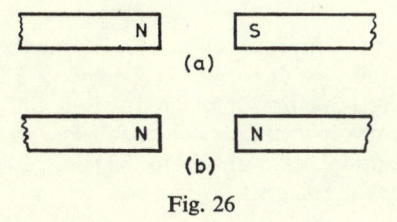

Fig. 26

Arrange one pair of magnets with their magnetic axes in line and a gap of about 20 mm between their ends which have unlike poles. The second pairs is to be similarly placed except that like poles are to be facing, see Fig. 26(*a*) and (*b*), for a plan view of the arrangements. Place a piece of cellophane paper over each pair of magnets and on this sprinkle a thin layer of iron filings. Make a sketch of the magnetic field patterns.

C

Are the patterns alike? What conclusions do you draw about the magnetic fields?

Repeat the procedure with two pairs of magnets arranged as shown in Fig. 27(a) and (b).

What do you notice about the magnetic fields? Make sketches of them.

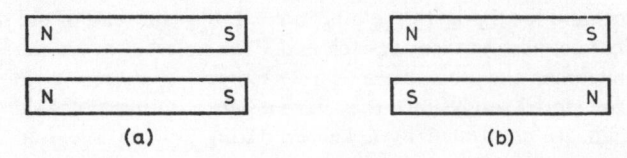

```
┌─────────────┐          ┌─────────────┐
│ N         S │          │ N         S │
└─────────────┘          └─────────────┘
┌─────────────┐          ┌─────────────┐
│ N         S │          │ S         N │
└─────────────┘          └─────────────┘
     (a)                      (b)
```

Fig. 27

Experiment. *To observe the magnetic fields around the poles of a horse-shoe magnet, without a pole piece, and with a pole piece.*

Place two horse-shoe magnets side by side, one having a keeper, the other without. Put a piece of cellophane paper over each magnet and around the poles sprinkle a thin layer of iron filings. Tap the papers gently if necessary to even out the filings, see Fig. 28.

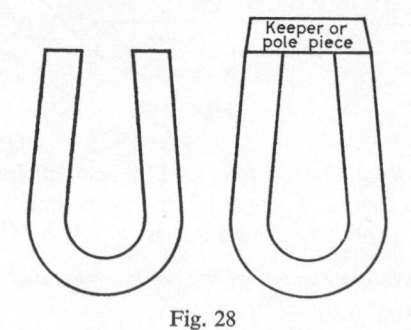

Fig. 28

What is the significant difference between the two magnetic fields?

When a piece of soft iron is placed on the end of a magnet, connecting the poles, the lines of force pass into the soft iron and are retained within the bar. Remember this is what happens when a keeper is fitted to magnets. The fact is that soft iron is less resistant to the passage lines of force than is the atmosphere around the magnet.

Experiment. *To find out what effects soft iron and brass bars have on the field around a bar magnet.*

Using any two poles of two bar magnets place a piece of soft iron adjacent to one and a piece of brass touching the other, as shown by Fig. 29(a) and (b). Place cellophane paper over each magnet and sprinkle a thin layer of iron filings around each pole.

MAGNETISM

How do the magnetic fields differ? Explain any differences that you may notice.

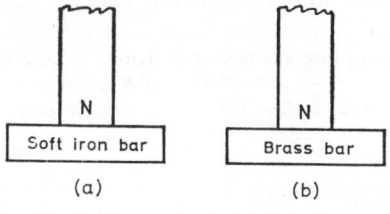

```
     N                    N
┌─────────────┐     ┌─────────────┐
│ Soft iron bar│     │  Brass bar  │
└─────────────┘     └─────────────┘
     (a)                  (b)
```

Fig. 29

Experiment. *Investigation of magnetic fields under different conditions.*

Arrange two pairs of bar magnets in the manner shown in Fig. 30(*a*) and (*b*), with their poles about 25 mm apart. Place a bar of soft iron between the poles of one of the pairs. As in previous experiments, place pieces of cellophane paper over the poles and sprinkle thin layers of iron filings on them.

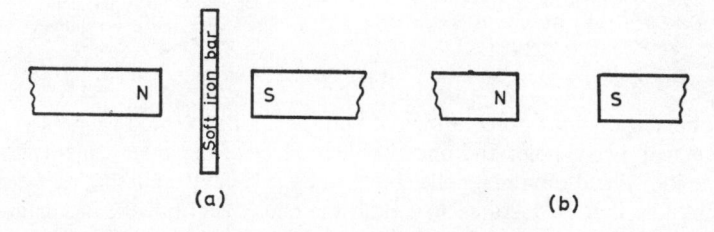

```
┤   N   ├ │Soft iron bar│ ┤ S   ├      ┤   N   ├    ┤   S   ├
              (a)                            (b)
```

Fig. 30

What significant difference is there between the fields? You could make sketches of the fields in your laboratory notebook.

Experiment. *To observe the effect of placing a soft-iron ring in a magnetic field.*

Place the opposite poles of a pair of bar magnets at a suitable distance apart and between them place a soft-iron ring, see Fig. 31. Place a piece of cellophane paper over the arrangement, and on it sprinkle a thin layer of iron filings.

```
┤   N   ├   (Soft iron ring)   ┤   S   ├
```

Fig. 31

What is it that is peculiar to this magnetic field? Place a compass inside the ring and observe its behaviour.

Why does a ship's compass have a soft-iron ring round it?

Experiment. *To demonstrate the working principle of the magnetic chuck that uses permanent magnets.*

Arrange three bar magnets in a wooden block as shown in Fig. 32(a) and fix three soft-iron pole pieces into suitably cut holes in a piece of plywood (cardboard would do). Slide the pole pieces, or the magnets, over so that the pole pieces act as keepers, see Fig. 32(b). Place a mild steel block (job) on the keepers whilst in position (b) and whilst keepers are in position (a).

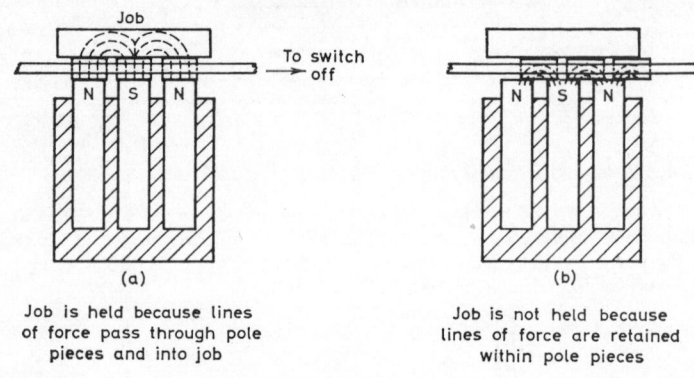

(a) Job is held because lines of force pass through pole pieces and into job

(b) Job is not held because lines of force are retained within pole pieces

Fig. 32

Comment on your observations of both positions. Would we get the same results with two S poles and one N pole in place of the arrangement shown?

The job is held on a magnetic chuck when it is in effect acting as a keeper. The handle that is operated to switch the chuck on and off, i.e. magnetise and de-magnetise it, is in fact sliding magnets under pole pieces. The 'on' position gives the conditions as shown by Fig. 32(a), and the 'off' position as shown by Fig. 32(b). The magnetic chuck has more than three magnets. This is realised when one counts the number of wide strips between the non-magnetic inserts in the top plate. The number will vary according to the size and holding capability of the chuck.

Electro-magnets

There is a magnetic field around a straight conductor when an electric current flows through the conductor. To check whether or not this statement is correct, we will carry out a simple experiment.

Experiment. Make a slot in a piece of cardboard and through it push one side of a coil of copper wire. Connect this coil in series with a switch and a 4-volt cell, see Fig. 33. Now pass a current through the coil and whilst it is flowing sprinkle iron filings around the conductor. Tap the cardboard gently if necessary in order to thin out the filings.

By using a conductor made up from strands of insulated wire the strength of the magnetic field is greater than that of a single wire. Because the current flowing is greater. The whole may be regarded as one conductor.

MAGNETISM

We should find that the filings arrange themselves into concentric circles (all having the same centre) around the conductor. If now we cut off the current, then tap the cardboard, we should find that the filings become disarranged. If we switch on the current again we should see a re-forming of the concentric circles. Clearly there is 'something' surrounding the conductor when it carries an electric current which is similar to the magnetic field which surrounds a permanent magnet. The 'something' is in fact an *electro-magnetic* field. Remember the magnetic field was not in existence when the electric current was switched off.

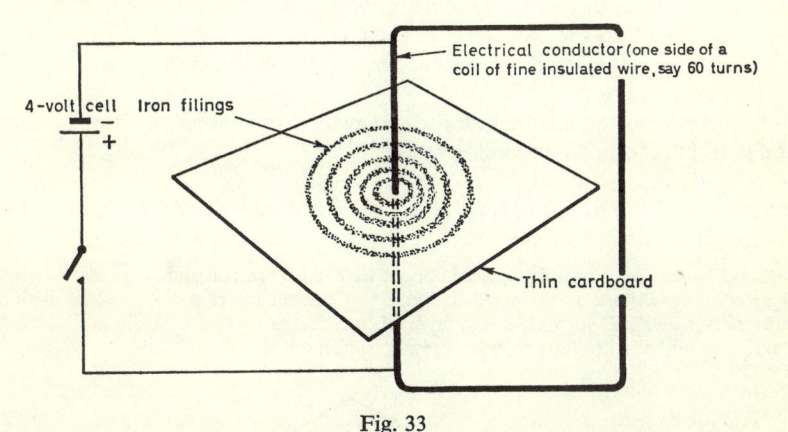

Fig. 33

Experiment. *To show that the magnetic field around an electrical conductor has a direction in a N–S sense, and that this direction is dependent upon the direction of the current.*

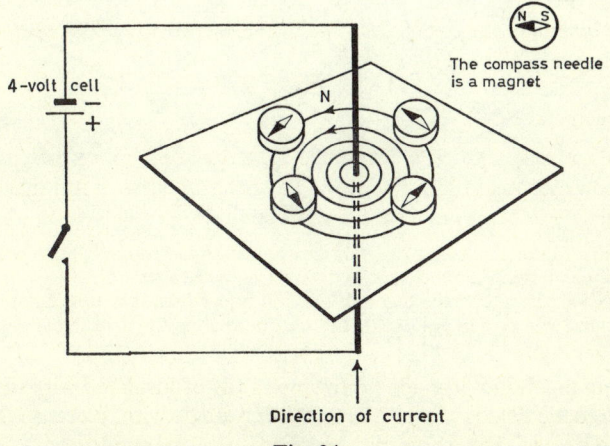

Fig. 34

60 ELECTRICITY IN THE WORKSHOP

Draw a number of concentric circles say about 15 mm difference in radius, on a piece of cardboard. Through the centre of these circles pierce a small hole and through it push a straight conductor connected to a cell as in the previous experiment. Place four compasses on any one of the circles in positions approximately at right-angles to each other, see Fig. 34, and note the direction in which the needles point when the current is switched off. Close the switch and observe the behaviour of the compass needles. Switch off the current and reverse its direction by changing over the battery terminal connectors. Close the switch again and observe the movements of the needles. Note which are the N-seeking ends of the needles.

Imagine your pen to be the conductor. Grasp it in your right hand, the pen being parallel to the conductor, with your thumb pointing in the direction in which the current is flowing in the wire. What do you notice about the direction of the magnetic field and the direction in which your fingers are pointing round the pen?

If the compasses are moved further away from the conductor do they indicate that the field is becoming weaker?

Experiment. *To find out whether or not there is a magnetic field around a bent conductor.*

Bend about 60 strands of insulated copper wire to form a coil and arrange it through a piece of cardboard as indicated in Fig. 35. Connect the ring coil in series with a battery and switch. Sprinkle a thin layer of iron filings on to the cardboard. Switch on the current and tap the cardboard gently to thin out the filings.

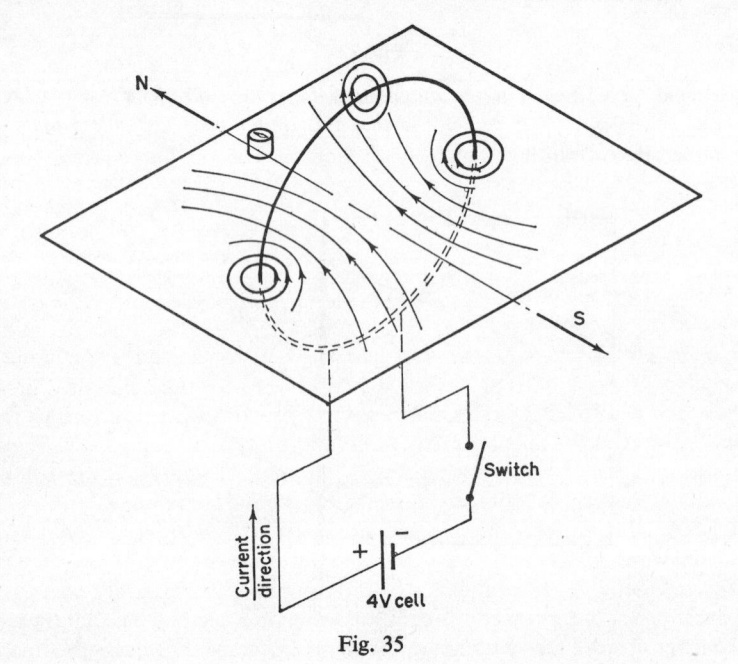

Fig. 35

MAGNETISM

Is there a magnetic field around the conductor when an electric current flows?

Using compasses determine whether or not the direction of the magnetic field is reversed when the current is reversed.

What do you notice about the lines of force around the centre of the ring when compared with those around the centre of the wire? You might move the cardboard round the ring and check the shape of the magnetic field in some other position.

If we have a series of rings and put them about the same axis, end on, then we have a coil, e.g. a continuous length of wire formed into a coil.

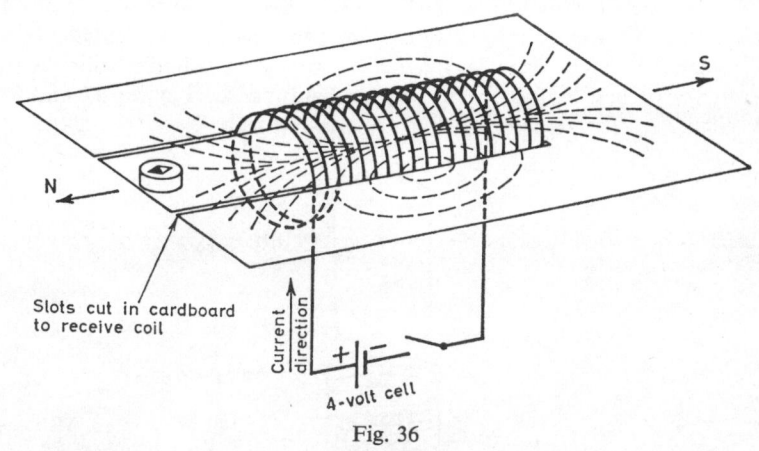

Fig. 36

Experiment. Fit a coil of current conducting material, say copper wire, into slots cut into a piece of cardboard as shown in Fig. 36. Connect the coil in series with a switch and battery. Sprinkle a thin layer of iron filings on to the cardboard in and around the coil. Switch on the current, tap the cardboard gently and observe the lines of force formed. Place compasses on the axis of the coil, at either end and note their readings. Feel for the magnetism with a bar magnet at either end of the coil using each pole of the magnet. Switch the current on and off whilst observing the compasses and the bar magnet. Reverse the current flow and again observe the compasses and bar magnet.

Is there a magnetic field around the coil when the current is flowing? Is there one when the current is switched off? Does the magnetic field have a N pole and a S pole? Does the current direction affect the positions of these poles? Compare the magnetic field around a permanent magnet with the one around an electro-magnet by placing a bar magnet alongside the coil and sprinkling filings on cellophane paper placed over the bar magnet.

Try the above experiment using un-insulated wire, then repeat using insulated wire.

The magnetic field around a straight conductor is relatively weak. We have seen when the conductor is bent to form a ring, see Fig. 35, that there are force rings around the conductor. At the ring centre, however, the lines of force are approximately straight and in planes which are normal to that in which the ring is situated. The magnetic effect is increased. If a number of rings are placed on a common axis end on, or, more easily, in the form of a coil, the magnetic effect is further increased. Such a coil, closely wound from insulated copper wire, and fitted with a soft iron core (or one made from other suitable metal) is known as a solenoid if the core reciprocates (moves to and fro) within the coil. If however, the core is fixed, the coil being tightly wound onto it, the system is known as an electro-magnet. By fitting a core of magnetic material to the coil its magnetic strength is greatly increased, i.e. it has more lines of magnetic force per square millimetre of cross-section than it would have if the core were made of air or brass. It is essential that the core material be capable of being magnetised instantly and demagnetised as quickly.

The solenoid is in effect a magnetic piston and cylinder suitable for operating switches, brakes, valves and mechanisms, see Figures 37 and 38.

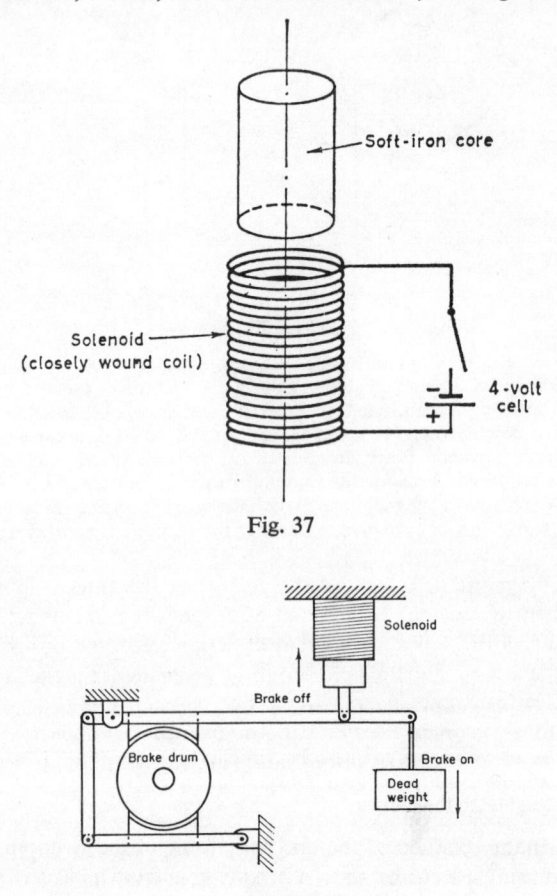

Fig. 37

Fig. 38

MAGNETISM

Whilst the electro-magnet attracts objects to it, e.g. the magnet on a crane for lifting scrap iron, or steel sheets and plate. It is also used for activating alarm bells and buzzers, telephone receivers, etc. and for lifting planing machine tool boxes clear of the work on the return stroke, see Figures 39 and 40.

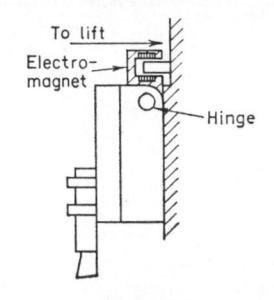

Fig. 39

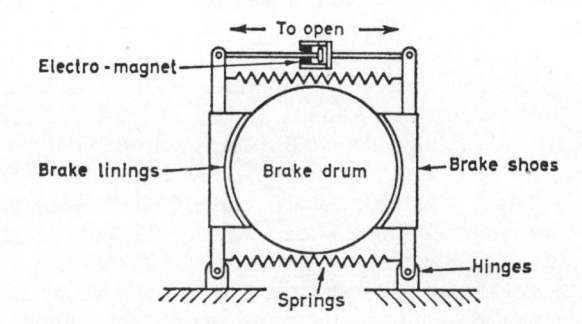

Diagrammatic arrangement of an electro-magnetic brake

Fig. 40

Experiment. *To find out whether or not the magnetic force is increased when a soft-iron core is used in preference to an air core in a solenoid; also to determine the magnitude of the magnetic pull of the solenoid.*

Suspend a metre rule in a balanced position by means of a piece of strong twine from the arm of a retort stand. Attach to a point somewhere near one end of the rule an inverted tin-plate cylindrical container, open at one end, say 25 mm diameter, and balance the beam by adding 'weight' to a hanger suitably placed, see Fig. 41. Wind a coil loosely round the ' tin can ' to make a simple solenoid. Connect the solenoid in series with a switch and a battery. Close the switch. The solenoid is now magnetised and the beam is thrown out of balance. Restore the balance of the beam by adding 'weights' to the hanger.

What is the magnitude of the magnetic force in the solenoid? Repeat the experiment using a soft-iron core, e.g. a core cut from a bar of dead mild steel,

64   ELECTRICITY IN THE WORKSHOP

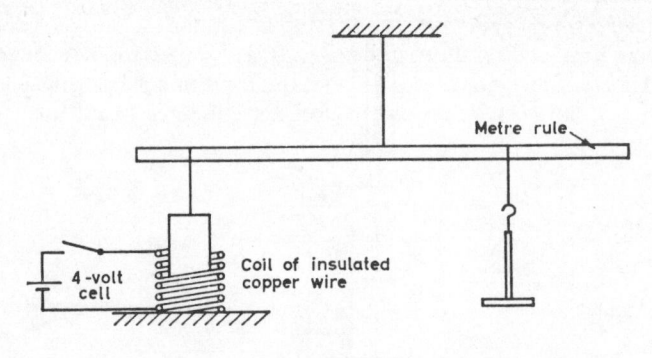

Fig. 41

in place of the air core. How does the magnetic force of the solenoid in this instance compare with that of the previous case? You could carry out this experiment using a battery with a higher voltage, or with a solenoid having twice as many turns of wire on it, then compare all results obtained.

As previously stated, when a coil is fitted with a soft-iron core which remains stationary with the coil, we have an electro-magnet that has many practical applications, from operating a door bell striker to lifting scrap ferrous metal in a steel works. The same principle is involved in the electro-magnetic chucks used on lathes, grinding machines, etc. By this means a very strong magnet can be made by flicking a switch. Equally as quickly the magnetism can be destroyed. Soft-iron is the ideal core because it intensifies the strength of the magnetic field and it is easily magnetised and de-magnetised. If a hard steel core were used it would take longer for the magnet to reach maximum strength. It would also take much longer for it to become de-magnetised. Clearly both are disadvantages from a practical point of view.

Experiment. *To find how the magnetic force of an electric magnet varies with the core material.*

Arrange an electro-magnet, having a soft iron core, under a mild steel disc attached to a spring balance as shown in Fig. 41(*a*). Connect the coil in series with a switch and battery. Note the spring balance reading then switch the current on, take the spring balance reading again, ensuring that there is about 3 mm gap between the electro-magnet and the disc attached to the spring balance. This relationship between disc and magnet core must be maintained at all times so as to ensure that the strength of the field is being measured in the same position each time. Any adjustment can be made through screw and nut, i.e. the turn buckle attachment between the spring balance and support point. Repeat the procedure using cores of brass, copper, aluminium and high-carbon steel—read the spring balance with current off, then with current switched on.

MAGNETISM

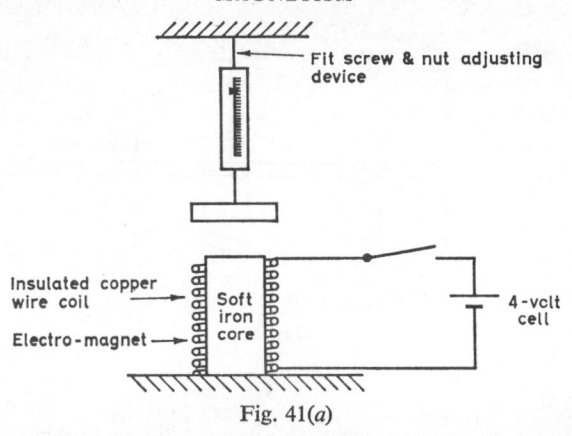

Fig. 41(a)

Carry out a similar experiment using only a soft-iron core and having an ammeter and rheostat in the solenoid circuit, the object being to find out how the magnetic force varies with the current supplied to the coil.

Another type of electro-magnet is the horse-shoe type. This is one which has a U-shaped soft-iron core round which a coil is wound, see Fig. 42. It is the type used in magnetic chucks. Unlike the bar type previously dealt with, it provides through the job a complete circuit of lines of force, i.e. a complete magnetic circuit. The straight bar form does not permit this. It allows the lines of force to pass into the atmosphere. Because of the complete magnetic circuit, the U-shaped electro-magnet has a much greater lifting

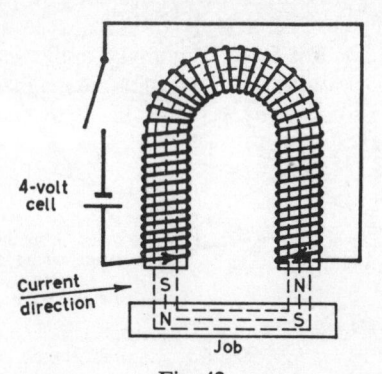

Fig. 42

capacity than that of the straight form. The coil need not cover the whole core, it need be only around the limbs; i.e. coil in two parts, one on each limb, see Fig. 43. The direction of the field in a coil may be determined in the manner similar to that used for the straight conductor. Grip the coil by the right-hand, whether it be wound round a straight or bent core, with the fingers pointing in the direction in which the current is flowing, the thumb

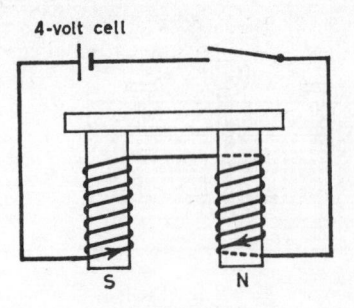

Fig. 43

being parallel to the axis of the coil. The N pole lies in the direction in which the thumb is pointing. The adaption of this type of electro-magnet to the magnetic chuck is shown in Fig. 44. This figure shows the inside of a chuck, the box and top cover plate not being shown. The strips of non-magnetic material fit into the top cover as permanent inserts. They direct the lines of force through the cover and into the job. If they were not there, the

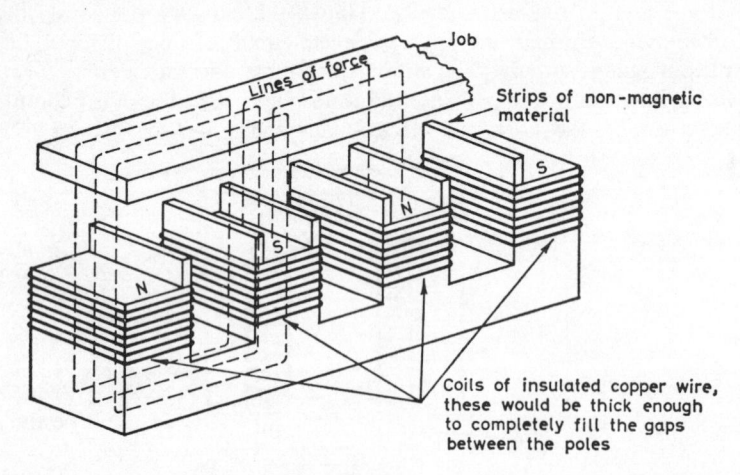

Fig. 44

magnetic lines would short circuit through the cover, resulting in the chuck being much less effective. The magnetism is switched on and off by simply switching the current supply to the coils on and off. In the case of a circular magnetic chuck, as used on a lathe, the working principle is similar, the magnets being arranged in a circular form. The electro-magnet used for lifting iron and steel works on the same principle, see Fig. 45. It may be circular or rectangular in shape.

MAGNETISM 67

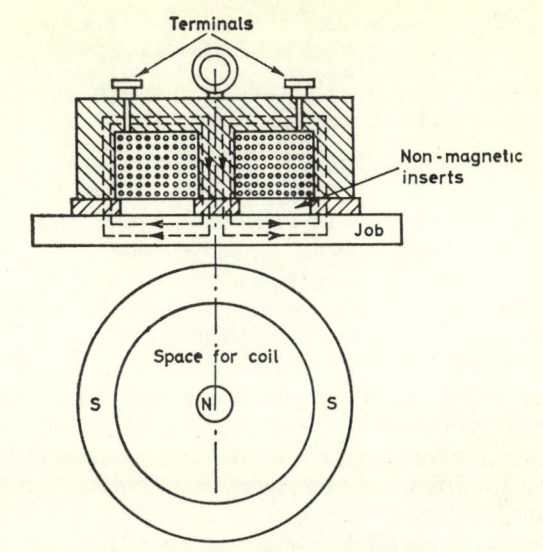

View on underside with cover & coils removed
Fig. 45

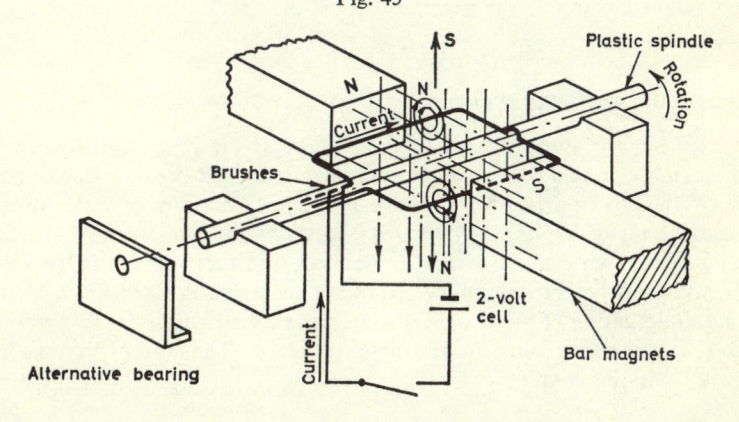

A simple electric motor
Fig. 46

The electric motor

The electric motor is dependent on electro-magnetism for its operation. A simple motor could be constructed using permanent as well as electro-magnets. Such a motor is shown in Fig. 46.

Experiment. *To construct a simple electric motor, see Fig.* 46.

Bend a coil of copper wire, about 60 turns, to the shape shown, and fix it, by glue or insulation tape, to a plastic spindle, i.e. a straight knitting needle 5 mm dia.

Support the spindle by wooden bearings, or bearings cut from sheet metal. Arrange two pieces of copper wire in a vertical position to act as brushes, touching the loose ends of the coil. Place two bar magnets in the positions shown. Connect the brushes into a series circuit with a switch and 2-volt cell. The whole may be arranged on a base board to which the bearings and brushes may be fixed.

When the switch is closed, a magnetic field will be set up around the coil of polarity as indicated by Fig. 46. The lines of force of this field will be normal to the lines of magnetic force between the poles of the two bar magnets. The coil like a compass needle will spin, the N pole of the bar magnet attracting the S pole of the coil, and in so doing the circuit will be broken. This will result in the breakdown of the coil magnetic field. However, the coil should have gained sufficient momentum to rotate it sufficiently to bring the next brush into contact with the other limb of the coil. The process will then be repeated and a continuous motion of the coil rotor (armature) maintained.

You might experiment further by introducing a second coil in a plane at right-angles to the first one, making sure that the two are insulated from each other and from the spindle.

An electric motor may have up to 10 such coils equally spaced and fitting neatly into slots cut in a suitable core, i.e. a splined core. Also instead of the fixed magnets being of the permanent type they are of the electro-magnetic form.

The dynamo (electricity generator)

Stated simply, the principle of the *dynamo* is the reverse of the electric motor, in so far as a coil is rotated in a magnetic field between unlike poles. The coil has to be driven; this is usually done by a steam or water turbine, or diesel engine. As the coil rotates it cuts the lines of force of the field. This generates an electric current in the coil, and then current is tapped off by brushes. Whether or not the electricity generated is d.c. (direct current) or a.c. (alternating current) depends upon the construction of the generator. *Note:* The current from cells and batteries is d.c. The current that we have referred to in this chapter is also d.c.

The ammeter

Many electrical instruments rely on the electro-magnet principle for their operation, see Fig. 47(*a*) and (*b*), for a simple illustration. Of the current to be measured, only a small amount passes through the instrument so as to enable the instrument to be of delicate and sensitive construction. Since this small amount of current is a known definite proportion of the total, the instrument scale is graduated to give the value of the total current. A process of scaling is used. When the small current passes through the coil it causes the coil to rotate in the same way as a motor coil is rotated. Since,

MAGNETISM

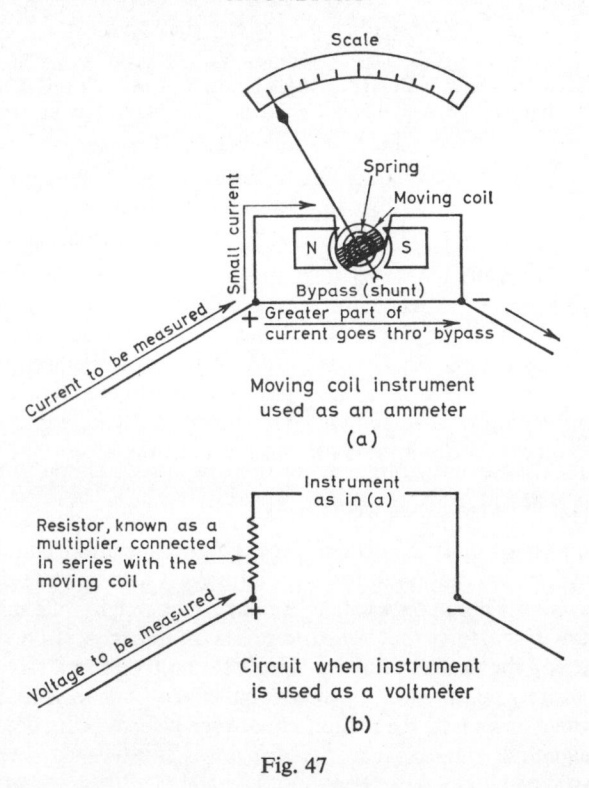

Fig. 47

Moving coil instrument used as an ammeter
(a)

Circuit when instrument is used as a voltmeter
(b)

however, the coil is controlled by hair springs its movement is limited. It will be proportional to the current flowing, since the strength of the coil field is dependent upon the current. This type of instrument, connected to a thermo-couple, is used for measuring temperature. The scale of the instrument is then graduated in degrees of temperature, although the instrument is measuring the value of the current put out by the thermo-couple. The current flowing from the thermo-couple is proportional to the temperature of the couple. A similar instrument is used on the optical pyrometer for recording the temperature measured by this instrument.

The voltmeter

If the moving coil instrument has a circuit as shown in Fig. 47(*b*), then it can be used for measuring the voltage of a current; i.e. it can be used as a voltmeter. The p.d. to be measured is applied across the moving coil and the resistor (multiplier). According to Ohm's law, the current which flows through the coil will be proportional to the p.d. applied; therefore the scale is calibrated in volts instead of amperes.

MATERIALS AND MACHINES IN THE WORKSHOP

Chapter 6. Force and Stress

Force

Force is used to cut, bend and shape metal, to press a bush into its housing, to operate the chain block, to turn the handle which operates the lathe carriage, to operate the gear lever on the lathe headstock. There is a force in the belt which drives the drilling-machine spindle. A force is used to tighten the bench vice. In fact, there are very few workshop processes in which a force is not used. The force may be caused by hand or by machine, e.g. an electric motor. A force may be the result of a load, e.g. the load on a set of rope blocks causes a gravitational force on the ropes that support it.

Gravitational force (so-called weight)

The unit of mass in the SI system is the kilogramme (kg). The earth exerts an attractive, or gravitational, force on all objects and because of this each object has what is commonly called 'weight'. The 'weight' of a mass of 1 kg, i.e. the gravitational force that the earth exerts on the mass, is $1 \times 9\cdot 81$ newton at this part of the earth's surface. The $9\cdot 81$ multiplier is a factor which is governed entirely by the earth's gravitational effect; it enables us to calculate the force effect of a mass, the magnitude of such a force being determined by using the equation:

$$\underset{\text{(newton)}}{\text{Force effect of mass}} = \underset{\text{(kg)}}{\text{mass}} \times 9\cdot 81$$

Measurement of force

Force is measured in *newton* in the SI system, or multiples of the newton, e.g. kilonewton, meganewton or giganewton.

$$1 \text{ kN (kilonewton)} = 10^3 \text{ N (newton)}$$
$$1 \text{ MN (meganewton)} = 10^6 \text{ N}$$
$$1 \text{ GN (giganewton)} = 10^9 \text{ N}$$

Application of force

A force may be applied in a number of ways. Three of the most common applications are shown in Figs. 48, 49 and 50.

Tensile force (Fig. 48)

The *driver* pulley revolves the *driven* pulley by pulling it round by a belt. This causes a *pulling force* in the belt. There is a *tensile force* in the tight

Driver   Driven

Tight Side (Tensile Force)

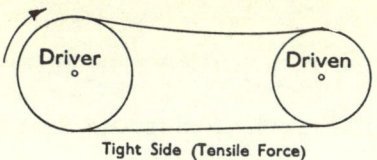

Fig. 48

side of the belt, and in the slack side. They are not of the same value. By the way, the slack side of the belt is at the top so as to increase the arc of contact between belt and pulley, producing a more effective drive.

A *tensile force* causes *tension*. A bolt is put in *tension* when the nut is tightened. The wire rope on a crane which is lifting a load is in *tension* because of the gravitational pull on the load mass.

Compressive force (Fig. 49)

Clamping Force (Compressive Force)

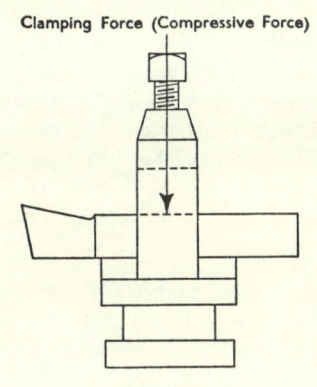

Fig. 49

The force clamping the tool in the lathe tool-post compresses that part of the tool that it contacts, i.e. it is a *compressive force*. A *compressive force* causes *compression*.

Other examples of a compressive force are: the force gripping a component between the jaws of a vice, the feed force acting on the drill as it is fed into the metal, the force acting on the screw that lifts the milling machine table (caused by the weight—gravitational force—of the table).

Shear force (Fig. 50)

The force which acts on a piece of tinplate as an effort is made to cut it with tin snips is a *shearing force*. Other examples of shear force are: the force at the blade of a guillotine shearing machine, the force at the cutting

Cutting Force (Shearing Force)

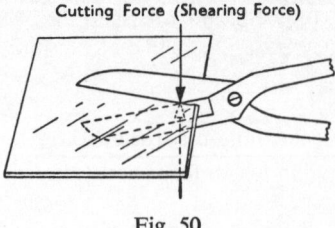

Fig. 50

edge of a chisel, the force acting on a key or taper pin fixing a gear wheel to a spindle as the gear wheel drives the spindle.

Stress

When a force acts on a material, e.g. the load (gravitational force) acting on the rope of a pulley block, an equal resisting force is set up within the material. The amount of this resisting force acting upon 1 square millimetre of the cross-section of the material, i.e. the load (gravitational force) per mm^2, is known as the *stress* in the material. It is a direct stress because it is the stress on a plane which is normal (at right-angles) to the direction of the load causing it, e.g. normal to the length of the rope.

All articles in the workshop which have to support a force of any kind are in a state of stress.

Calculation of stress

Consider a square bar of steel 50 mm × 50 mm, one end of which is fixed to the roof of a workshop in such a way that it hangs vertically downwards, see Fig. 51. A force of 15 kN is applied to the lower end of the bar. This force

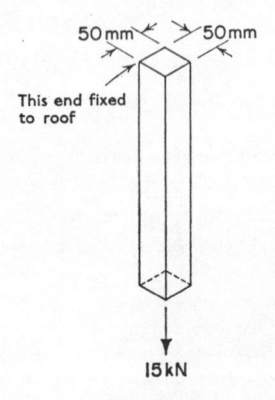

Fig. 51

can be obtained by attaching a 1530 kg mass to the bar since the gravitational pull on such a mass is $1530 \times 9 \cdot 81 \simeq 15\,000$ N, i.e. 15 kN.

STRESS in bar (N/mm²)

$$= \frac{\text{Force acting on bar (newton)}}{\text{AREA upon which force acts (mm}^2\text{)}} \text{ or } \frac{\text{FORCE}}{\text{AREA}}$$

$$= \frac{15 \times 10^3}{50 \times 50}$$

$$= 6 \text{ N/mm}^2 \ (6 \times 10^6 \text{ N/m}^2 \text{ or } 6 \text{ MN/m}^2)$$

Units of stress

Using SI units *stress* is measured in newton per square metre. This is the basic unit, multiples and sub-multiples of the N/m² are used.

A giganewton/m², $\text{GN/m}^2 = 10^9 \text{ N/m}^2 = \dfrac{10^9 \text{ N}}{\text{m} \times 10^3 \times \text{m} \times 10^3}$

i.e. $\quad 1 \text{ GN/m}^2 = \dfrac{10^3 \text{ N}}{\text{mm} \times \text{mm}} = 1 \text{ kN/mm}^2$

Types of stress

Tensile stress is caused by a *tensile force* (as in Fig. 51); e.g. a tightened bolt and the screw of a tightened bench-vice have a tensile stress.

Compressive stress is caused by a *compressive force*; e.g. in a centre punch as the hammer strikes it, in a press tool at the instant it shapes the pressing, in the screw supporting the table of a milling machine.

Shear stress is caused by a *shear force*; e.g. a piece of sheet metal being cut on the guillotine shears has a shear stress in it the instant the blade touches it. The driving force acting on the key that fixes a pulley to its shaft causes a shear stress in the key. Metal that is being cut on a blanking press has a shear stress in it the instant the punch touches it.

Examples

1. A mild steel tie-bolt 40 mm dia. that is being used on a small furnace, has to carry a tensile load of 60 kN. Calculate the tensile stress in the bar because of this force.

    First make a sketch of part of the bolt, see Fig. 52.

    AREA on which 60 kN force acts

    $$= \frac{\pi \times \text{dia.}^2}{4} = \frac{\pi \times (40)^2}{4} = 400\pi \text{ mm}^2$$

74   MATERIALS AND MACHINES IN THE WORKSHOP

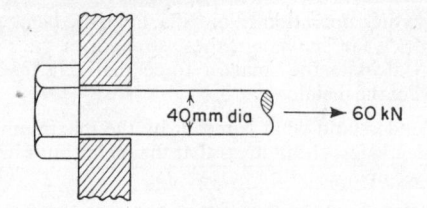

Fig. 52

TENSILE STRESS in the tie-bolt

$$= \frac{\text{LOAD}}{\text{AREA}} = \frac{60 \times 10^3}{400\pi}$$
$$= 48 \text{ N/mm}^2 \text{ (48 MN/m}^2\text{)}$$

2. The punch that is fitted to a certain punching machine has a shank 20 mm dia. Just as the punch begins to cut the metal it has a compressive force of 25 kN acting along its axis. Find the compressive stress in the punch shank.

In this problem we have a force of 25 kN tending to compress the shank of a punch of 20 mm dia., as shown in Fig. 53.

AREA on which the 25 kN force acts

$$= \frac{\pi \times \text{dia.}^2}{4} = \frac{\pi \times (20)^2}{4} = 100\pi \text{ mm}^2$$

COMPRESSIVE STRESS in punch shank

$$= \frac{\text{LOAD}}{\text{AREA}} = \frac{25 \times 10^3}{100\pi}$$
$$= 80 \text{ N/mm}^2 \text{ (80 MN/m}^2\text{)}$$

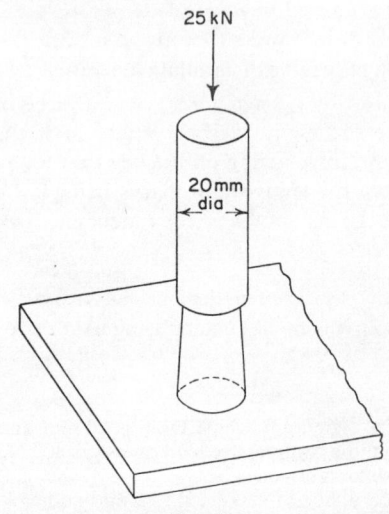

Fig. 53

FORCE AND STRESS

3. In a particular punching operation 18 mm dia. holes are being punched in a sheet of brass 3 mm thick. The maximum shear strength of the sheet may be taken as 200 N/mm². Calculate the smallest force that has to be applied to the punch in order to cut the metal.

First we must understand what is meant by the maximum shear strength of the plate. It is the greatest shear stress that the metal can carry before it is cut. Fig. 54 shows the set-up.

Note that the area of metal that the punch cuts through is given by: the circumference of the piece cut out multiplied by the thickness of the sheet (shown shaded in Fig. 54).

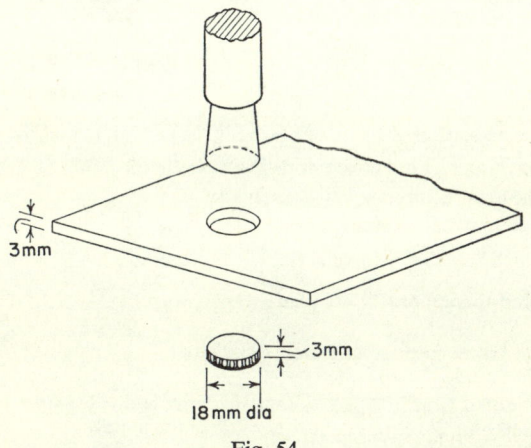

Fig. 54

AREA cut out by punch

$$= \text{circumference of hole} \times \text{thickness of sheet}$$
$$= \pi \times \text{dia.} \times \text{thickness}$$
$$= \pi \times 18 \times 3 = 54\pi \text{ mm}^2$$

Now $\text{STRESS} = \dfrac{\text{FORCE}}{\text{AREA}}$, and in this problem we have to find a value for shear FORCE

Hence we have

$$\text{Maximum shear strength of metal} \times \text{Area cut} = \text{Punch force}$$

FORCE required on punch $= 200 \times 54\pi$ newton

$$= 34 \text{ kN}$$

4. The support leg of a shaping machine table is 20 mm dia. During a certain machining operation, the compressive load on it was 800 N. Find the compressive stress in the leg under this load.

It is always a good policy to make a sketch and add to it all the given data, see Fig. 55.

AREA on which load acts

$$= \frac{\pi \times \text{dia.}^2}{4} = \frac{\pi \times (20)^2}{4} = 100\pi \text{ mm}^2$$

COMPRESSIVE STRESS in leg

$$= \frac{\text{FORCE}}{\text{AREA}} = \frac{800}{100\pi} = 2 \cdot 5 \text{ N/mm}^2 \; (2 \cdot 5 \times 10^6 \text{ N/m}^2)$$

5. The axle in the lower block of a set of Weston chain blocks is to be of such diameter that the shear stress in it never exceeds 40 N/mm². The maximum mass that the blocks have to carry is 2 040 kg. What is the smallest permissible diameter for the axle?

$$\text{Gravitational force on a 2 040 kg mass} = 2\,040 \times 9 \cdot 81$$
$$= 20 \text{ kN}$$

Draw a cross-section of the block (see Fig. 56) and note that the area of metal supporting the force of 20 kN is in two parts, one on either side of the pulley.

First find the area required to carry this load.

$$\text{We know that STRESS} = \frac{\text{FORCE}}{\text{AREA}}, \;\therefore\; \text{AREA} = \frac{\text{FORCE}}{\text{STRESS}}$$

AREA required to support 20 kN load

$$= \frac{20 \times 10^3}{40} = 500 \text{ mm}^2 \; (0 \cdot 005 \text{ m}^2)$$

We have noted from the sketch that this area is divided into two parts, i.e. an area of 250 mm² is required on either side of the pulley.

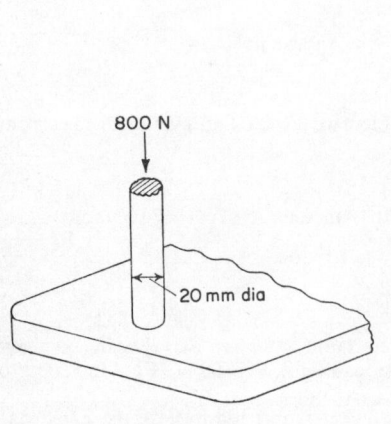

Fig. 55

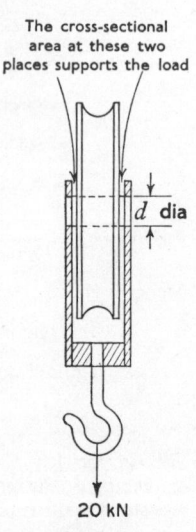

Fig. 56

FORCE AND STRESS 77

We know that the cross-sectional area of the axle could be given in the form $(\pi/4) \times \text{dia.}^2$. Therefore we can state:

$$\text{Cross-sectional AREA of axle} = \frac{\pi \times \text{dia}^2.}{4} = 250 \text{ mm}^2$$

From this we can find the diameter of the axle,

$$\pi d^2 = 4 \times 250$$

$$d = \sqrt{\frac{1000}{\pi}}$$

and diameter of axle = 18 mm

6. In a broaching operation the force pulling the broach through the component is 15 kN. The core diameter of the broach is 15 mm. What is the tensile stress on the core of the broach?

We will again make a sketch to help us understand the problem more easily (see Fig. 57).

AREA on which 15 kN force acts

$$= \frac{\pi \times \text{dia.}^2}{4} = \frac{\pi (15)^2}{4} \text{ mm}^2$$

TENSILE STRESS on core dia. of broach

$$= \frac{\text{FORCE}}{\text{AREA}} = \frac{15 \times 10^3}{225\pi/4}$$

$$= \frac{60 \times 10^3}{225\pi}$$

$$= \underline{85 \text{ N/mm}^2 \text{ (85 MN/m}^2\text{)}}$$

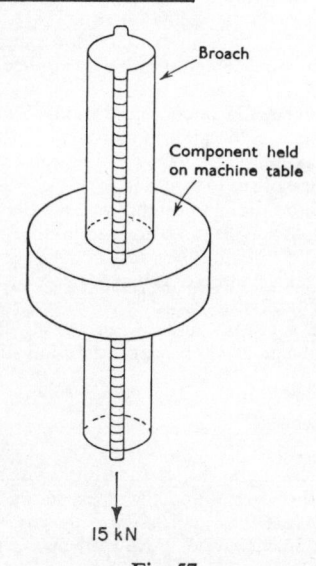

Fig. 57

The foregoing examples show how the *stress* in certain workshop tools, and machines, can be calculated. We must note the use that has been made of sketches to assist us in understanding the problems. A simple sketch of that part of the set-up with which we are concerned is sufficient.

$$1 \text{ N/mm}^2 = 1 \times 10^6 \text{ N/m}^2 = 1 \times 10^3 \text{ kN/m}^2 = 1 \text{ MN/m}^2$$

Exercise 5

Section A

1. A square bar 20 mm by 20 mm cross-section has one of its ends fixed to a roof girder in such a way that the bar hangs vertically downwards. On the lower end of the bar a set of chain blocks is hung. The blocks are to be used to lift a centre lathe 'weighing' 5 kN. Calculate the stress in the bar and state what type it is.
2. A 12 mm dia. punch has a punch force of 15 kN acting on it. Calculate the stress in the punch and state what type of stress it is.
3. (*a*) When a hacksaw blade is tightened in its frame the tensile stress set up in it must not exceed 15 N/mm². If the size of the blade is 12 mm wide by 0·6 mm thick, find the greatest permissible tightening force that can be applied to the blade. Pin holes are 3 mm dia.
   (*b*) Make a sketch to show the weakest part of the blade.
4. A piece of metal strip 50 mm wide by 3 mm thick is to be sheared on the guillotine shears. If the maximum shear strength of the strip is 320 N/mm², calculate the maximum force required at the blade of the shears.
5. A bar of steel 40 mm diameter is to be turned between centres. The force applied to the end of the bar when the tailstock centre is tightened is 70 N. Calculate the stress in the bar and state what type it is.
6. A certain mechanical saw blade is 25 mm wide by 0·75 mm thick. When it is tightened in its frame the stress in it must not exceed 15 N/mm². What is the maximum safe tightening force that may be applied to the blade? Pin holes are 5 mm dia.
7. In a particular belt drive a flat open belt is used. The belt is 5 mm thick and 100 mm wide, whilst the safe load that it can carry is 10×10^3 N per metre of its width. Find (*a*) the maximum safe load that the belt can carry and (*b*) the stress at this load, (*c*) what type of stress would it be?
8. A brass bush 50 mm bore dia. and 60 mm outside dia. is to be pressed into its pulley by means of an Arbor press. If the minimum force required to push the bush into the pulley is 1·5 kN, find the stress in the brass under this load.
9. If a certain tightened bolt, 25 mm dia., has a stress of 32 N/mm² in it, find the force that is causing this stress.
10. A stub axle 18 mm dia. is gripped endways in a vice. The gripping force applied by the vice jaws to the axle is 400 N. Calculate the stress in the pin, and state what type it is.

Section B

1. Fig. 58 shows part of clamping device for a jig. The rivet shown has to withstand a maximum load of 10 kN and the stress set up in it must not be greater than 50 N/mm². Calculate the smallest diameter rivet that may be used. What type of stress is caused in the rivet? Give rivet dia. to nearest millimetre above the calculated value.

FORCE AND STRESS

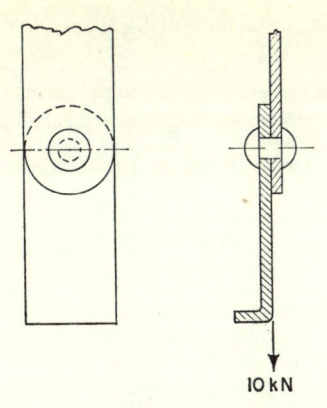

Fig. 58

2. A spindle bearing cap is held by 2 bolts. The maximum upward force acting on the cap is 2·5 kN, whilst the greatest permissible stress in the bolts must not exceed 30 N/mm². Calculate a suitable core diameter for the bolts.
3. Calculate the stress set up in the parallel shank of a 5 mm dia. drill when a feed force of 400 N is applied to the spindle.
4. The greatest force available at the blades of a set of bar shears is 20 kN. Find the largest dia. bar of metal than can be cut by the shears. (Maximum shear strength of metal being sheared is 150 N/mm²).
5. A set of rope blocks has 6 falls of 18 mm dia. rope, i.e. there are 6 ropes to support the load. Calculate the stress in each rope when the blocks lift a load of 400 kg (100 × 9·81 N) if the nett cross-sectional area of a rope is 0·8 of the gross value.
6. In a certain punching operation 20 mm dia. holes are punched in metal plate 5 mm thick, maximum shear strength 350 N/mm². Find the force that must be applied to the punch and the compressive stress in the punch.
7. Fig. 59 shows part of a drive to a machine. Pulley, shaft and bearings have a mass of 60 kg (60 × 9·81 N) and the max. pull in the belt is 1500 N acting

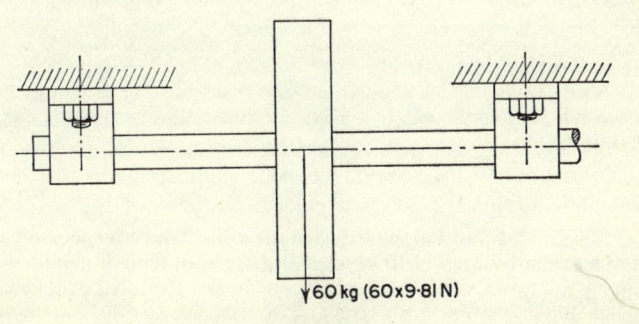

Fig. 59

vertically downwards. Calculate suitable diameters for the fixing bolts, 2 per bearing, if the stress in them is not to exceed 4·5 N/mm².

8. (a) A 450 N/mm² bolt has a core dia. of 7.5 mm. It is made from mild steel, maximum tensile strength 450 N/mm². Calculate the breaking load of the bolt.

(b) What tensile force acting along the axis of the screw would be required to break a bolt of the same material, its core dia. being 15 mm?

9. Fig. 60 shows a safety coupling fitted to the lead-screw of a lathe. If the carriage of the lathe is obstructed in its travel whilst screw cutting is in progress, one of the pins will shear to prevent major damage. The pins are slightly tapered, but for the purpose of calculation you may use an average dia. of 3.5 mm.

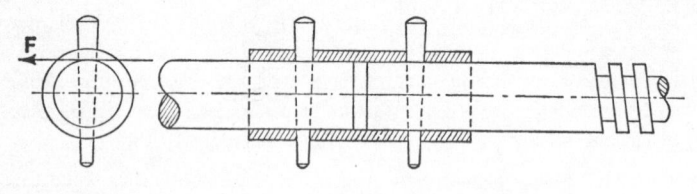

Fig. 60

Assuming that the coupling should fail when a shear force of 4·5 kN acts on one of the pins, state from what material you would make the pin, given the following materials and their maximum shear strengths (select the one nearest, on the low side, to the value that you calculate):

Mild steel	Aluminium	Brass
290 N/mm²	145 N/mm²	216 N/mm²

10. A hydraulic cylinder has a 300 mm bore dia. and the greatest pressure within it is 4·5 N/mm². The front cylinder cover is held by 12 studs 20 mm dia. Find the stress on the cross-section of the outside dia. of the studs, and the stress on the cross-section of the core (core dia. of thread is 15.5 mm).

Strain

A material which is stressed is slightly deformed; e.g. the wire rope on a crane carrying a load stretches; at the same time its diameter is reduced slightly. A simple but convincing example of this deformation can be seen by stretching a rubber band gripped between the thumb and finger of each hand. It can be seen too when a piece of plasticine is similarly stretched. The deformation of a stressed component is known as *strain*. Thus all components that carry stress are at the same time strained. Just how much a component is strained depends upon the strength of the material from which it is made; e.g. under the same load, and therefore the same stress, a 5 mm diameter lead wire would be strained far more than a mild steel wire of the same diameter.

Calculation of Strain. Suppose that the bar shown in Fig. 51 is 2 m long and that it stretches 0·03 mm under the 15 kN load. The longitudinal strain on the bar, i.e. the strain along the length of the bar is given by:

$$\text{LONGITUDINAL STRAIN on bar} = \frac{\text{EXTENSION of bar (mm)}}{\text{ORIGINAL LENGTH of bar (mm)}};$$

$$\text{LONGITUDINAL STRAIN on bar} = \frac{0\cdot03}{2 \times 10^3}$$

$$= 0\cdot015 \times 10^{-3} \text{ (no units, it is a ratio)}$$

The longitudinal strain on a component will always be greater than the strain on its transverse dimensions. The transverse strain is approximately a quarter of the longitudinal strain. It is with the longitudinal strain that we are concerned. Under a compressive stress a specimen will be shortened in length and its transverse dimensions will be increased.

Units of Strain. Strain is just a number; i.e. it has not units since the units of the dividend and the divisor cancel.

Elasticity

Earlier we cited an example using a piece of rubber and a piece of plasticine to show that a material is deformed under a load. There was a very significant difference between the behaviour of the rubber and the plasticine when the load was removed. When the load was removed from the rubber, it recovered its original shape and dimensions, whereas with the plasticine the material remained in its stretched condition. Obviously the rubber has some property in its makeup that the plasticine does not have. This property is known as *elasticity*. Hence the reason for calling a rubber band an elastic band. The plasticine clearly does not have elasticity but it does have *plasticity*, the property which enables a material to take any shape which a force may cause. It may of course be a desired shape in many instances. Rubber does not have plasticity in normal conditions of temperature and pressure. We are concerned with the property of elasticity at this point.

The metals with which we work have elasticity to some degree. None will have the same degree of elasticity as the rubber, nevertheless they do have it. For instance a tightened mild steel bolt is stretched slightly but when it is slackened it reverts to its original dimensions.

Experiment. *To find out whether, or not, metals do have elasticity and if so whether they have the same degree of elasticity.*

Suspend copper, aluminium and mild steel wires, say 1 mm diameter, from a ceiling joist, or wall bracket, and arrange vernier scales, to measure 0·1 mm, at their lower ends so that the lengths of wire between the gripping points are 2·5 m. Attach load hangers to the wires and apply loads up to a total of 10 kg (10 × 9·81 N) per wire in increments of 1 kg (1 × 9·81 N). Note the extensions during the loading.

Remove the loads in decrements equal to the increments and note extension values for comparison with those taken when load was increasing.

Did the metals stretch under the loads?
What happened when the loads were removed?
Did all the wires stretch by the same amount?
Make a list of the materials in the order of greatest to least amount of elasticity.

We should by now appreciate that metals do have elasticity.

Experiment. *To find out whether or not there is a limit to this elasticity.*

As in the previous test, fix three wires to a ceiling joist, or wall bracket. Apply a small load to each wire in turn; observe the extension on the vernier scale. Remove the load and observe whether, or not, the wire returns to its original length. Repeat this procedure, gradually increasing the load each time, until you see that the wire has not returned to its original length, if this is possible.

Does the wire have a limit to its elasticity? If so, did the limit occur at the same load for each wire?

I think by now we shall have reached the conclusion that there is a limit to the amount of elasticity a metal may have.

We know that when a force is applied to a component, it is in a state of stress and accompanying this stress is a strain. When the force is removed stress and strain will not be present and the component will return to its original shape providing it has not exceeded its elastic limit.

Hooke's law

Robert Hooke, who lived between the years 1635–1703, found out by experiment that the stress in a component was directly proportional to the strain whilst the metal was elastic. What he did was to apply a gradually increasing load to a specimen and measure the deflections at certain load values. He then calculated the stress values at these loads and the corresponding strain values. A graph of stress against strain was then plotted. This graph turned out to be a straight line which meant that by reading off any stress value and dividing it by the corresponding strain value, he always achieved the same result. Stated mathematically, he found out that

$$\frac{\text{Stress}}{\text{Strain}} = \text{constant}; \text{ i.e. it is a constant ratio.}$$

This was only true, he said, within the elastic limit of the material.

Thus he stated the law:

Within the elastic limit the stress (tensile, compressive or shear) *in a material is directly proportional to the strain that accompanies it.*

Modulus of elasticity

This is given as follows. It indicates the relationship between the amount of physical distortion and the force producing it:

$$\frac{\text{Stress (N/m}^2)}{\text{Strain (number)}} = \text{Modulus of Elasticity (N/m}^2)$$

The modulus of elasticity is a constant value for any one material but has different values for different materials.

Experiment. *To determine the modulus of elasticity for a selection of metals.*

Fix a number of wires, 20 gauge (0·900 mm), mild steel, brass, copper and aluminium, to a ceiling joist or wall bracket, their lengths being about 2·5 m. Apply suitable loads to each wire and by means of a vernier scale measure the extension at each load. Obtain about 8 to 10 readings and tabulate these, then proceed to complete a table of stress-strain values. Plot graphs of load against extension and stress against strain. Use the latter graph to determine the value of the modulus of elasticity for each metal.

How do the values obtained compare with accepted standard values?

Are the two graphs alike for each metal? Would you expect them to be?

Modulus of elasticity values for some of the common engineering metals are shown in the following table.

Material	Modulus of Elasticity N/m^2 (average values)
Mild steel	210×10^9
Cast iron	100×10^9
Brass	90×10^9
Aluminium alloy	90×10^9

Note: The value will be dependent upon the composition of the material.

Stress–Strain diagrams

Every steelworks and most other metal producing plants have a materials testing laboratory and in this laboratory machined specimens from billets, bars, and sheets, produced from every cast of metal, are subjected to various tests depending upon the use to which the metal is to be applied. For example, specimens machined from bars that are to be used for reinforcing concrete beams would be subjected to tensile and compressive tests, whilst steel for king pins and track rods would be subjected to an impact load test. Steel for rivets would be tested for shear strength.

Most of the modern tensile and compressive testing machines are fitted with autographic recorders which produce a stress–strain graph, on the specimen under test, whilst the test is actually taking place. Figure 61 shows such a graph for a mild steel specimen under tensile test. Mild steel is chosen because it is one of the most used metals. The graph shows various important points and stages during the testing of a specimen to destruction.

The graph is divided into two stages. First there is the *elastic stage* in which the load increases at a much greater rate than that of the extension. The straight line indicates that the stress is directly proportional to the strain. The diameter of the specimen is gradually being reduced all along the gauge length, whilst the length is increasing, see Fig. 62a. A point is then reached

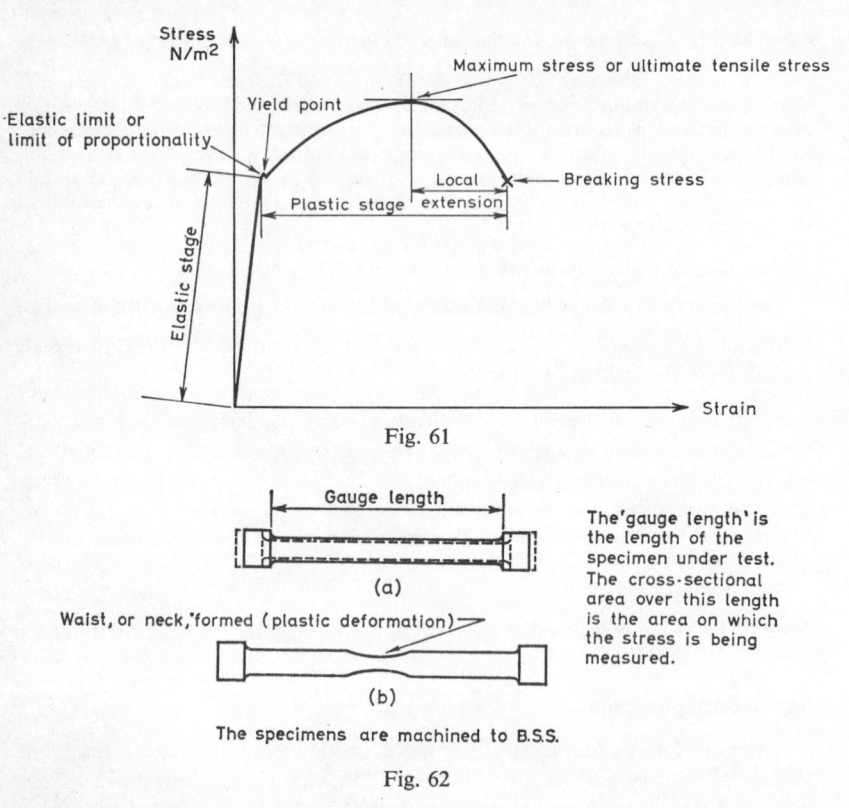

Fig. 61

Fig. 62

The 'gauge length' is the length of the specimen under test. The cross-sectional area over this length is the area on which the stress is being measured.

The specimens are machined to B.S.S.

at which the stress is no longer proportional to the strain; this point is the *elastic limit*, or *limit of proportionality*.

The specimen now moves into the second stage, the *plastic stage*, and it suddenly gives when it reaches the *yield point*. This is denoted by a slight dip in the graph. The dip exists until the slack has been taken up in the specimen, i.e. until the load catches up with the extension. The elastic limit and yield point are so close to each other that they are often taken as one point for practical purposes. When the load does catch up with the extension its rate of increase is much slower than it was in the elastic stage, whilst the extension rate is greatly increased. The diameter of the specimen is still reduced all along the gauge length. This continues until the maximum load is reached. At this point the maximum stress occurs, the U.T.S. (*ultimate tensile stress*). After this point, the specimen begins to extend so rapidly that the load cannot keep pace with it. The graph shows a falling-off of the stress (and the load, too), though it is still great enough to cause extension. Over this latter period, the specimen 'necks' rapidly, see Fig. 62*b*, usually in the centre of the gauge length; i.e. there is a rapid local extension. Finally the specimen fractures. Over this period of necking the stresses shown on the graph are based on the original cross-sectional area of the specimen even though the actual cross-sectional area obviously will be less. These stresses, the chief being the breaking stress, are apparent, not true, stresses. To obtain the true breaking stress, by calculation, the cross-sectional area at fracture must be obtained. This means that the diameter at fracture has to be measured, a task often performed with the aid of a micrometer having conical measuring points. In practice, stresses based on the original cross-sectional area are usually sufficient for most purposes.

It may be useful to point out before we proceed further with this topic why it is essential to know the degree of elasticity and plasticity possessed by a metal. We will do this by citing certain examples. In wire and bar drawing, the metal being worked must be sufficiently elastic to withstand the force which pulls it through the die, yet plastic enough to permit a change in its cross-sectional dimensions. In cold pressing operations, the metals used must be plastic enough to allow the shape of the component to be formed from the flat sheet. In hot pressing the degree of plasticity required is obtained by heating the sheet. If the heating stage can be avoided then clearly time and expense can be saved. Deep drawing depends on plasticity too for a successful operation. We might pose the question, 'Why is cast iron never drawn or pressed'? Clearly the process used for shaping a metal depends upon the properties of the metal. These are determined by metal testing.

To continue with stress–strain graph forms, these are different for different materials, and differ according to the physical condition of the metal, each graph having its own particular shape and characteristics. The stress–strain graphs for brass, copper and aluminium alloys, for instance, do not have clearly defined yield points, neither does the one for cast iron. In fact we could state that they are completely different from that for mild steel, see Fig. 63. Cast iron, incidentally, does not even have a straight line portion. As can be seen from the graphs that do have straight line portions it is not easy to determine the yield points. It is a common feature with aluminium alloys, brasses and bronzes, not to be able to determine a yield point stress. The yield point stress is important because it is a guide to the amount of

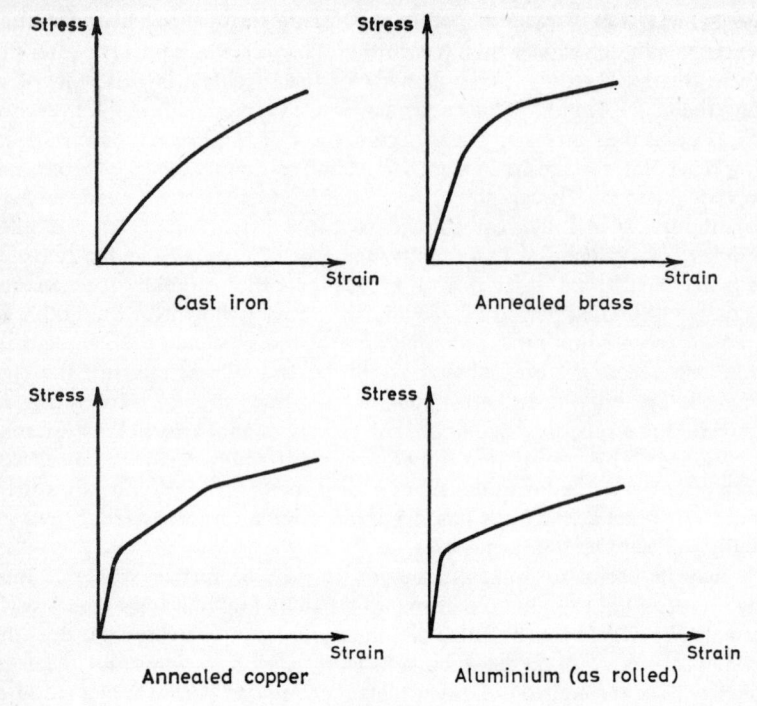

Fig. 63

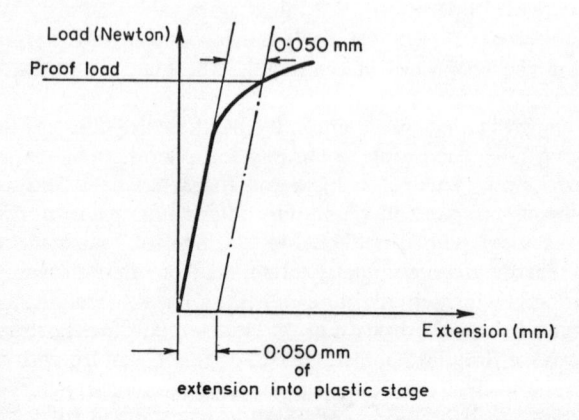

Fig. 64

stress a metal can withstand before the elastic limit is reached. In such cases of no definite yield point, a *proof stress* is used.

Proof stress

This is a stress given by a point on the stress–strain graph at which a certain amount of plastic deformation has taken place. The amount of plastic deformation allowed to take place before the proof stress is measured is usually an extension into the plastic stage equivalent to 0·1 % of the gauge length of the specimen under test. On a specimen of 50 mm, gauge length, this would give a plastic deformation (extension into plastic stage) of 0·1 % of 50 mm = $10^{-3} \times 50$ = 0·050 mm; see Fig. 64.

If the gauge length of the specimen were 200 mm, the proof stress would be taken at a point at which 0·1 % of 200 mm extension into the plastic stage had taken place. This is in fact a procedure for estimating a yield-point stress.

From a stress–strain graph we obtain certain values which have a practical application, namely: Yield Stress (or Proof Stress), Maximum Stress (U.T.S.) and Breaking Stress. These could be calculated using the formulae:

$$\text{Yield Stress} = \frac{\text{Yield Load}}{\text{Original Cross-Sectional Area}}$$

$$\text{Proof Stress} = \frac{\text{Proof Load}}{\text{Original Cross-Sectional Area}}$$

$$(\text{U.T.S.}) \text{ Maximum Stress} = \frac{\text{Maximum Load}}{\text{Original Cross-Sectional Area}}$$

$$\text{Apparent Breaking Stress} = \frac{\text{Fracture Load}}{\text{Original Cross-Sectional Area}}$$

$$\text{True Breaking Stress} = \frac{\text{Fracture Load}}{\text{Cross-Sectional Area at Fracture}}$$

Percentage elongation and reduction in area

Two other values which are obtained from a tensile test, either by calculations or direct from gauges, are: the *percentage elongation* of the specimen and the *percentage reduction in area*. These values are a guide to the ductility of the material being tested. If we consult our Workshop Technology notes, we shall find that this is an essential property of metals which are drawn through dies as in wire drawing. These values may be calculated from the formulae:

$$\text{Percentage Elongation } (\% \text{ El.}) = \frac{\text{Total Extension}}{\text{Original Length}} \times \frac{100}{1}$$

This is really an expression of the *total strain* as a percentage. Since the value of the % El. will vary according to the gauge length over which it is measured, it is usual to state the gauge length on which it is calculated.

$$\text{Percentage Reduction in Area} \atop (\% \text{ Red. in Area}) = \frac{\text{Reduction in Cross-Sectional Area}}{\text{Original Cross-Sectional Area}} \times \frac{100}{1}$$

Experiment. *To obtain stress–strain graphs for mild steel, brass and copper specimens using a Hounsfield tensometer.*

Note the condition of the specimens; e.g. annealed, work hardened, as rolled or cast, specimens machined to Hounsfield specification. Compare the yield (or proof) stresses, the ultimate stresses and breaking stresses, % El. and % Red. in area.

The modulus of elasticity cannot be obtained by the Hounsfield tensometer unless the machine factor is taken into consideration. This is because the extension values recorded are not true values. The stresses are true values, so too are the % El. and % Red. in area obtained from the gauges. However, the modulus of elasticity can be determined using the tensometer along with an extensometer. You might repeat this experiment with the same metals but in different conditions of physical state.

Experiment. *To determine the modulus of elasticity for mild steel, copper, brass and aluminium on a Hounsfield tensometer machine using a Hounsfield extensometer.*

Obtain specimens of the materials machined to the Hounsfield specifications and note their physical conditions. Fix specimens 'in turn' in the Hounsfield tensometer which is to be used for applying the load. Fit the extensometer to the specimen, so that its gripping points are spaced to grip over a 50 mm gauge length. Zero the tensometer and extensometer, then apply a load and read off the corresponding extension. Repeat this procedure for at least eight readings. Tabulate the results, plot appropriate graphs and use them to determine modulus of elasticity and yield (or proof) stress for each material. Make reference to the current Hounsfield publications.

How do the values obtained compare with each other and with the accepted standard values?

Experiment. *To observe the behaviour of various metals under a compressive load.*

Machine specimens of copper, mild steel, aluminium, cast iron and brass to 15 mm diameter 15 mm long. Place each in turn centrally between the platens of a universal testing machine, behind safety guards. Gradually apply the load until the specimen is flattened or fractured. In those instances where fracture takes place you might record the fracture load and use it to obtain the fracture stress. Comment on the final condition of each specimen and suggest what properties are indicated.

Do the specimens which do not fracture have stress lines showing on their surfaces?

Is the fracture inclined to the axis of the specimen? If so, what does this indicate?

Safe working stress—Factor of safety

Clearly the components that make up the workshop crane, or railway bridge spanning a river, or wings of an aircraft, or shackle pins of a road vehicle must not be stressed beyond their elastic limit. This of course applies to a host of other components; the highest stress to which a component is subjected is well below the yield stress and is known as the *safe working stress*. The value of this stress is determined by dividing the ultimate tensile stress by a number known as the *factor of safety*.

$$\text{Safe Working Stress} = \frac{\text{Ultimate Tensile Strength}}{\text{Factor of Safety}}$$

Or we can say that the factor of safety is the ratio between the ultimate strength (stress) and safe working stress,

$$\text{Factor of Safety} = \frac{\text{Ultimate Strength}}{\text{Safe Working Stress}}$$

Suppose a certain component is made from steel which has an ultimate tensile stress of 480 N/mm² and in designing it a factor of safety of 3 is used,

$$\text{Safe Working Stress for component} = \frac{480}{3} = 160 \text{ N/mm}^2$$

This means that it would be working well within its maximum load capabilities and elastic limit stress.

If a component from the same cast of steel were used on a machine, or structure, where the safety requirements demanded a factor of safety of 6

$$\text{Safe Working Stress for component} = \frac{480}{6} = 80 \text{ N/mm}^2$$

We could say that the component in the latter case is twice as strong as that in the former since the stress it is permitted to carry is half as much. A bridge whose design was based on a factor of safety of 4 would be only half as strong as it would be if a factor of safety of 8 were used.

In deciding upon a value for the *factor of safety* for a particular job, the following points are taken into consideration:
1. The nature of the product; e.g. is personal safety affected as in an aircraft?
2. The nature of the material; e.g. the behaviour of cast iron is unpredictable in tension.
3. The consequences of breakdown; e.g. a burst in a steam boiler drum as against a burst in a watering can.
4. The type of loading on the component; e.g. is load applied gradually, or suddenly, or is it impact loading? Is the load constant, or intermittent?

A load gradually applied will cause half the stress that would be caused if load were suddenly applied.
5. The climatic conditions; e.g. are these of a corrosive nature? Do tornado conditions prevail? This is especially important in the case of bridges and similar structures.
6. Errors in workmanship; the dimensions of a component decide the area of a metal resisting stress. A bar could be made on the small side whilst its mating hole might be on the large side.
7. The accuracy of any assumptions made; e.g. in behaviour of material, or loading.

In certain cases, for instance steam boilers and aircraft, the factor of safety will be laid down by regulation of the appropriate Government department.

Although we have generally referred to tensile loading in the preceding notes and formulae, the details are equally true for compressive and shear loading.

Examples

1. The support leg of a shaping machine table is 20 mm dia. and the accuracy standards of the machine demand that it must not shorten more than 0·008 mm on its maximum length of 375 mm. If the leg is made of steel which has a modulus of elasticity of 210×10^9 N/m², find the compressive load that it may withstand.

The arrangement is similar to that shown in Fig. 55 on page 76, the compressive load on the leg being F newton and the leg diameter 20 mm.

Cross-sectional area of leg $= \dfrac{\pi}{4} \times (20)^2 = 100\pi$ mm²

Max. compressive strain on leg $= \dfrac{0\cdot008}{375}$

Max. compressive stress on leg $= \dfrac{F}{100\pi}$ N/mm²

$\dfrac{\text{Compressive stress}}{\text{Compressive strain}} = $ Modulus of Elasticity

$$\dfrac{F}{100\pi} \div \dfrac{0\cdot008}{375} = \dfrac{210 \times 10^9}{10^6} \text{ N/mm}^2$$

$$\dfrac{F}{100\pi} \times \dfrac{375}{0\cdot008} = 210 \times 10^3$$

$$F = 210 \times 10^3 \times \dfrac{0\cdot008 \times 100\pi}{375}$$

$$W = \underline{1\cdot4 \text{ kN}}$$

FORCE AND STRESS

2. On a universal testing machine there are four vertical tie bars of equal diameter fixed to the machine bed and supporting the cylinder. They have to carry a maximum load of 500 kN without extending more than 0·20 mm in 2·5 m. Find the diameter of these bars if they are made from steel having a modulus of elasticity of 210×10^6 kN/m². See Fig. 65 for an illustration of the problem.

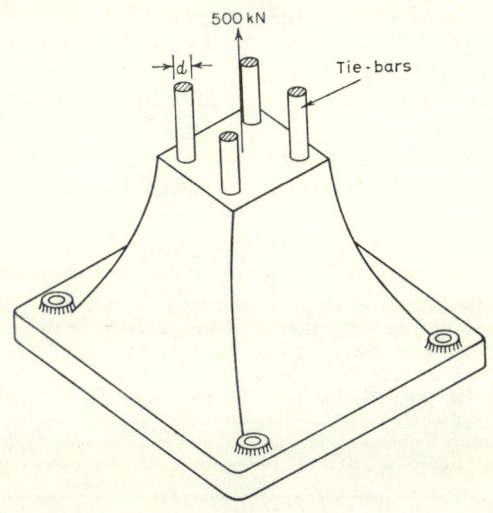

Fig. 65

Let d mm be the diameter of the tie bars.

$$\text{Tensile strain on tie-bars} = \frac{0 \cdot 2}{2 \cdot 5 \times 10^3}$$

$$\frac{\text{Tensile stress}}{\text{Tensile strain}} = \text{Modulus of Elasticity}$$

$$\frac{\text{Tensile stress}}{0 \cdot 2/(2 \cdot 5 \times 10^3)} = \frac{210 \times 10^6 \times 10^3}{10^6} \text{ N/mm}^2$$

$$\text{Tensile stress} = \frac{210 \times 10^3 \times 0 \cdot 2}{2 \cdot 5 \times 10^3}$$

$$\text{Tensile stress} = \frac{500 \times 10^3}{\text{Cross-sectional area of four tie-bars}}$$

$$\frac{42}{2 \cdot 5} = \frac{500 \times 10^3}{\text{Cross-sectional area of four tie-bars}}$$

Cross-sectional area of four tie-bars $= \dfrac{500 \times 10^3 \times 2\cdot 5}{42}$

Cross-sectional area of one tie-bar $= \dfrac{1250 \times 10^3}{42 \times 4}$

$$\dfrac{\pi d^2}{4} = \dfrac{1250 \times 10^3}{42 \times 4}$$

$$d^2 = \dfrac{1250 \times 10^3}{42\pi}$$

$$d = \sqrt{\dfrac{1250 \times 10^3}{42\pi}}$$

$$d = 97\cdot 5 \text{ mm}$$

3. In a tensile test on an aluminium alloy specimen 14 mm diameter, gauge length 150 mm, the following extensions were obtained for the given loads.
   The maximum load was 48·5 kN, the total extension 31·1 mm, the final diameter 11·48 mm, and the load at fracture 40 kN. Find the modulus of elasticity for the material, the proof stress based on 0·1% gauge length of plastic deformation, the limit of proportionality stress, maximum stress, the percentage elongation, the percentage reduction in area and the true and apparent stresses at fracture.

Load, kN	2·0	4·0	6·0	8·0	10·0	12·0	14·0
Extension mm	0·075	0·100	0·125	0·150	0·163	0·200	0·228

Load, kN	16	18	20	22	24	26
Extension, mm	0·250	0·275	0·300	0·312	0·325	0·338

Load, kN	28	30	32	36	40	44
Extension, mm	0·350	0·387	0·425	0·562	1·125	1·800

Load, kN	46	48	46	44	42	40
Extension, mm	2·650	4·000	4·280	5·750	5·970	6·120

We will begin by plotting load–extension graphs, see Figs. 66 and 67.

$$\text{Modulus of Elasticity} = \dfrac{\text{Stress}}{\text{Strain}} = \dfrac{\text{Load}}{\text{X-Sect. Area}} \times \dfrac{\text{Orig. length}}{\text{Extension}}$$

$$= \dfrac{\text{Load}}{\text{Extension}} \times \dfrac{\text{Orig. length}}{\text{X-Sect. Area}}$$

The value for $\dfrac{\text{Load}}{\text{Extension}}$ is given by the gradient of the load-extension graph.

Initial cross-sectional area $= \dfrac{\pi}{4}(14)^2 \qquad = 154 \text{ mm}^2$

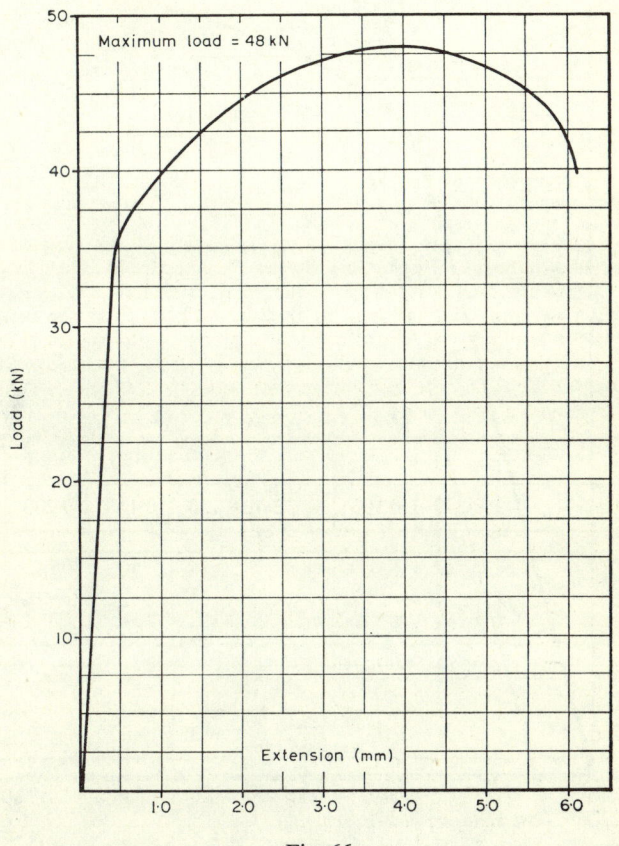

[Graph: Load (kN) vs Extension (mm); Maximum load = 48 kN]

Fig. 66

Final cross-sectional area $= \dfrac{\pi}{4}(11\cdot 48)^2 \qquad = 103 \text{ mm}^2$

Modulus of Elasticity $= 76\cdot 4 \times 10^3 \times \dfrac{150}{154} = \underline{74\cdot 5 \times 10^3 \text{ N/mm}^2}$

Plastic deformation

for proof load = 0·1% of 150 = 0·150 mm
Proof load = 36·4 kN (see Fig. 67)

$$\text{Proof stress} = \frac{36 \cdot 4 \times 10^3}{154} = \underline{236 \text{ N/mm}^2}$$

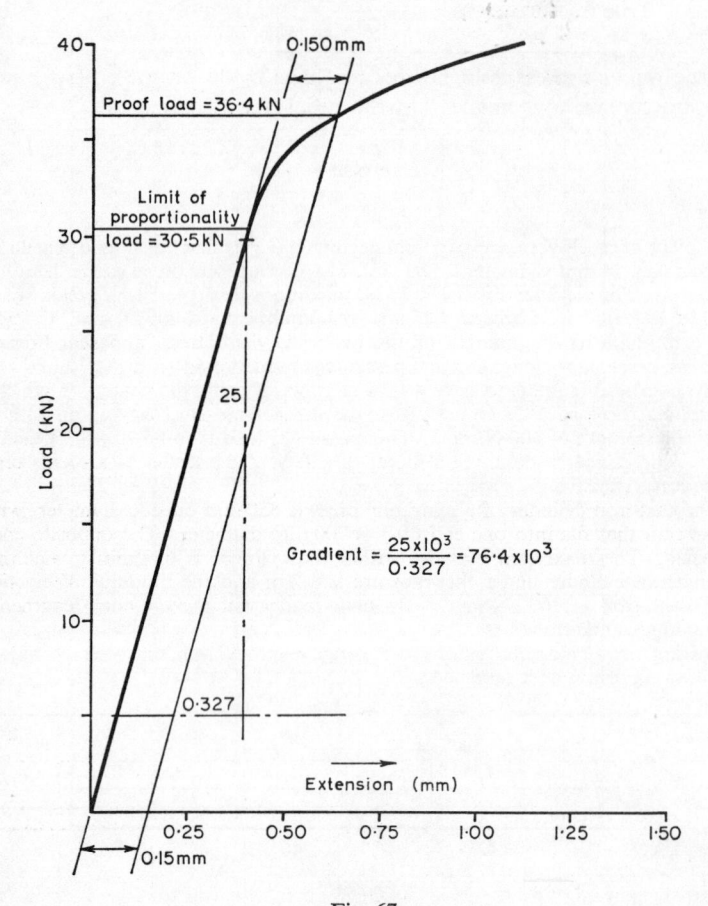

Fig. 67

$$\text{Limit of proportionality stress} = \frac{30 \cdot 5 \times 10^3}{154} = \underline{198 \text{ N/mm}^2}$$

Whichever one was used would be a matter of choice for the designer.

$$\text{Percentage elongation} = \frac{31 \cdot 1}{150} \times 100 = \underline{21\%}$$

$$\text{Percentage reduction in area} = \frac{(154-103)}{154} \times 100 = \underline{33\%}$$

$$\text{Apparent fracture stress} = \frac{40 \times 10^3}{154} = \underline{260 \text{ N/mm}^2}$$

$$\text{True fracture stress} = \frac{40 \times 10^3}{103} = \underline{390 \text{ N/mm}^2}$$

The true fracture stress is higher because it is based on the cross-sectional area at fracture which is smaller than the initial cross-sectional area.

Exercise 6

Section B

1. A piece of steel from a broken connecting rod was machined as a tensile test specimen 14 mm diameter. The test was carried out on a gauge length of 50 mm. The yield load was 112·7 kN, maximum load 143·5 kN, breaking load 111·5 kN, fracture diameter 9·25 mm, maximum extension 9·50 mm. Calculate the ultimate tensile strength of the material, yield stress, apparent breaking stress, percentage elongation and percentage reduction in area.
2. An overhead crane is to have 4 falls of rope, of equal diameters, on which to support the load. The ropes are to be made from steel having an ultimate tensile strength of 600 N/mm². The maximum load is to be 20 Mg. Calculate the rope diameter assuming a factor of safety of 6 and 0·8 of a rope's cross-section is effective.
3. The cast-iron cylinder of a hydraulic press is 850 mm outside diameter, whilst the ram that fits into one end of it is 700 mm diameter. The opposite end is sealed. The maximum pressure within the cylinder is 6 N/mm²; maximum length of cylinder under this pressure is 1·2 m and the modulus of elasticity of cast iron is 100 kN/mm². By how much will the cylinder lengthen at maximum conditions?
4. During a test on an aluminium wire, gauge length 50 mm, diameter 14 mm, the following results were obtained:

Load, kN	0	10	20	30	35	40
Extension, mm	0	0·020	0·040	0·057	0·067	0·080

Load, kN	42·5	45	46			
Extension, mm	0·095	0·125	0·162			

Determine the proof stress of the material for a plastic deformation equivalent to 0·1% of the gauge length.
5. The chain link coupler between the locomotive and a train of wagons is made of wrought iron having an ultimate tensile strength of 450 N/mm². There are 35 wagons, each of 10 Mg mass, and the tractive resistance is 200 N/Mg. Using a factor of safety of 6, determine the least permissible diameter of the link material.

6. On a small light-alloy press the two tie-bars are made from aluminium alloy and have to withstand a total load of 16 kN. The diameter of each bar is 12 mm and length 225 mm. If the modulus of elasticity of the metal is 90×10^9 N/m^2, calculate the extension of the bars under maximum load assuming it to be the same on both.
7. Two specimens, 18 mm diameter by 20 mm long, one copper and the other phosphor bronze, are subjected to a compressive load of 72 kN. Determine by how much each shortens if their modulii of elasticity are 115 kN/mm^2 and 108 kN/mm respectively.
8. A 20 mm diameter hole is to be punched in mild steel plate 6 mm thick and ultimate shear strength 340 N/mm^2. If the stress in the punch shank is not to exceed 120 N/mm^2, what must its diameter be?
9. In a tube drawing operation the drawn sizes of the mild steel tube on one particular pass are 60 mm outside diameter and 50 mm inside diameter. The tensile stress within the steel is not to exceed 250 N/mm^2. Calculate the maximum permissible drawing force.
10. Mild steel bar of 20 mm diameter is being cold drawn in lengths of 6 m. The yield stress of the steel is 250 N/mm^2, and its modulus of elasticity is 200 kN/mm^2. What is the maximum drawing force that may be applied to the bar before it yields and the extension at this load on a 5 m length?

Chapter 7. Moment of a Force

We have learned from Chapter 6 some of the consequences of force in the workshop; in this chapter we will learn yet another.

In order to find out what is meant by the statement 'Moment of a force about a point', we will consider the use of an open jaw spanner.

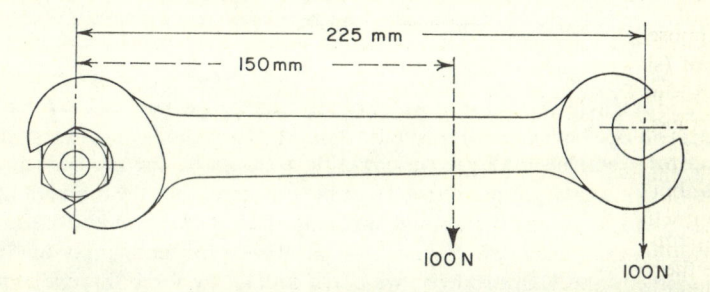

Fig. 68

Figure 68 shows an open jaw spanner which has an effective length of 225 mm. We will assume that a force of 100 N is applied in the position shown in full line, at a distance, 225 mm from the centre of the nut being tightened.

We know from our workshop experience that as the nut is tightened the 100 N force moves in a circular path, the point about which the force turns being the centre of the nut. The nut will reach a certain degree of tightness beyond which it cannot be tightened any more by the 100 N force acting at a distance, 225 mm from the point about which it was turning.

We therefore have a situation in which a force is either turning, or tending to turn, about a point. In such a circumstance we say that the force has a *turning moment about the point*, or that there is a *moment of force about a point*. Actually *turning moment* and *moment of a force* about a point are the same thing.

Calculation of the moment of a force about a point

The *moment of a force* is calculated as follows:

moment of a force (Nm) about a point

= force (newton) × distance (metre) of force from the point, measured at 90° to the direction of the force.

The force could be measured in some multiple of the newton, e.g. kilonewton, meganewton, whilst the distance could be measured in millimetres. The unit of the moment would be modified accordingly.

The point about which the turning takes place is sometimes called a *pivot* or *fulcrum*. We shall see this later in the worked examples. Let us calculate the size of the *moment* on our spanner.

TURNING MOMENT on spanner = 100 (N) × 225 (mm) = <u>22·5 Nm</u>

Suppose this 100 N force was applied at a point 150 mm from the centre of the nut (shown in dotted line in Fig. 68), then the *turning moment* on the spanner in this instance would be:

TURNING MOMENT on spanner = 100 (N) × 150 (mm) = <u>15 Nm</u>

This means that by shortening the distance between the force and the point about which it is turning we have reduced the *moment* of the *force* and the nut will not be tightened as much. On the other hand, we can see that if our spanner had been longer we could have increased the moment and so made the nut tighter, although we must bear in mind that the moment could be made too large and we should find ourselves twisting the bolt end off, or stripping the thread. If, however, we did have a longer spanner and used our judgment and skill properly, we could reduce the force that we applied to the spanner and so reduce the strain on our body. The distance between the force and the point about which it acts has been taken into consideration in the designing of spanners, so that under normal conditions a moment large enough to twist the bolt end off cannot be caused. Spanners of different sizes are made in different lengths. The spanners for small nuts, which do not require such large *turning moments* to tighten them, are made shorter than the spanners for larger nuts. In this practice, it is assumed that the average pull that we can cause normally with our hands will remain about the same. Even so, we must beware of our strength when tightening small nuts and set-screws, etc. Some spanners are made to slip at a specific torque, e.g. a torque wrench. Such spanners are used increasingly in the assembly of components, e.g. motor vehicle engines, food mixers, vacuum cleaners, to mention but a few.

Types of moments

So far we have only thought of tightening the nut, and said nothing of slackening it. The main difference between tightening and slackening the nut is that the moment moves in an exactly opposite direction when slackening. There are only two possible directions in which the moment can move. These are clockwise and anti-clockwise. This applies to all turning moments, since they move in a circular path.

In the engineering workshop we can see numerous operations in which a turning moment is used, e.g., tightening and slackening a vice, using a tommy bar, operating the gear lever on a lathe headstock, or clutch lever on a lathe, hand-feed on a drilling machine, using a crow-bar, or lever, on guillotine shears, etc., in fact all types of levers.

A number of varied experiments should be carried out to verify the principle of moments; e.g.,

1. Pivot a bar at one end and support it, in a horizontal position, by a spring balance at the other end. Attach loads to the bar at various points along its length, then bring it into a horizontal position again by adjusting the screw at spring balance support. Check the spring balance reading by calculation.

2. Determine the reactions at the bearings of a lathe spindle by supporting the spindle, on spring balances, at the journals.

Examples

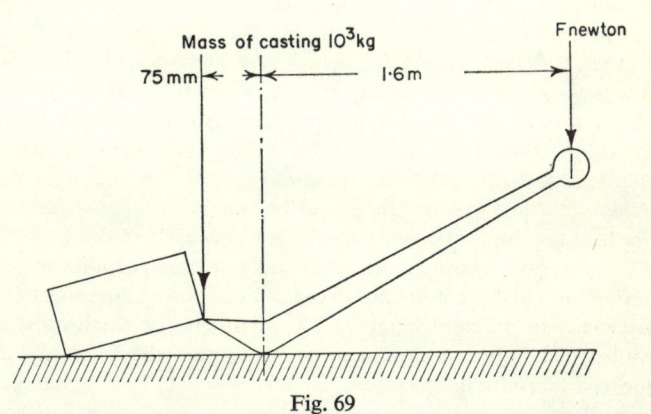

Fig. 69

1. Figure 69 shows a crow-bar being used to partly lift a casting so that a chain sling can be put round it. What force, F newton, must be applied to the upper end of the crow-bar in order to just lift the casting? The load on the lower end of the bar may be taken as 10^3 kg.

    Note: The gravitational force acting on a 1 kg mass is 9·81 newton. Therefore the gravitational force acting on 10^3 kg (1 Mg) is $10^3 \times 9·81$ newton.

    MOMENT of F about pivot
    $= F \times 1·6$ Nm (this is a CLOCKWISE MOMENT)

    MOMENT caused by the gravitational force on the 10^3 kg mass about pivot
    $= (10^3$ kg$) \times 9·81 \times 75 \times 10^{-3}$ Nm (this is an ANTI-CLOCKWISE MOMENT)

    When casting is just lifted, and the system is in balance,

    CLOCKWISE MOMENT = ANTI-CLOCKWISE MOMENT

    MOMENT of force F balances MOMENT of $10^3 \times 9·81$ N

    $$F \times 1·6 = 10^3 \times 9·81 \times 75 \times 10^{-3}$$

    $$F = \frac{9·81 \times 75}{1·6} = \underline{460 \text{ N}}$$

2. A plan view of a clutch lever fitted to a lathe headstock is shown in Fig. 70. The sliding member of the clutch requires a force of 100 N to be applied to it in order to operate it. Calculate the force F newton that has to be applied to the clutch lever in order to operate the clutch.

In this problem we must note that, if the sliding member requires a force of 100 N to move it into engagement with the other half of the clutch, it must be offering a resistance of 100 N, and it is this *resisting force* which we have to consider in our calculations. As can be seen from Fig. 70, this *resisting force* has an *anti-clockwise moment* about the pivot. This has to be balanced by the *clockwise moment* of the force F newton applied by the operator.

Therefore by taking moments about the pivot we have:

CLOCKWISE MOMENT = 0·3(m) × F(newton) = 0·3 F newton m

ANTI-CLOCKWISE MOMENT = 0·09(m) × 100(N) = 9 Nm

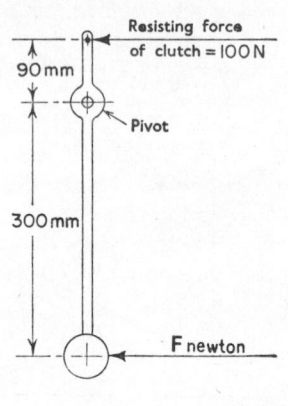

Fig. 70

At the instant that the *clockwise moment* balances the *anti-clockwise moment*, the sliding member of the clutch is on the point of moving into its driving position. Hence we have:

CLOCKWISE MOMENT = ANTI-CLOCKWISE MOMENT

0·3 F = 9

$$F = \frac{9}{0·3} = \underline{30\ N}$$

A Reaction

Before we work out Example 3, let us find out what is meant by a *reaction*.

When we stand at our lathe, or shaping machine, or bench, what supports us? Of course it is the floor we stand on. Yes, ' of course ', but let us look a little more closely. The mass of the body attracted to the earth by the earth's gravitational pull constitutes a downward force, ' weight ', and this force has to be supported. It is supported by an equal and directly opposite force which is part of the floor. This force is part of the strength of the floor and adjusts itself to balance the force that it is opposing. For instance, if it is a 70 kg man

MOMENT OF A FORCE

who is standing by the bench, then the balancing force in the floor is $70 \times 9 \cdot 81$ N. If the man has a mass of 85 kg, then the balancing force is $85 \times 9 \cdot 81$ N, and so on. There is a limit to the resistance of the floor. That limit is reached when the floor has not got sufficient strength to cause a balancing force equal to the force that is acting on it. The result is that the floor will break if it is hollow underneath, or it may be deformed if it is more solid. This balancing force is known as a *reaction*. We note from chapter six that when we apply a force to a piece of metal a resisting force is set up within the metal. We could quite rightly call this a *reaction*.

As the metal in a turning operation is turned on to the tool point with a specific force, a reaction is set up in the tool point equal and opposite to this force. When the shaper tool is driven into the metal with a certain force, there is a reaction within the metal to this force. This applies in all metal cutting operations and when the driving force becomes too great for the reaction to resist, the metal is cut. The maximum value of the reaction can only be as great as the strength of the metal.

The important fact to learn from this is that to every applied force there is always a reacting force which is called a *reaction*. When the reaction cannot sustain the applied force the material constituting the reaction breaks or is cut (as in metal cutting), or movement takes place, e.g., as in levers. The applied force need only be slightly larger than the reaction in order to overcome it. To every *action* there is an equal and opposite *reaction*.

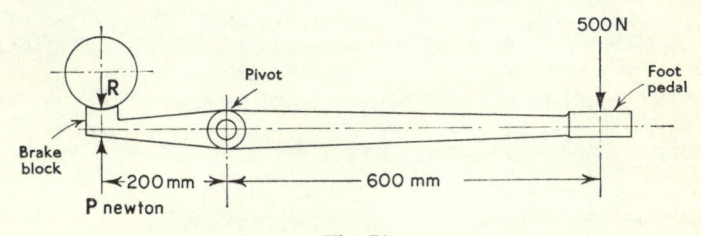

Fig. 71

3. In Fig. 71 an arrangement of a foot brake fitted to a winch is shown. Calculate the force P newton that is acting on the brake block when a force of 500 N is applied to the foot pedal.

For the purpose of solving this problem, we must consider the reaction R instead of the applied force P. By this means we can consider the moment that opposes the moment of the force applied to the foot pedal.

By taking moments about the pivot we have:

CLOCKWISE MOMENT $= 500 \times 0 \cdot 6$ Nm

ANTI-CLOCKWISE MOMENT $= R \times 0 \cdot 2$ Nm

When the *clockwise moment* balances the *anti-clockwise moment*,

$$0 \cdot 2 R = 500 \times 0 \cdot 6 \qquad R = \frac{500 \times 0 \cdot 6}{0 \cdot 2} = 1500 \text{ N}$$

Since R is the *reaction* to P it has the same value as P. Therefore, the force, P newton, acting on the brake block is

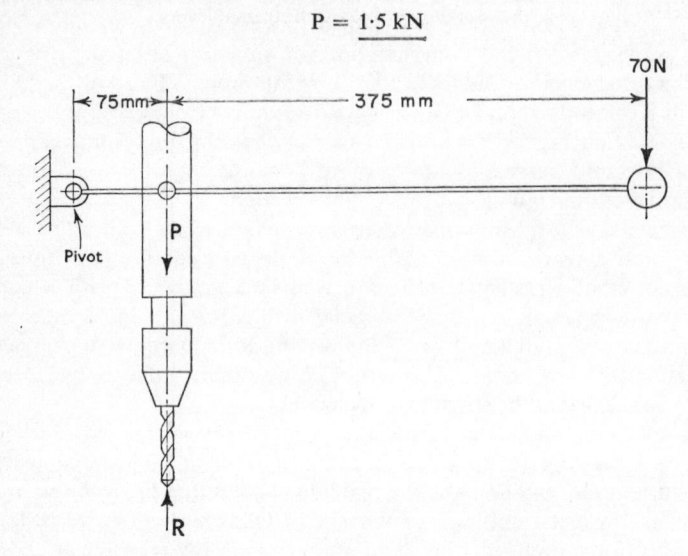

$$P = \underline{1\cdot5 \text{ kN}}$$

Fig. 72

4. Figure 72 shows the hand feed mechanism of a small drilling machine. What is the feed force acting on the drill when a force of 70 N is applied to the feed handle by the operator?

As in the previous problem we must consider the *reaction* R when taking moments.

By taking *moments* about the *pivot* we have:

CLOCKWISE MOMENT = 70 × 0·45 Nm

Note: The line of action of the 70 N force is (375 mm + 75 mm = 450 mm), 0·45 m from the pivot.

ANTI-CLOCKWISE MOMENT = R × 0·075 Nm

When the *clockwise moment* balances the *anti-clockwise moment*,

$$R \times 0\cdot075 = = 70 \times 0\cdot45$$

$$R = \frac{70 \times 0\cdot45}{0\cdot075} = 420 \text{ N}$$

Since R is the *reaction* to P, it is equal to P.

The *force* acting on the drill = $\underline{420 \text{ N}}$

5. A bell crank lever that is part of a belt shifting mechanism is shown in Fig. 73. The belt must have a 110 N force applied to it to cause it to slide from the loose pulley to the fast pulley. What force P newton must be applied to the other arm of the lever by the operator?

We note that, if the belt requires a force of 110 N to push it from one pulley to the other, it has a resistance of 110 N, and it is the resisting force that we

MOMENT OF A FORCE   103

must use in our calculation of moments. The next point to appreciate is that the method for solving this problem, where we have a cranked lever, is exactly the same as that used when solving problems on straight levers.

By taking *moments* about pivot we have:

CLOCKWISE MOMENT $= P \times 0.115$ Nm

ANTI-CLOCKWISE MOMENT $= 110 \times 0.08$ Nm

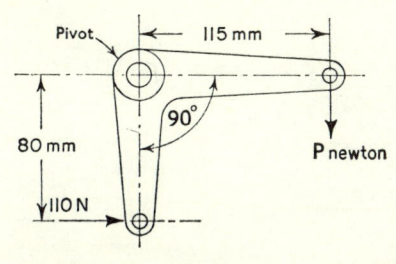

Fig. 73

At the instant the belt begins to slide from the loose pulley to the fastened pulley, the *clockwise moment* balances the *anti-clockwise moment* and

$$P \times 0.115 = 110 \times 0.08$$

$$P = \frac{110 \times 0.08}{0.115}$$

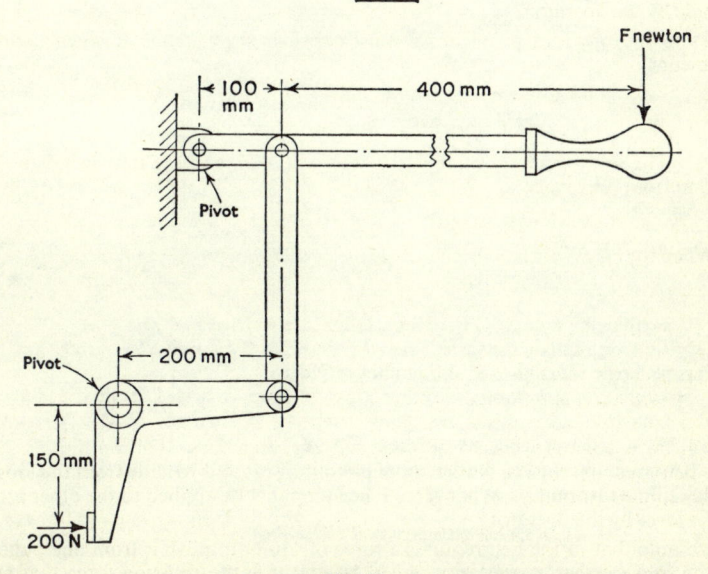

$$P = \underline{76.5 \text{ N}}$$

Fig. 74

6. Figure 74 shows a hand-operated knock-off mechanism for knocking off components from a conveyor. A force of 200 N must be applied to a component to knock it off. What force F newton must be applied to the handle of the mechanism in order to push off each component?

This is really two problems in one, for we have a straight lever connected to a cranked lever.

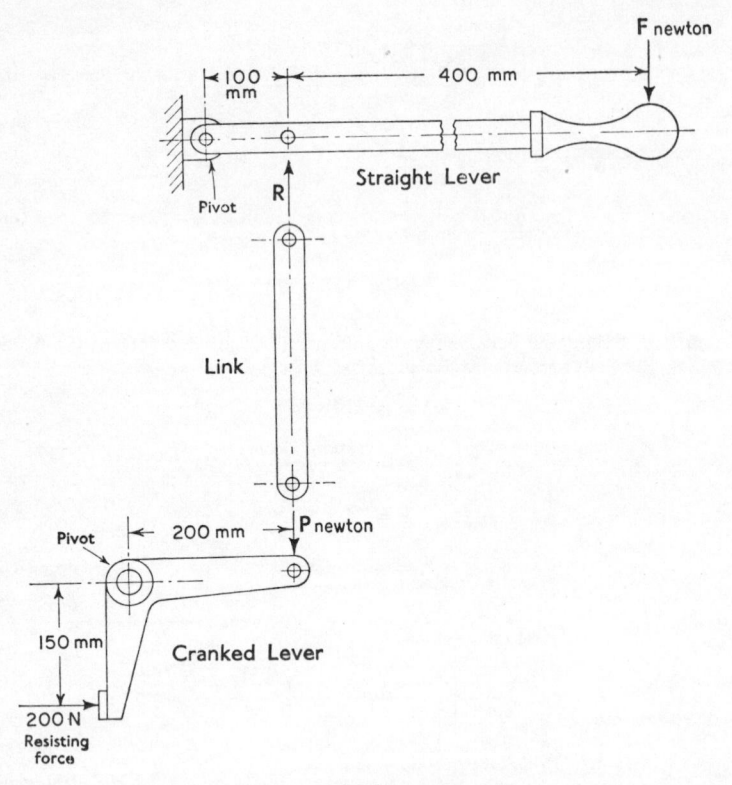

Fig. 75

We will begin by drawing each part separately. (See Fig. 75.)

Next, we calculate the force P newton that must be applied to the cranked lever in order to produce the required knock-off force.

Now if each component requires a 200 N force to knock it off the conveyor, it means that each has a resistance to being pushed off of 200 N, so there is a 200 N resisting force, as shown in Fig. 75.

Therefore by taking moments about the pivot of the cranked lever to find force, P, we have:

CLOCKWISE MOMENT $= P \times 0.2$ Nm

and ANTI-CLOCKWISE MOMENT $= 200 \times 0.15$ Nm

MOMENT OF A FORCE

At the instant a component is pushed off, the *clockwise moment* balances the *anti-clockwise moment*,

$$0.2 \, P = 200 \times 0.15$$

$$P = \frac{30}{0.2} = 150 \text{ N}$$

Now this force P is caused by applying a force F newton to the handle of the straight lever, F being carried to the cranked lever by the link.

Next, in order to calculate F we must consider the reaction to P instead of P itself. We will call this reaction, R, and R = P.

Then, by taking moments about the pivot of the straight lever to find F we have:

CLOCKWISE MOMENT = $F \times 0.5$ Nm

and ANTI-CLOCKWISE MOMENT = $R \times 0.1 = 150 \times 0.1$ Nm

since R = P, and P = 150 N

At the same instant as the two moments balance each other on the cranked lever, they do so on the straight lever. Hence we have:

$$F \times 0.5 = 150 \times 0.1 \qquad F = \frac{15}{0.5} = \underline{30 \text{ N}}$$

The force that must be applied to the handle of the knock-off mechanism in order to push off each component is 30 N.

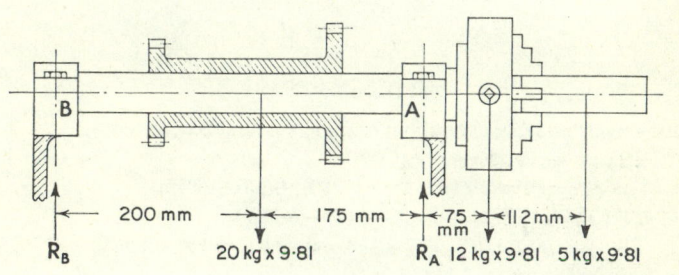

Fig. 76

7. A lathe spindle supported on two bearings and which has a four jaw chuck fitted to it is shown in Fig. 76. The total mass of the spindle and the gears that it carries is 20 kg. The chuck mass is 12 kg and that of the job it grips 10 kg. The gravitational effect of these masses may be taken as acting from the positions shown. Find the reactions R_A and R_B in each bearing and the directions in which they act.

This example appears to be different from the ones that we have so far tackled. The previous problems concerned levers. We can use the same method for solving this problem as we used for solving those.

We should note that each bearing supports a certain load because of the gravitational effects on the masses that it is helping to support. These loads are not necessarily equal. In fact, in our example they are not. We can see by inspection of the diagram that the load on bearing A is likely to be greater than that on bearing B. The positioning of the chuck, job and gears governs this. It is not the load on each bearing we are asked to calculate, it is the *reaction*, R_A and R_B, to each load. The reactions of the bearings are equal to the loads on the bearings.

To solve a problem by *moments* we must have a *pivot*, and in this example there does not appear to be one. In fact, we could take the pivot at any point along the centre line of the spindle. It will be convenient to take it at either A or B, so that we eliminate one of the unknown *reactions*. Thus we find only one *unknown* at a time. Begin by taking moments about bearing A in order to determine the magnitude of reaction B. (We could have begun by taking moments about bearing B and found the reaction in bearing A.)

Note: Gravitational force on a mass = Mass (kg) × 9·81 = newton

CLOCKWISE MOMENTS

$$= (12 \times 9·81 \times 0·075) + (5 \times 9·81(0·075 + 0·112)) + R_B(0·2 + 0·175)$$
$$= (12 \times 9·81 \times 0·075) + (5 \times 9·81 \times 0·187) + 0·375 \, R_B$$
$$= 8·83 + 9·17 + 0·375 \, R_B$$

ANTI-CLOCKWISE MOMENT

$$= 20 \times 9·81 \times 0·175$$

Now, the lathe spindle is in a horizontal position and is never moved out of this position irrespective of what mass of job is put in the chuck, i.e., it remains in a balanced position. It is said to be in *equilibrium* and, when a system of *moments* is in *equilibrium* or *balance*, the sum of the clockwise moments equals the sum of the anti-clockwise moments,

$$8·83 + 9·17 + 0·375 \, R_B = 20 \times 9·81 \times 0·175$$
$$0·375 \, R_B = 34·3 - 18$$
$$R_B = \frac{16·3}{0·375}$$
$$\text{and } R_B = \underline{43·3 \text{N}}$$

Next we will take moments about bearing B to find the reaction A.

CLOCKWISE MOMENTS

$$= (20 \times 9·81 \times 0·2) + (12 \times 9·81(0·2 + 0·175 + 0·075))$$
$$\quad + (5 \times 9·81(0·2 + 0·175 + 0·075 + 0·112))$$
$$= (20 \times 9·81 \times 0·2) + (12 \times 9·81 \times 0·45) + (5 \times 9·81 \times 0·562)$$
$$= 9·81(4 + 5·4 + 2·81)$$

ANTI-CLOCKWISE MOMENT

$$= R_A(0·2 + 0·175)$$
$$= 0·375 \, R_A$$

Remembering that the spindle is in *equilibrium*,

Sum of ANTI-CLOCKWISE MOMENTS = Sum of CLOCKWISE MOMENTS

$$0·375 \, R_A = 9·81(4 + 5·4 + 2·81)$$
$$R_A = \frac{9·81 \times 12·21}{0·375} = \underline{320 \text{ N}}$$

Since the signs of R_A and R_B are positive their directions are as shown in Fig. 76.

When a system of forces is in equilibrium, besides their moments being in balance the forces must be in balance, both in vertical and horizontal directions. In this case,

TOTAL UPWARD FORCES = TOTAL DOWNWARD FORCES

$$R_A + R_B = 9·81(20 + 12 + 5)$$
$$320 + 43·3 = 363·1$$
$$363 \text{ N} = 363 \text{ N}$$

MOMENT OF A FORCE

Having determined one of the reactions by the principle of moments, the other could have been determined by this latter method,

$$R_A + 43 \cdot 5 = 363$$
$$R_A = 363 - 43$$
$$R_A = \underline{320 \text{ N}}$$

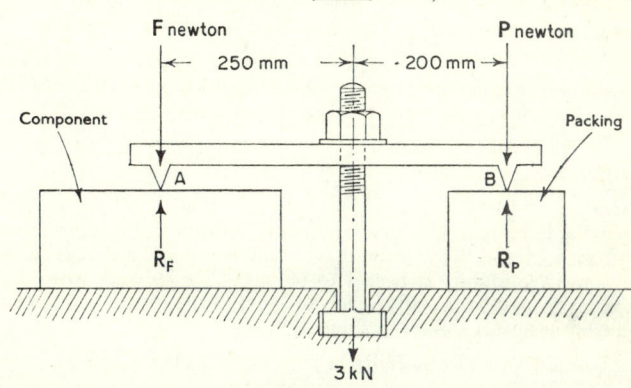

Fig. 77

8. Figure 77 shows how a component is clamped to the table of a drilling machine. Calculate the clamping force, F, if the pull in the bolt is 3 kN.

Because of the clamping force in the bolt, clamping forces are caused at points A and B. These are F newton and P newton respectively. F has a reaction R_F and P a reaction R_P. It is F that we have to find and we can do this by first finding R_F.

To find R_F we will take moments about point B.

CLOCKWISE MOMENT $= R_F(0 \cdot 25 + 0 \cdot 20) = 0 \cdot 45 \, R_F$

ANTI-CLOCKWISE MOMENT $= 3 \times 10^3 \times 0 \cdot 200 = 600$ Nm

Since the clamping set-up remains stationary, the clockwise moment must be balancing the anti-clockwise moment, and the opposing forces equal, i.e., the set-up is in equilibrium.

$$0 \cdot 45 \, R_F = 600$$
$$R_F = \frac{600}{0 \cdot 45}$$
$$\text{and } R_F = \underline{1 \cdot 33 \text{ kN}}$$

We already have learned that an *applied force* is equal to its *reaction*. Hence the clamping force on the component is 1·33 kN.

A point worth noting here, is that if we had taken point A nearer the clamping bolt, F would have been greater. On the other hand, if we had taken it further away, the clamping force would have been less.

108   MATERIALS AND MACHINES IN THE WORKSHOP

9. An adjustable tap wrench 240 mm long has a single hole cut in its centre to fit a range of taps. In a particular tapping operation equal forces of 15 N are applied at each end of the wrench. What is the turning moment on the tap?

Let us make a sketch as shown in Fig. 78, so that we understand the problem more easily. A plan view will show the most detail.

We will assume that a right-hand thread is being cut. The pivot is the axis of the tap.

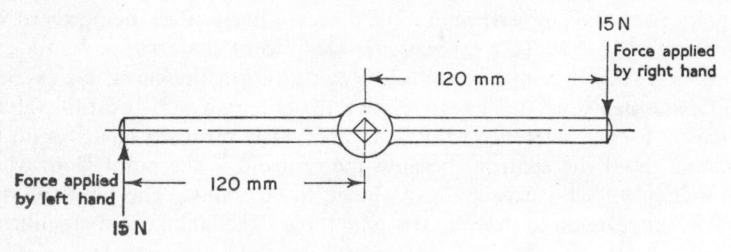

Fig. 78

By taking moments about the centre of the tap,

$$\text{TURNING MOMENT} = (15 \times 0.120) + (15 \times 0.120) \text{ Nm}$$
$$= 1.8 + 1.8$$
$$= \underline{3.6 \text{ Nm}}$$

The moment is a clockwise moment.

When two forces are acting in the manner shown in Fig. 78, i.e., two equal forces with their lines of action parallel, but their directions—in the 'up and down' sense—directly opposite, they are known as a *couple*.

A couple

When two equal forces acting in opposite directions have their lines of action parallel they constitute a *couple*, see Fig. 79.

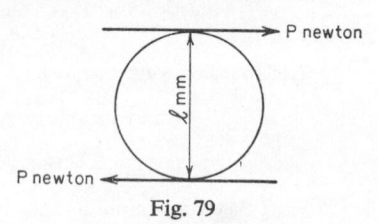

Fig. 79

Moment of a couple

The moment of a couple is given by the product of one of the forces and the normal distance between them. Referring to Fig. 79, by moments about a point midway between the two forces (or any other point) we have,

$$\text{Moment of couple} = \left(P \times \frac{l}{2}\right) + \left(P \times \frac{l}{2}\right), \text{ in a clockwise sense.}$$

$$\text{Moment of couple} = P\left(\frac{l}{2} + \frac{l}{2}\right)$$
$$= P \times l,$$

as stated.

Centre of Area

The centre of area of a surface is sometimes referred to as the *centroid*. If we experiment with a postcard, there is a point on it, which if placed on a pin-point (with the pin vertical and the postcard horizontal) the postcard will balance, see Fig. 80. This point is the centroid of the area, or lamina. A lamina is a sheet having a relatively small uniform thickness, e.g. a sheet of tin plate has small thickness relative to its length and breadth. Hence the reason for the reference to it as an area. The postcard balances on the pin-point about the centroid because the centroid is the point from which the ' weight ' of the card may be assumed to be acting. The pin as it were provides the reaction to this gravitational force. The lamina is in equilibrium about this point.

It is easy to obtain the centroid of a rectangular lamina because it is in fact at its geometric centre, i.e. the point at which the diagonals intersect. The centroid of a circular lamina is the point from which the lamina is described, i.e. the centre of the circle. Try balancing a square and a circular lamina on a pin point about their geometric centres. Whilst the centroid of a regular shaped lamina can be obtained readily without calculation, calculations have to be used to find the centroid position of more complicated shapes. The principle of moments is used to solve such problems.

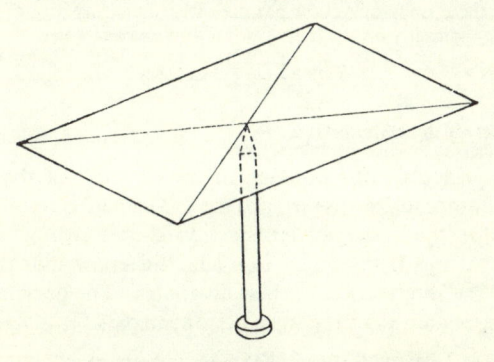

Fig. 80

Experiment. *To find the position of the centroid of the lamina shown in Fig. 81.*

Cut out a lamina from cardboard, or tinplate, to the dimensions given, or to your own dimensions, and suspend it by means of a pin as shown in Fig. 82(*a*), from a forces board. Attach a plumb-bob to the pin. Swing the lamina slightly on the pin and wait for it to become stationary. When this is achieved, carefully trace the plumb-line on to the lamina. Repeat the whole procedure with the lamina suspended from some other point as indicated in Fig. 82(*b*).

The two plumb-line traces should intersect, the point of intersection being the centroid of the lamina. You might swing the lamina from a third position as shown by Fig. 82(c), merely as a check. Try to balance the lamina at this point on a pin point. What do you conclude? Does the centroid lie on the centre line XX? Give reasons for your answer.

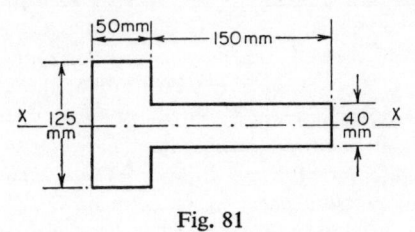

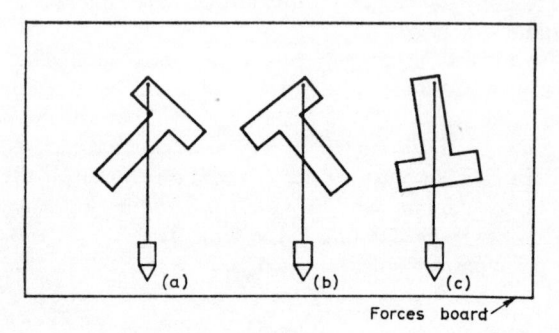

Fig. 81

Fig. 82 (a) (b) and (c)

Position of centroid by calculation

We will now calculate the position of the centroid of the lamina shown in Fig. 81, and compare the answer with the experimental result.

First we divide the T-shaped lamina into simple regular shapes, i.e. the two rectangles A and B shown in Fig. 83. We know that the centroids of A and B are at the intersection of their diagonals. The gravitational pulls on the rectangles are taken as acting from these centroids.

$$\text{Weight of A} = 6{\cdot}25 \times 10^3 w \text{ newton}$$

$$\text{Weight of B} = 6 \times 10^3 w \text{ newton}$$

where w is the ' weight ' of the lamina in newton per mm^2.

If we were to put a single support under the lamina so as to ' balance ' it, the line of this support would have to pass through the centroid of the lamina, i.e. through the point whose position we have to determine. The centroid of the lamina lies on the line XX because the lamina is symmetrical about this line, i.e. of equal area on either side. The support that we would

MOMENT OF A FORCE 111

balance the lamina on would have to carry a load of $6·25 \times 10^3 w + 6 \times 10^3 w$ since, for equilibrium, upward forces equal downward forces, see page 106. Suppose the centroid of the lamina is distance $\bar{x}$ from the edge YY and that we take moments about this edge.

$$\text{Clockwise moment} = (6·25 \times 10^3 w \times 25) + (6 \times 10^3 w \times 125)$$
$$= 156·25 \times 10^3 w + 750 \times 10^3 w$$
$$= 906 \times 10^3 w$$
$$\text{Anti-clockwise moment} = 12·25 \times 10^3 w \times \bar{x}$$
$$= 12·25 \times 10^3 w \bar{x}$$

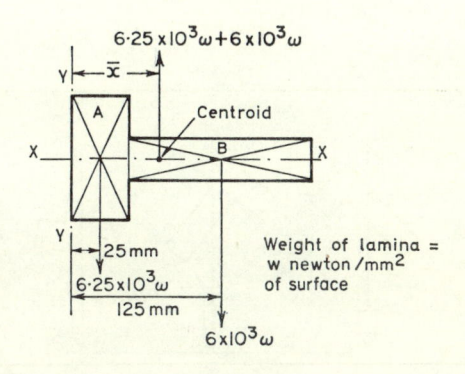

Fig. 83

For equilibrium,

$$\text{Anti-clockwise moment} = \text{Clockwise moment}$$
$$12·25 \times 10^3 w \bar{x} = 906 \times 10^3 w$$
$$\bar{x} = \frac{906 \times 10^3 w}{12·25 \times 10^3 w}$$
$$= 74 \text{ mm.}$$

The w cancels, so we could have ignored it in the first place. In other words we could have found the moment of an area (correct name, *first moment of area*) instead of the moment of a force. We must remember this when solving problems of this type in the future.

Experiment. *To find the centroid of a triangular lamina.*

Cut out from cardboard, or tinplate, a number of different shaped triangular laminae, suspend them from a pin on a forces board and, using a plumb-bob repeat the procedure used in the previous experiment.

Now check the experimental results by geometric constructions. Using odd-leg calipers determine the mid-points of two sides of each lamina. Next draw straight lines from these mid-points to the opposite angles, see Fig. 84. The point of intersection of these lines is the centroid of the lamina. You could carry out the same procedure on the third side of the lamina.

Do the centroids obtained by geometric construction coincide with the ones obtained by experimental means?

How far does each centroid lie along the lines drawn from a mid-point in a side to the opposite angle? Check each one. What are your conclusions?

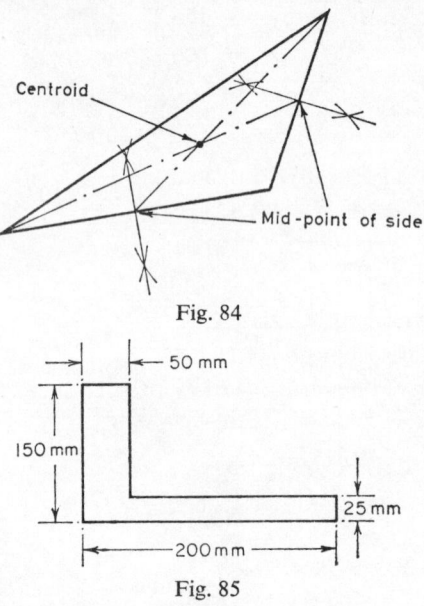

Fig. 84

Fig. 85

Example. Find the position of the centroid of the L-shaped lamina shown in Fig. 85.

In this case the centroid will lie outside the lamina since it is asymmetrical.

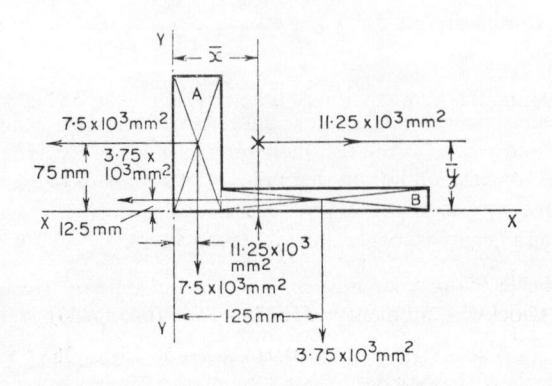

Fig. 86

MOMENT OF A FORCE

We shall have to calculate its position from the 150 mm and 200 mm edges, i.e. we shall have to calculate $\bar{x}$ and $\bar{y}$, see Fig. 86.

$$\text{Area of limb A} = 150 \times 50 = 7 \cdot 5 \times 10^3 \text{ mm}^2$$
$$\text{Area of limb B} = 150 \times 25 = 3 \cdot 75 \times 10^3 \text{ mm}^2$$

By moments about edge YY to find $\bar{x}$

$$\text{Clockwise moment} = (7 \cdot 5 \times 10^3 \times 25) + (3 \cdot 75 \times 10^3 \times 125)$$
$$= 125 \times 10^3 (1 \cdot 5 + 3 \cdot 75)$$

$$\text{Anti-clockwise moment} = 11 \cdot 25 \times 10^3 \times \bar{x}$$

For equilibrium,
$$11 \cdot 25 \times 10^3 \bar{x} = 125 \times 10^3 \times 5 \cdot 25$$
$$\bar{x} = \frac{125 \times 5 \cdot 25}{11 \cdot 25}$$
$$= \underline{58 \cdot 3 \text{ mm}}$$

Take moments about edge XX to find $\bar{y}$.

$$\text{Clockwise moment} = 11 \cdot 25 \bar{y}$$
$$\text{Anti-clockwise moment} = (3 \cdot 75 \times 10^3 \times 12 \cdot 5) + (7 \cdot 5 \times 10^3 \times 75)$$
$$= 10^3 (46 \cdot 9 + 562 \cdot 5)$$

For equilibrium
$$11 \cdot 25 \bar{y} = 609 \cdot 4$$
$$\bar{y} = \frac{609 \cdot 4}{11 \cdot 25}$$
$$= \underline{54 \cdot 1 \text{ mm}}$$

We could use the same method to find the centroid of an I-shaped lamina.

Example. Find the centroid of a circular lamina 150 mm diameter which has a 50 mm diameter hole cut in it. The hole is 15 mm off centre, see Fig. 87.

Arranging the lamina as shown, we have one calculation to make, that for $\bar{x}$.

$$\text{Area of 150 mm dia. lamina without hole} = (150)^2 \frac{\pi}{4} = 5625\pi \text{ mm}^2$$

$$\text{Area of 50 mm dia. hole} = (50)^2 \frac{\pi}{4} = 625\pi$$

This is taken as a negative area because it is cut away.

By moments about A,
$$\text{Clockwise moment} = (5625\pi \times 75) - (625\pi \times 90)$$
$$= 10^3 \pi (421 \cdot 9 - 56 \cdot 25) = 365 \cdot 6 \times 10^3 \pi$$
$$\text{Anti-clockwise moment} = 5000\pi \times \bar{x}$$

For equilibrium,
$$5000\pi \cdot \bar{x} = 365 \cdot 6\pi \times 10^3$$
$$\bar{x} = \frac{365 \cdot 6 \times 10^3}{5000} = \underline{73 \cdot 1 \text{ mm}}$$

You should check this by experiment.

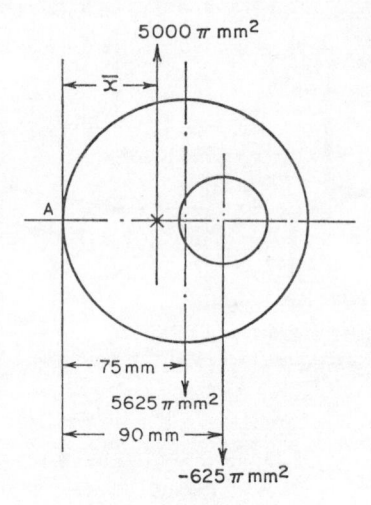

Fig. 87

Centre of Gravity

When dealing with *centre of gravity* we are concerned with three-dimensional objects, and volume, as against an assumed two dimensions and area as with a lamina. The centre of gravity of an object, or structure, is the point from which the whole 'weight' of the object or structure may be assumed to be acting. All objects are made up of atoms, as we noted on page 1. Atoms have mass and consequently gravitational attraction. We express the total gravitational force, i.e. 'weight' of the atoms as a single value, i.e. as a resultant 'weight' of an object, which is a more convenient form than a host of smaller 'weights.' This 'weight' is taken as acting from the centre of gravity of the object. In certain instances it is important to know accurately where the centre of gravity of a component, machine or structure is, e.g. of an aircraft or motor vehicle, so as to ensure future stability. For instance a single-deck bus is more stable than a double-deck bus when cornering, because the centre of gravity of the 'single-decker' is nearer the ground level. Efforts are made when designing vehicles to improve stability by keeping the centre of gravity low. Why do some of the motor cars, which have been made light in order to improve the power-weight ratio, have small diameter wheels? It is in order to shorten the distance of the centre of gravity above ground level so improving the stability of the vehicle. The position of the centre of gravity of a ship is important in obtaining overall balance. The balance and accuracy of a lathe spindle will to some extent be influenced by its centre of gravity position. In other instances the approximate position of the centre of gravity is sufficiently accurate for practical purposes. It is necessary for a crane slinger to judge well the position of the

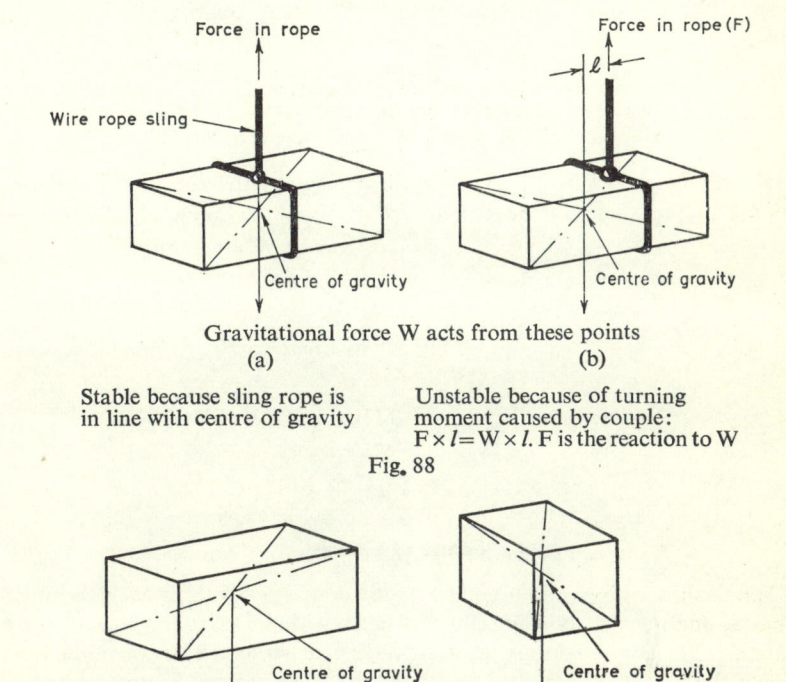

Gravitational force W acts from these points
(a)   (b)

Stable because sling rope is   Unstable because of turning
in line with centre of gravity   moment caused by couple:
                                 $F \times l = W \times l$. F is the reaction to W
Fig. 88

Gravitational force W acts from these points
(a)   (b)
Fig. 89

centre of gravity of a component to be lifted by the crane; this is especially true when putting a shaft between the lathe centres, or lifting a turbine rotor from its bearings.

The importance of the centre of gravity position in relation to a sling is illustrated in Fig. 88(a) which shows the sling in the correct position, and Fig. 88(b) which shows the effect of it not being correctly placed. The out-of-balance couple in this case is anti-clockwise. Even when a double-sling chain is used, the two chains meeting at the crane hook, if the vertical line through the centre of the crane hook does not pass through the centre of gravity of the load there will be an out-of-balance couple. This will mean that the component is not horizontal when lifted and may result in it slipping from the sling. This could be highly dangerous. A good slinger is capable of making an accurate assessment of the position of the centre of gravity of the load to be lifted.

A designer may need to calculate the c. of g. position accurately. For simple regular shaped objects such as rectangular and square prisms and spheres, the centre of gravity is at their geometric centres, see Fig. 89(a) and (b).

Example. Calculate the position of the centre of gravity of a shaft which is 150 mm diameter for 0·6 m of its length, 110 mm diameter for the next 1 m of length and 100 mm diameter for the remaining 0·8 m of its length, see Fig. 90.

We may ignore ' weight ', as when calculating a centroid position, and use volume instead, since volume is directly proportional to ' weight '.

$$\text{Volume of cylinder 1} = (0\cdot15)^2 \frac{\pi}{4} \times 0\cdot6$$

$$= 3\cdot375\pi \times 10^{-3} \text{ m}^3$$

$$\text{Volume of cylinder 2} = (0\cdot110)^2 \frac{\pi}{4} \times 1$$

$$= 3\cdot025\pi \times 10^{-3} \text{ m}^3$$

$$\text{Volume of cylinder 3} = (0\cdot1)^2 \frac{\pi}{4} \times 0\cdot8 = 2\pi \times 10^{-3} \text{ m}^3$$

$$\text{Total volume} = 8\cdot4\pi \times 10^{-3} \text{ m}^3$$

By moments about the end YY,

$$\text{Clockwise moment} = \pi \times 10^3[(3\cdot375 \times 0\cdot3)+(3\cdot025 \times 1\cdot1)+(2 \times 2)]$$

$$= \pi \times 10^{-3}[1\cdot012+3\cdot327+4]$$

$$= 8\cdot339\pi \times 10^{-3}$$

$$\text{Anti-clockwise moment} = 8\cdot4\pi \times 10^{-3}\bar{x}$$

For equilibrium

$$8\cdot4\pi \times 10^{-3}\bar{x} = 8\cdot34\pi \times 10^{-3}$$

$$\bar{x} = \frac{8\cdot34}{8\cdot4}$$

$$\bar{x} = 0\cdot99 \text{ m}$$

MOMENT OF A FORCE

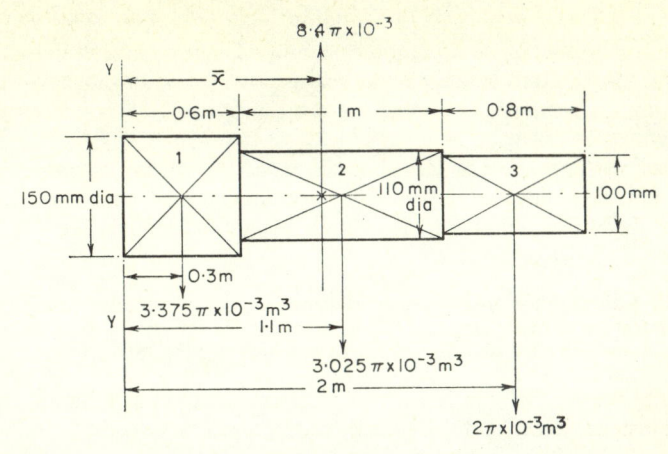

Fig. 90

Hence if this shaft is to be slung horizontal in a one- or two-sling chain the resultant pull in the slinging chain must pass through a point on the shaft axis 1 metre from the left-hand end. Had we taken the density of the shaft material into consideration in our calculation, it would have cancelled out.

Exercise 7(a)

Section A

1. In a particular hand screwing operation a stock and die require a couple of magnitude 13 Nm. The stock is 0·35 m long and the die situated in the centre of it. Calculate the force required at each end of the stock.
2. To slacken a nut, a turning moment of 50 Nm is required. If a spanner having an effective length of 0·4 m is to be used, find what force must be applied to the end of it.
   What force would be required to do the same operation at the end of a 0·32 m long spanner?
3. Figure 91 shows the working principle of a set of guillotine shears. The shears are to be used to cut a piece of metal strip 50 mm wide by 1 mm thick, maximum shear strength 200 N/mm². Calculate the force that must be applied to the handle of the shears.
4. A shaft, clamped in vee-blocks, to the table of a planing machine is shown in Fig. 92. What will be the reaction in each vee-block due to the clamping forces and the 'weight' of the shaft? The 'weight' of the shaft may be taken as acting from its centre of gravity as shown.
5. The sketch in Fig. 93 shows a shaft supported between the centres of a lathe. If the 'weight' of the shaft is 1·25 kN, calculate the normal reactions in the lathe centres.
6. The foot lever for operating the clutch on a small shearing machine is shown in Fig. 94. A spring is attached to the lever to assist in the disengaging of the clutch when the foot is removed from the lever. When the clutch is engaged, the pull in the spring is 70 N. Find what force, F newton, must be applied to the foot pedal in order to keep the clutch engaged. Resistance of clutch is 225 N.

118 MATERIALS AND MACHINES IN THE WORKSHOP

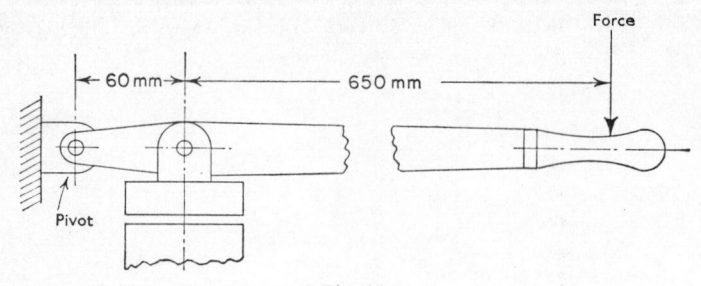

Fig. 91

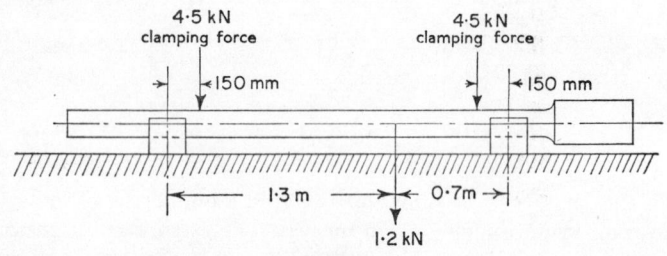

Fig. 92

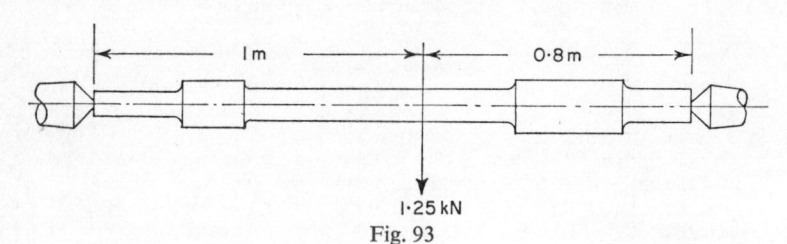

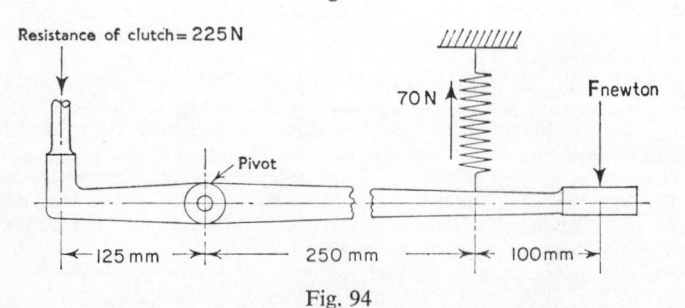

Fig. 93

Fig. 94

7. (*a*) A sketch of a tool-post for a 150 mm centre lathe is shown in Fig. 95. The force on the tool point due to the cutting of metal is 700 N. A setscrew having a core diameter of 7·5 mm clamps the tool in the post. Calculate the greatest stress in the setscrew and state what type of stress it is.

(*b*) Why is it a good policy to keep the tool overhang as small as possible?

MOMENT OF A FORCE

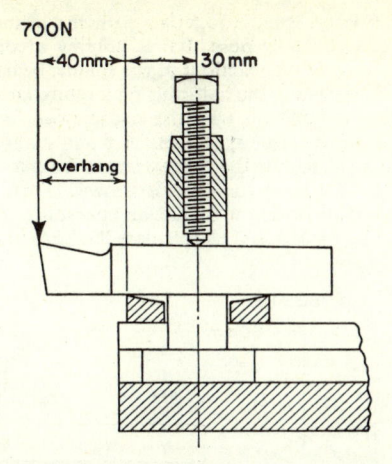

Fig. 95

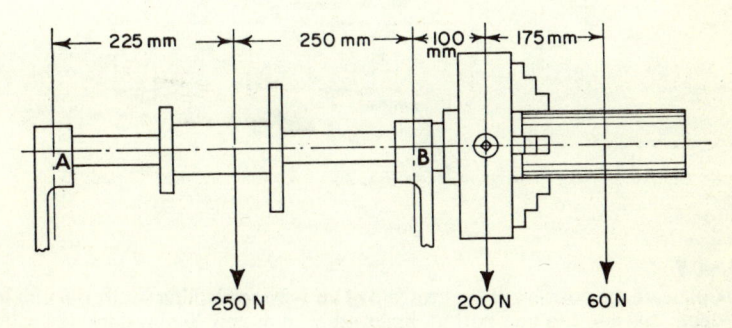

Fig. 96

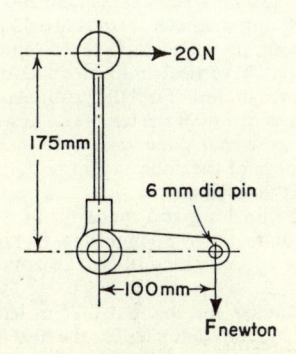

Fig. 97

120   MATERIALS AND MACHINES IN THE WORKSHOP

8. Figure 96 shows a lathe spindle together with the gears that are fixed to it. It has a chuck screwed to its nose that is holding a component. Determine the sizes and directions of the reactions at the spindle bearings A and B.
9. A cranked lever for operating the half-nuts on a centre lathe is shown in Fig. 97. In order to engage the half-nuts with the leadscrew, a force of 20 N has to be applied to one arm of the lever as shown. A 6 mm dia. pin connects the other arm to the half-nuts. Calculate the shear stress in the pin.
10. In Fig. 98 a sketch of a lathe tool-post is shown. There is a cutting force of 500 N on the tool point during a particular operation. If the force caused in screw A by the cutting force is 700 N, calculate the force in screw B.

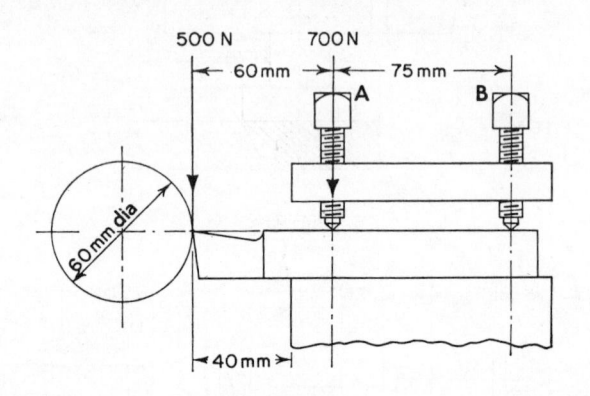

Fig. 98

Section B

1. Calculate the position of the centroid of an I-shaped lamina width 100 mm total depth 200 mm top and bottom limbs are 24 mm and 20 mm thick respectively and central limb 15 mm thick.
2. Find the position of the centroid of a T-shaped lamina having dimensions: cross-member 150 mm × 20 mm, vertical member 200 mm × 15 mm.
3. A circular template 150 mm diameter is to have a 45 mm diameter hole cut in it, 50 mm off centre. Determine the position of the centroid of the template.
4. An L-shaped lamina has a vertical limb of dimensions 250 mm × 30 mm and horizontal limb 200 mm × 20 mm. Find the position of its centroid.
5. Find the position of the centroid of the template shown in Fig. 99.
6. A casting comprises a 75 mm cube and a 60 mm diameter, 150 mm, long cylinder attached to a face of the cube so that both lie on a common axis. Find the centre of gravity of the casting.
7. A drop forging is 175 mm long and made up of 3 cylinders, which lie on a common axis, 45 mm diameter by 50 mm long, 60 mm diameter by 50 mm long and 100 mm diameter by 75 mm long. Find the position of the centre of gravity of the forging.
8. A shaft is 150 mm diameter for the first 0·25 m length, 170 mm diameter for the next 0·2 m length, 160 mm diameter for the next 2 m length, 140 mm for the next 0·22 m length and 120 mm diameter for the remaing 0·22 m length. Determine the position of its centre of gravity.

MOMENT OF A FORCE 121

9. A hydraulic cylinder 2·4 m long is 1 m diameter for the first 1·2 m of its length and 0·4 m diameter for the remaining 1·2 m. It is bored 0·76 m dia. to a depth of 1·15 m at which point the bore diameter is reduced to 0·22 m. Find the position of the centre of gravity of the cylinder.
10. Find the position of the centre of gravity of the casting shown in Fig. 100.

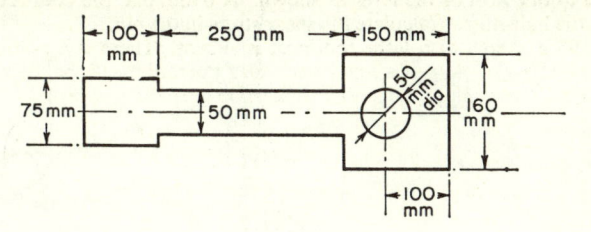

Fig. 99

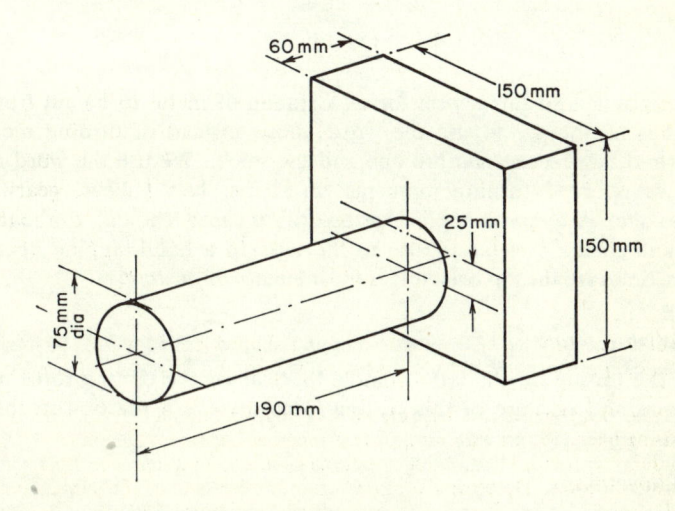

Fig. 100

Torque

In order to understand what is meant by *torque*, we will consider the operation of machining a bar of metal in a lathe as shown in Fig. 101. Once the component is gripped properly in the chuck it becomes, like the chuck, part of the lathe spindle. The job having been set up, we would select the correct speed, and then switch on the power and engage the clutch. Next we would apply the cut. The metal is cut because the turning moment on the lathe spindle compels the bar to be twisted against the resistance set up by

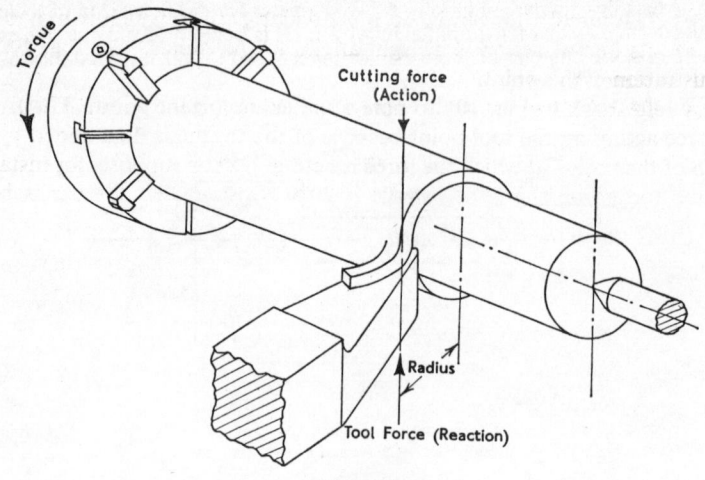

Fig. 101

the tool point, and this action forces a ribbon of metal to be cut from the bar. It is customary to use the word *torque* instead of turning moment. *Torque* and *turning moment* are one and the same. We use the word *torque* when we refer to turning moments on shafts, belt pulleys, gearwheels, spindles, etc. A *torque* is applied to the drill to cause it to cut, to the milling cutter and emery wheel, and also to the tap. In a hand-tapping operation we sometimes see the tap twist under the influence of the *torque*.

Calculation of torque

It is the torque on the lathe spindle that causes the cutting force on the tool point and because of this cutting force there is a reaction in the tool point.

We know that:

TURNING MOMENT = FORCE × DISTANCE of the force from the pivot.

Therefore, since torque and turning moment are the same thing:

TORQUE = FORCE × DISTANCE of the force from the pivot.

In the case of the bar being turned in a lathe, or in the use of a drill, tap milling cutter, pulley, gear wheel, etc., the pivot is the centre of rotation. The distance of the force from this point will be a *radius*.

Hence, we can make the following statement:

TORQUE (Nm) = FORCE (newton) × RADIUS (metre).

MOMENT OF A FORCE

Note: The force is tangential to the circular path in which it moves and therefore its line of action is at right-angles to the radius. See Fig. 102 for an illustration of this point.

Now refer back to Fig. 101 to note another important point. The size of the force acting on the tool point because of the torque is determined by the length of the radius at which the force is acting. Let us suppose, for instance, that the torque on the lathe spindle is 2000 Nm and that the bar is being

Fig. 102

machined to a radius of 50 mm, i.e. the force on the tool point due to cutting metal is at 50 mm radius. Therefore, as

TORQUE = FORCE × RADIUS

we have 2000 (Nm) = force (N) × 0·05 (m)

$$\frac{2000\ (Nm)}{0·05\ (m)} = \text{force (N)}$$

40 kN = Force.

Note: Had we been given a depth of cut, we should have used a mean radius instead of the finished bar radius; e.g., if depth of cut were 4 mm, mean radius would be 52 mm. This radius would have been used in place of the 50 mm.

If the radius of the bar were 100 mm, the force on the tool point due to cutting metal would be as follows:

2000 (Nm) = force (N) × 0·1 (m)

$$\frac{2000\ (Nm)}{0·1\ (m)} = \text{force (N)}$$

20 kN = Force.

We have a similar situation with milling cutters of different radii working under the same conditions of cut, etc., as is shown by the following example.

Examples

1. In a certain milling operation, there is a torque of 50 Nm on the arbor.

(*a*) If a 150 mm dia. cutter is being used, find the force acting on the cutter's teeth.

(*b*) If a 200 mm dia. cutter were used, what would be the force on the teeth?

We will make a sketch to help us see the problem more clearly (see Fig. 103(*a*) and (*b*)).

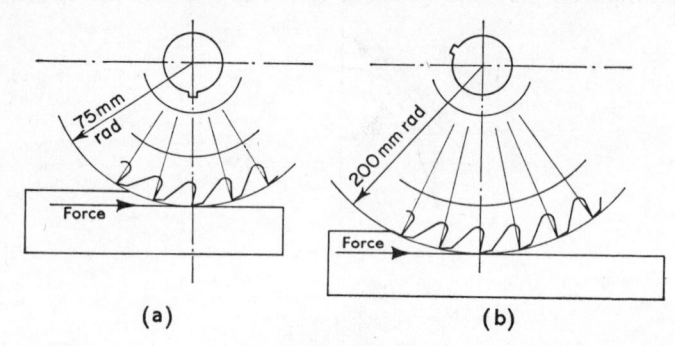

Fig. 103

(*a*)

$$\text{TORQUE} = \text{FORCE} \times \text{RADIUS}$$

$$50 = \text{force} \times 0{\cdot}075$$

$$\frac{50}{0{\cdot}075} = \text{force}$$

$$667 \text{ N} = \text{Force}$$

The force acting on the cutter's teeth is 667 N and this will be shared by the number of teeth cutting at any instant.

(*b*) In this case we have:

$$50 = \text{force} \times 0{\cdot}1$$

$$\frac{50}{0{\cdot}1} = \text{force}$$

$$500 \text{ N} = \text{Force}$$

The force acting on the teeth when the larger diameter cutter is used is 500 N and will be shared by the teeth in contact with the component.

2. The operator on a certain lathe applies a force of 20 N to the handle on the carriage handwheel in order to traverse the carriage. The handle is at a radius of 90 mm. What is the torque on the handwheel spindle?

In this problem we have a force of 20 N acting at a radius of 90 mm and we are asked to find a torque,

$$\text{TORQUE on handwheel} = 20 \times 0{\cdot}09$$

$$= \underline{1{\cdot}8 \text{ Nm}}$$

MOMENT OF A FORCE 125

3. A belt pulley of 300 mm dia. is keyed to its 46 mm dia. shaft by a key 50 mm long by 12 mm wide. The belt which drives the pulley causes an effective pull of 300 N at its rim. Calculate (a) the torque on the pulley shaft, (b) the force acting on the key because of this torque, and (c) the shear stress in the key.

First we will make a sketch of the pulley (see Fig. 104).

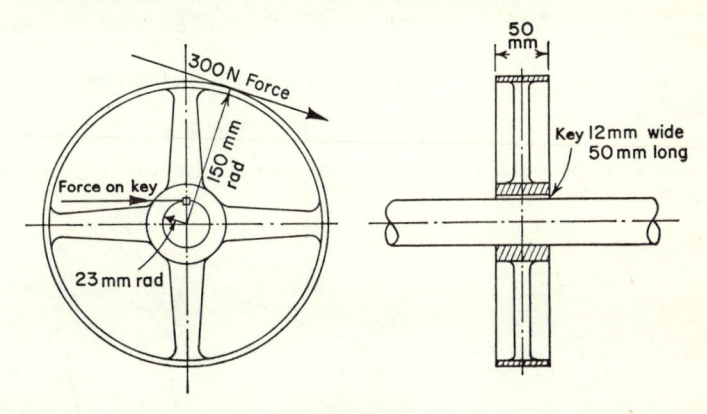

Fig. 104

(a)   TORQUE on pulley shaft = effective belt pull × pulley radius

$$= 300 \times 0.150 = 45 \text{ Nm}$$

(b) Now the force that acts on the key due to this torque is at a radius of 23 mm.

$$45 \times 10^3 = \text{FORCE on key} \times 23$$

$$\frac{45 \times 10^3}{23} = \text{force on key}$$

$$1.96 \times 10^3 \text{ N} = \text{force on key}$$

$$\underline{\text{Force on key} = 1.96 \text{ kN}}$$

(c) This force of 1.96 kN is tending to shear the key on an area of 12 mm × 50 mm

$$\text{SHEAR STRESS in key} = \frac{\text{FORCE}}{\text{AREA}}$$

$$= \frac{1.96 \times 10^3}{12 \times 50} = \underline{3.3 \text{ N/mm}^2}$$

4. Figure 105 shows a hole being tapped using a tap wrench which is 250 mm long, the tap being in the centre of the wrench. The force applied to each end of the wrench during the operation is 40 N. Determine (a) the torque on the tap (b) the force acting at the crest of the thread i.e. at a 5 mm radius.

(a)   TORQUE on tap $= (40 \times 0.125) + (40 \times 0.125)$ Nm

$\qquad\qquad\qquad\qquad = 40 \times 0.25$ (torque created by a couple)

$\qquad\qquad\qquad\qquad = \underline{10 \text{ Nm}}$

(b)   TORQUE on tap = CUTTING FORCE at crest of thread $\times 0.005$

$\qquad\qquad\qquad 10 =$ cutting force at crest of thread $\times 0.005$

$\qquad\qquad\qquad \dfrac{10}{0.005} =$ cutting force at crest of thread

$\qquad\qquad\qquad \underline{2 \text{ kN} = \text{Cutting force at crest of thread.}}$

Therefore, the cutting force acting at the crest of the thread, i.e., the outside surface, is 2 kN.

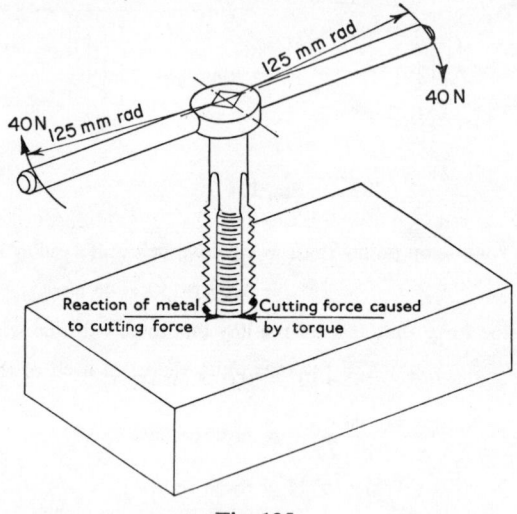

Fig. 105

This, incidentally, is the total of a number of small forces acting at the cutting edges in contact with the metal.

We might ask if the force at the root of the thread, i.e., at the core diameter, would be the same. The answer is no, it would be greater. Assume the core radius is 4 mm, the torque the same and no cutting at crest.

$\qquad\qquad$ TORQUE on tap = CUTTING FORCE at root of thread $\times$ core radius

$\qquad\qquad\qquad 10 =$ cutting force at root of thread $\times 0.004$

$\qquad\qquad\qquad \dfrac{10}{0.004} =$ cutting force at root of thread

$\qquad\qquad\qquad \underline{2.5 \text{ kN} = \text{Cutting force at root of thread.}}$

Therefore the cutting force at the root of the thread is (2·5–2·0), 0·5 kN greater than at the crest.

MOMENT OF A FORCE   127

5. A sensitive drilling machine has a torque of 8·6 Nm on its spindle when a 10 mm dia. hole is being drilled. A two fluted drill is used and it may be assumed that the forces on the two cutting edges are equal. What will be the value of these forces at the full radius of 5 mm?

Let the forces on the cutting edges be represented by F as indicated in Fig. 106, which shows a plan view of the drill point.

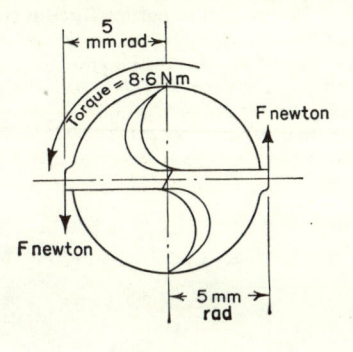

Fig. 106

Now,

TORQUE on drill point $= F \times 10 \times 10^{-3}$ (moment of a *couple*)

$$8 \cdot 6 = 10 \times 10^{-3}$$

$$F = \frac{8 \cdot 6}{10 \times 10^{-3}} = 860 \text{ N}$$

The cutting force acting at the maximum radius of each of the two cutting edges is 860 N.

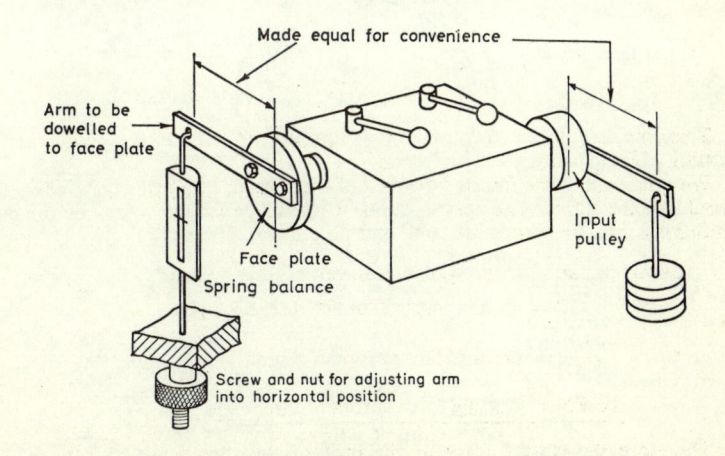

Fig. 107

Experiment. *To observe the effect of a gear-box on a torque.*

Using a lathe gear-box, say from a 150 mm centre lathe all-geared head-stock, fit as indicated by the diagram in Fig. 107. Arrange the two arms in a horizontal position, select first gear and add a mass to the arm on input pulley of sufficient magnitude to create the desired force. Restore the arms to a horizontal position. Note the spring balance reading before and after adding the mass. Calculate the input and output torques. Repeat the procedure for all the remaining gears and tabulate the results. Plot a graph of spindle torque against gear number.

Note: The gravitational effect on a 1 kg mass is 9·81 N.

How does the spindle torque in top gear compare with that of the bottom gear?

Determine the gear ratios and calculate the faceplate torques for a given input torque. Compare these values with values obtained experimentally for the same input torque.

Variable torque

We should at this point distinguish between a constant torque and a variable torque. The examples and problems which we have dealt with so

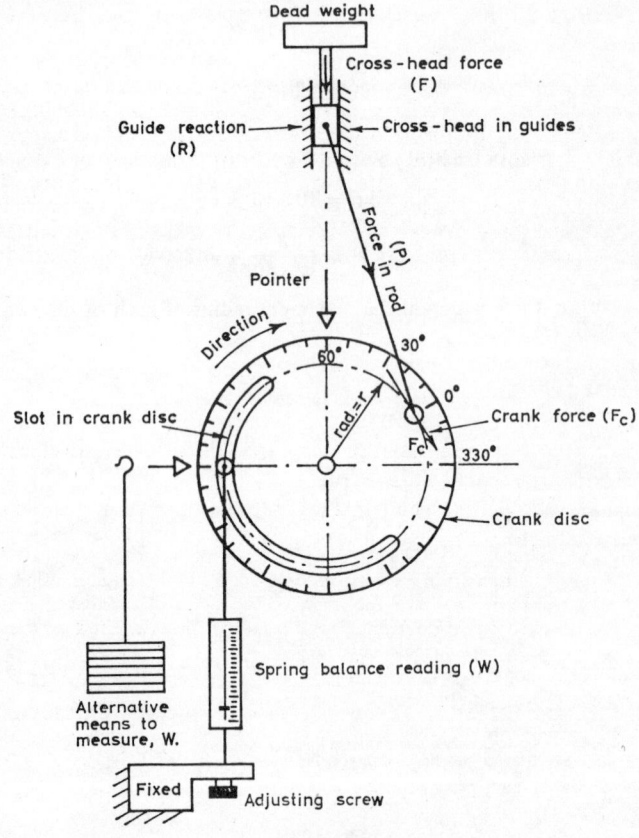

Fig. 108

MOMENT OF A FORCE

far have been concerned solely with a torque that remained constant for the given conditions. For instance, the torque on a lathe spindle remains constant, for a given component diameter, once the depth of cut and rate of feed have been fixed. The frictional resistances will remain at a virtually steady value.

The torque on the rope drum of a crane hoist, or a lift, will vary since the load being lifted will vary; e.g., the load will be reduced as the hoist rope is wound on to the drum. The maximum torque will be required when the maximum length of rope, or ropes, is paid out. The minimum torque is applied to the drum when the load is in its highest position, a greater part of the rope being wound in.

There are some mechanisms which by their very construction produce a variable torque. The engine mechanism will produce a variable torque on the crankshaft even if the cross-head force is constant.

Experiment. *To determine the crank disc torque on an engine mechanism which has a constant cross-head force.*

Arrange an engine mechanism—one can easily be constructed in the workshop—as shown in Fig. 108. Apply a force, a 5 kg mass will result in a gravitational force of 5×9.81 newton, to the cross-head and let the crank disc move through 30°. Lock the crank disc in this position by a thumb screw fitted to the disc spindle. Adjust the pin in the crank disc slot so that it is in the position shown. By means of a spring balance, or weight hanger and mass, apply a force to balance the crank pin force, F_C. As the equilibrium point is approached slacken the nut which is locking the disc. Repeat this procedure for crank pin positions at 30° intervals. Calculate the corresponding crank torques.

$$\text{Crank torque} = F_C \times r = W \times r$$

Plot a graph of crank pin position (horizontal axis) against crank torque (vertical axis).

Is the crank torque variable? If so explain why it is.

The engine mechanism is used on some machine tools in reverse; i.e., the crank pin drives the cross-head.

If the crank were driven by a constant torque motor, would the cross-head force be constant?

Example. A goods lift carrying a capacity load of 10 kN is hoisted 40 m at a slow uniform speed by a wire rope of 6 N/m mass being wound on to a drum, mean dia. 1 m. Find the maximum, minimum and mean torques required on the drum.

The maximum torque occurs at the beginning of the lift.

$$\begin{aligned}
\text{Maximum torque} &= \text{Torque to lift load} + \text{Torque to lift rope.} \\
&= (10 \times 10^3 \times 0.5) + (40 \times 6 \times 0.5) \\
&= 5000 + 120 \\
&= \underline{5.12 \text{ kN m}}
\end{aligned}$$

The minimum torque occurs when all the rope is wound onto the drum and the torque required is that to support the load.

$$\text{Minimum torque} = 10 \times 10^3 \times 0{\cdot}5 = 5 \text{ kN m}$$

$$\text{Mean torque} = \frac{5{\cdot}12 + 5}{2}$$

$$= \underline{5{\cdot}06 \text{ kN m}}$$

Exercise 7(b)

Section A

1. An electric motor that drives a lathe by means of a flat open belt has a pulley 250 mm dia. The lathe pulley is 300 mm dia. If the torque on the motor pulley is 50 Nm, find the effective pull in the belt, and the torque on the lathe pulley.
2. A pinion in the gear-box of a certain lathe has a pitch circle, 150 mm dia, and the torque on it is 30 Nm. It meshes with a gear-wheel which has a pitch circle, 250 mm dia. (see Fig. 109). Determine the force between the teeth, i.e., the tooth force, and the torque on the gear-wheel.

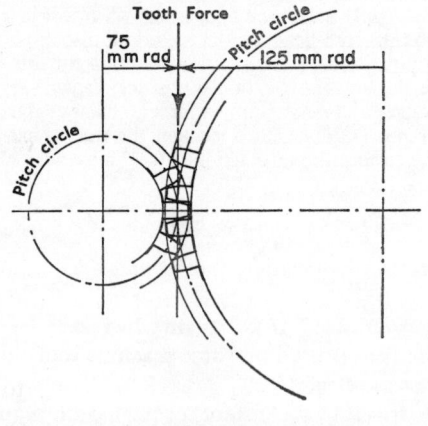

Fig. 109

3. (a) On a vertical boring machine, steel tyres for locomotives are being rough bored to a diameter of 1·2 m. The torque on the machine spindle during the operation is 900 Nm. Calculate the cutting force on the tool point.
   (b) If the tyres were being bored 1·3 m dia. with the torque on the spindle remaining at 900 Nm, what would be the cutting force on the tool point?
4. (a) Figure 110 shows a diagram of the feed mechanism of a shaping machine. During a certain operation the torque on the crank disc is 8 Nm and the crank pin is at a radius of 16 mm. Find the force acting on the push rod due to this torque for the position shown.
   (b) What would be the stress in the push rod due to this torque if the crank pin was at a radius of 20 mm? State what types of stresses the push rod has to withstand. Dia. of push rod = 15 mm.

5. The sketch in Fig. 111 shows a push-off mechanism, used on a conveyor, for pushing the components off the conveyor. Find the torque that must be applied to the pinion in order to produce a 100 N push-off force.

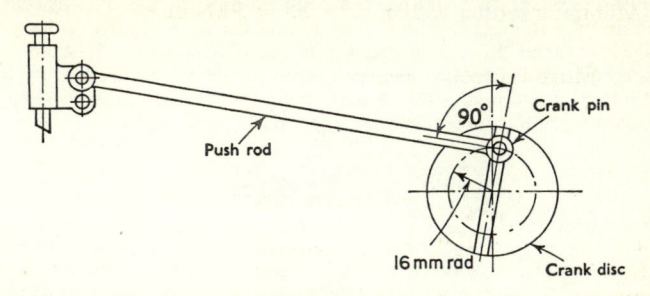

Fig. 110

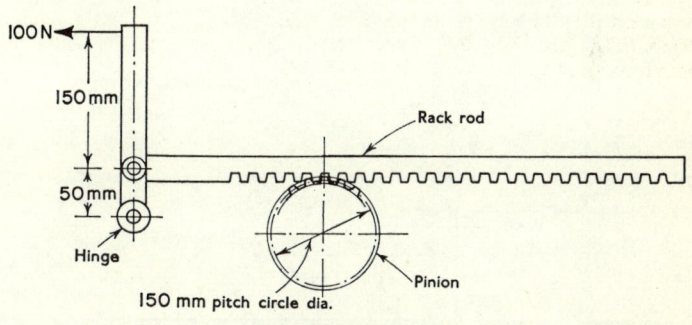

Fig. 111

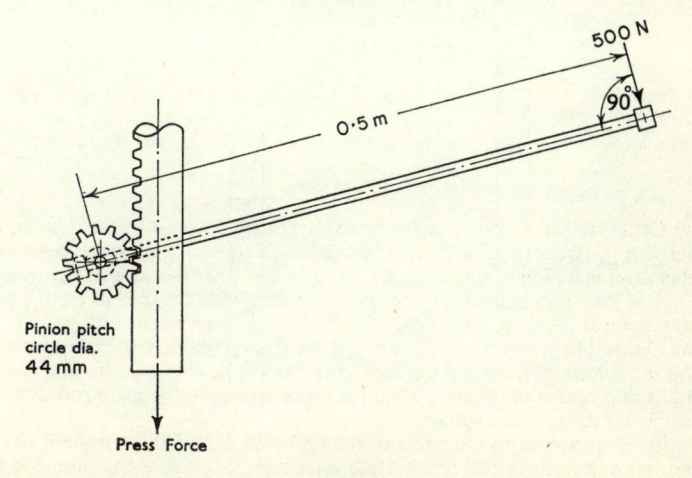

Fig. 112

6. The sketch in Fig. 112 shows a side elevation of the mechanism that operates the spindle of an arbor press. Calculate the press force at the spindle, when the operator applies a force of 500 N to the handle.
7. In Fig. 113 is shown a simplified line diagram of the mechanism which drives the ram of a shaping machine. The pinion drives the bull wheel and fixed to this wheel is an adjustable crank pin which operates the rocker arm. The upper end of the rocker arm is connected to the ram. During a certain machining operation, the torque on the pinion is 44 Nm. Under these conditions, calculate the ram force for the rocker arm position shown.

Fig. 113

8. If in question 7 the ram force that you have calculated is required on a longer stroke, which means that the crank pin has to be at a radius of 125 mm instead of 110 mm, what torque would be required on the pinion?

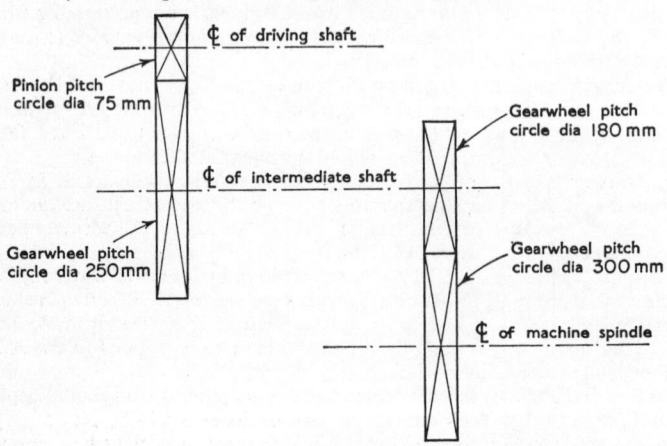

Fig. 114

MOMENT OF A FORCE

9. A line diagram of a train of gears from a machine tool gear-box is shown in Fig. 114. The torque on the driving shaft is 54 Nm. Calculate the torque on the machine spindle.
10. Figure 115 shows the belt and pulley drive to the spindle of a sensitive drilling machine. When 12 mm dia. holes are being drilled, the cutting forces on each of the two cutting edges at points, 6 mm radius from the centre line of the drill are 300 N each. An electric motor which has a 130 mm dia. pulley drives the 250 mm dia. spindle pulley. Calculate the torque required at the motor.

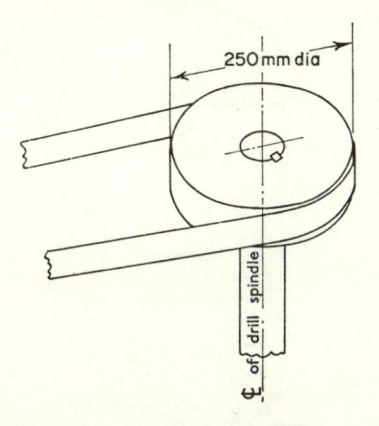

Fig. 115

Section B

1. A 110 mm diameter milling cutter has an average load of 200 N per tooth. When cutting under maximum conditions 5 teeth are in contact with the job. Find the minimum and maximum torques on the arbor during the operation.
2. In a drilling operation with a 20 mm diameter two-fluted drill the cutting force on each edge at the maximum radius is 500 N. The feed rate is 0·2 mm per rev. and the torsional load is increased by 1·5 N at the maximum radius for each such feed movement. Find the maximum and minimum torques required on the drill if the hole depth is 40 mm.
3. A machine for hauling a wagon up an inclined track, by means of a wire rope 25 mm dia. has a rope drum 1.6 m core dia., 6 layers of rope are wound onto the drum in the process. If the pull in the rope is constant at 1·2 kN, find the average torque on the drum. What would the maximum torque be?
4. On a 'moments disc' place a mass at some suitable radius and in such a position that it will cause a maximum torque on the disc. Calculate the torque, also state the position of zero torque and the values of the torques between the maximum and zero values at 10° intervals, see Fig. 116.
5. A goods lift, laden 'weight' 10 kN, is raised 40 m by means of wire ropes being wound onto a drum 1·2 m diameter. The ropes 'weigh' 12 N per metre length. Calculate the maximum and mean torques required on the drum to lift the load. What minimum braking torque would have to be applied to the drum to hold the load in its highest position?
6. A 150 mm diameter bar is to be turned to 78 mm diameter using equal depths of cut and feed rates for each cut. If the cutting force is 500 N, plot a graph to show how the spindle torque varies with bar diameter over the whole operation. Use at least six torque values.

7. An electric motor has a 4-step cone pulley having diameters 150, 200, 250 and 300 mm and torque, 16 Nm. Calculate the effective pull in the belt when each pulley is in use.
8. A mine shaft is 800 metre deep. A fully laden cage weighs 60 kN, and is hoisted up the shaft by means of a 40 mm diameter wire rope, weighing 50 N per metre length, which is wound onto a drum, 4 m mean diameter. Plot a graph showing how the drum torque varies over the 800 m.

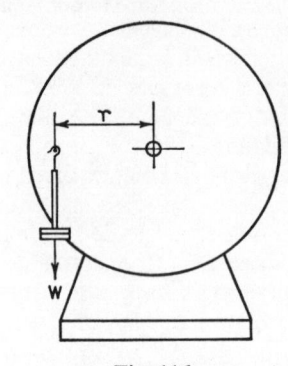

Fig. 116

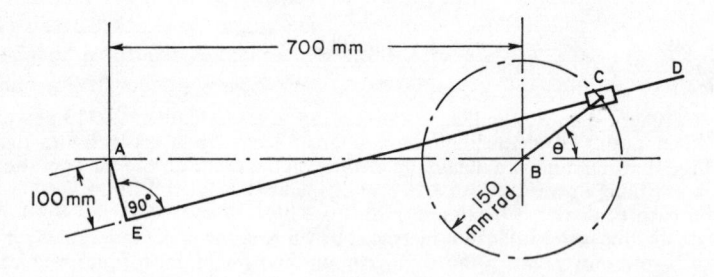

Fig. 117

9. The feed mechanism on some shaping machines is based on the engine mechanism. Explain, using a sketch, why the torque on the crank disc will vary, assuming that the torque requirements at the table screw remain constant for given conditions.
10. In the mechanism shown in Fig. 117, A and B are fixed points, i.e. bearings, about which the links AE and BC respectively can rotate. The block fitted to the crank pin C slides on the link ED and the angle AED is 90° permanently. By drawing the mechanism accurately to scale in order to obtain the required details, find the torque on the link ED for crank positions of $\theta = 30°$ and 60° when the torque on BC is constant at 54 Nm.

Chapter 8. Work and Power

In the previous two chapters we have learnt some of the uses of force. We have also found out what effects it can have when it acts on a material. This chapter will inform us that a force can do *work*. For instance, when metal is cut on a lathe, or slotting machine, or milling machine, *work* is done, just as when metal is drawn through a die, or rolls *work* is done. Also, when a casting or forging is lifted by a crane, or block and tackle, *work* is done.

In some workshop processes the force used causes movement, e.g., the force that acts at the cutting edge of the shaping machine tool as it cuts through the metal, or the force that causes the drill to feed into the metal during a drilling operation. The force in the belt that drives a machine tool causes motion. We must note, however, that there are forces in the engineering workshop which do not cause movement. Examples of these forces are, the gravitational force on a machine tool causing a downward force on the foundation that supports it, the force gripping the tool in a lathe tool-post, the force holding a component to a magnetic chuck, and all similar clamping and gripping forces. Another example is the force required to support a component suspended from a rope block, or similar lifting machine.

Work done by a constant force

When a force causes movement, work is done. The amount of work done depends upon the size of the force and the distance through which the force moves, and it is stated in the following way:

$$\text{WORK DONE (joule)} = \text{FORCE (newton)} \times \text{DISTANCE MOVED BY FORCE (metre)}$$

The multiples gigajoule and megajoule are used, also the sub-multiple millijoule.

Examples

1. The cutting force on a shaper tool during the machining of a particular component is 250 N and the length of the job being cut is 400 mm. How much work is done on one cutting stroke?

$$\text{WORK DONE on 1 cutting stroke (joule)} = \text{cutting force (newton)} \times \text{length of cut (metre)}$$

$$= 250 \times 0{\cdot}4$$

$$= \underline{100 \text{ J}}$$

MATERIALS AND MACHINES IN THE WORKSHOP

2. In a turning operation, the cutting force acting on the tool point is 500 N and the mean diameter of the component being turned is 75 mm. Calculate the amount of work done during one revolution of the component.

In one revolution of the component the cutting force moves through a distance equal to the circumference of the surface being turned and the circumference of the bar is given by $\pi \times$ mean diameter,

DISTANCE moved by cutting force in 1 revolution $= \pi \times 0.075$ metre

WORK DONE in 1 revolution of component $= 0.075\pi \times 500$ joule

$$= \underline{117.8 \text{ J}}$$

3. The drive from an electric motor to a drilling machine is through a 75 mm wide belt. The safe working load for the belt is 15 N per millimetre width. If the belt drives a pulley of 375 mm diameter, how much work would be done in 1 revolution of the pulley?

First calculate the greatest safe pull that can be carried by the belt.

A 1 mm wide belt can carry a safe load of 15 N

A 75 mm wide belt can carry a safe load of 15×75 N

Maximum safe pull that belt can carry $= 1125$ N

DISTANCE moved by belt in 1 rev of pulley $= 0.375\pi$ m

WORK DONE for 1 rev of pulley $= 1125 \times 0.375\pi$

$$= \underline{1.32 \text{ kJ}}$$

4. A casting of mass 5 Mg is lifted 2 m from the ground by means of a crane. Calculate the work done by the crane.

The crane has to create an upward force equal to the gravitational pull on the casting and it is this force that is doing *work*.

5 Mg $= 5 \times 10^3 \times 9.81$ N since the gravitational pull on a 1 kg mass is 9.81 newton.

'Weight' of casting $= 5 \times 9.81 \times 10^3$ N

WORK DONE lifting casting $= 5 \times 9.81 \times 10^3 \times 2$

$$= \underline{98.1 \text{ kJ}}$$

5. When turning a 80 mm dia. bar, the cutting force on the tool point is 1.5 kN and the feed force, which moves 0.5 mm in every revolution of the bar, is 200 N. How much *work* is done per revolution of the bar?

In this problem we have a double calculation. First find the work done per rev by the cutting force, then determine the work done by the feed force, and finally add the two together to give the total work done in 1 rev of the bar.

DISTANCE moved by cutting force in 1 rev of bar $= 0.08\pi$ m

WORK DONE by cutting force in 1 rev of bar $= 0.08\pi \times 1.5 \times 10^3$ J

DISTANCE moved by feed force in 1 rev of bar $= 0.5 \times 10^{-3}$ m

WORK DONE by feed force in 1 rev of bar $= 0.5 \times 10^{-3} \times 200$ J

TOTAL WORK DONE in 1 rev of bar $= 120\pi + 0.1$

$$= \underline{376.1 \text{ J}}$$

Work done by a variable force

So far we have thought only of work done by a constant force, having assumed that the force doing the work had a constant value over the whole period of its action. This may be true in some cases, e.g., the cutting forces in turning and drilling operations may be assumed constant, so may the pulling force in wire and tube drawing passes. It is not true in all cases. For instance, if we study the graph in Fig. 66 on page 93, we shall appreciate that the force causing the tensile stress is a variable one. This is also true of the compressive force in a compression test. Incidentally the work done in each instance is given by the area under the load-extension (or load-compression) graph; see the following experiment. The force acting on the piston of a motor vehicle engine is variable over the working stroke, since the pressure on the piston decreases due to expansion of the gas as the piston moves down the cylinder. At the head of a punching machine a variable force is at work when a hole is punched. A fraction of a second before the punch touches the metal to be cut, the shearing force on the punch will be zero. From the instant the punch touches the plate the shearing force on the punch will increase uniformly at a relatively rapid rate until the elastic limit of the metal is reached. The force will then continue to increase at a slower rate until the shear stress caused in the metal, by the punch force, just exceeds the maximum shear strength of the metal, see Fig. 118. The metal fails in shear, and the force on the punch falls rapidly to that value required to push the blank from the hole.

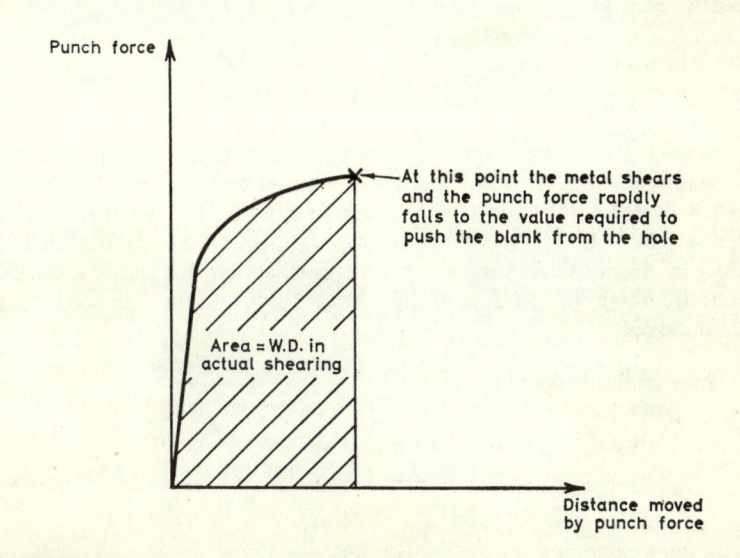

Fig. 118

When finding the *work done by a variable force*, it is usual to use an average value for the force, then proceed as in the case of a constant force. If the rate of doing work at a particular instant is required, then we would use the force acting at that instant.

Experiment. *To determine the work done by a variable force.*

Attach a helical spring to a wall bracket by one of its ends so that it hangs vertically; fix a weight hanger to the other end. Add loads to the hanger in suitable increments and measure the extension after each addition. Record about eight values, then plot a load-extension graph as shown in Fig. 119.

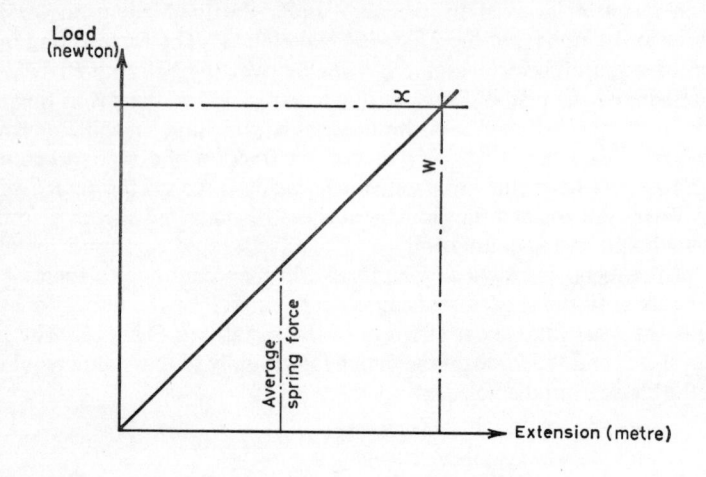

Fig. 119

Here we have a gravitational force (weight is a force) varying from zero to a maximum value W. The extension, x, caused by W is really the distance travelled by the force causing the spring to extend. Not all the force has travelled this distance. It will be an average force that has travelled the distance x. Hence to find the work done in extending the spring a distance x metre we obtain the product of the average force and the distance through which it moves.

$$\text{Average force on spring} = \frac{\text{Final Force} + \text{Initial Force}}{2}$$

$$= \frac{W + 0}{2} = \frac{W}{2} \text{ newton}$$

$$\text{Work done extending spring} = \frac{W}{2} \times x = \frac{Wx}{2} \text{ joule}$$

Note: The area under the graph represents this work.

WORK AND POWER

Would this theory apply if the spring had been compressed? Give reasons for your answer. If a mild steel wire were subjected to a similar test, within the elastic limit, how could the work done in extending the wire be found?

Incidentally the work done in stretching the spring is stored in the spring as energy—*elastic strain energy*.

There are many industrial examples of the uses of springs doing work; valve springs closing valves on motor vehicle engines; clutch springs maintaining the contact of the clutch surfaces; springs on press tool assemblies.

A variable force does not always follow a straight line when plotted against displacement (distance traversed). When it does follow a straight line it is directly proportional to the displacement.

Another example of a variable force is the hoisting force at the rope drum of a crane referred to earlier (page 129).

Examples

1. A mine cage 'weighing' 5 kN is to be hauled through a vertical height of 250 m by means of a winding engine. The rope 'weighs' 20 N/m length. How much work is done in lifting the cage?

$$\text{Force to lift cage} = 5 \times 10^3 \text{ N}$$

$$\text{Average force to lift rope} = \frac{(250 \times 20) + 0}{2} = 2500 \text{ N}$$

$$\text{Work done in hoisting} = (5 \times 10^3 + 2{\cdot}5 \times 10^3) \times 250$$
$$= \underline{1{\cdot}875 \text{ MJ}}$$

If we look at Fig. 120 we shall see a graphical representation of this problem.

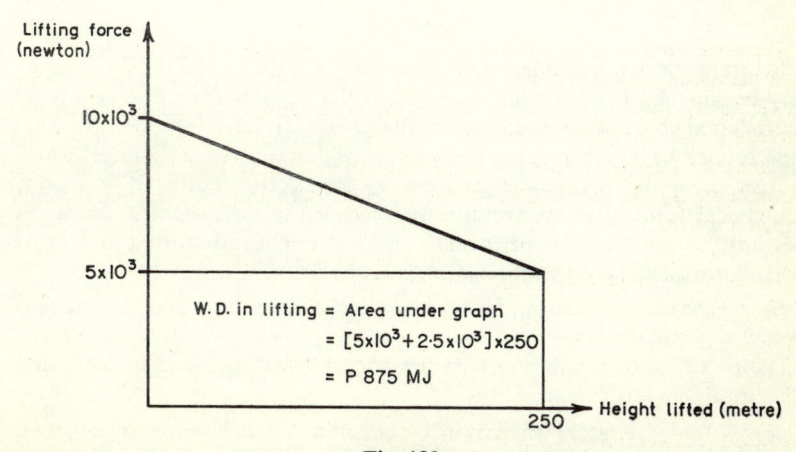

Fig. 120

140　MATERIALS AND MACHINES IN THE WORKSHOP

2. A 30 m length of wire rope 'weight' 24 N/m hangs vertically and has to be wound onto a drum. Calculate how much work is done, firstly, winding the first 10 m and secondly the whole length.

For a graphical representation of this problem see Fig. 121.

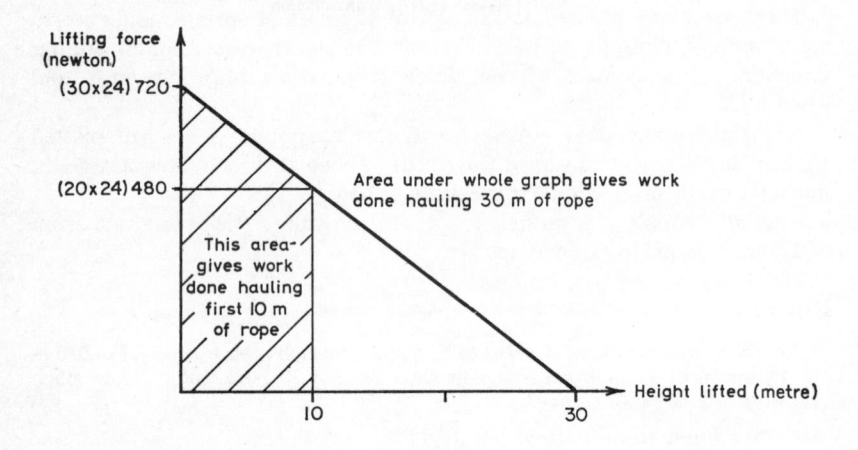

Fig. 121

$$\text{Work done hauling first 10 m (Area of a trapezium)} = \frac{720+480}{2} \times 10 = \underline{6 \text{ kJ}}$$

$$\text{Work done hauling 30 m rope (Area of a triangle)} = \frac{720 \times 30}{2} = \underline{10\cdot 8 \text{ kJ}}$$

Energy

Before work can be done, *energy* must be available to do it. This *energy* may be supplied from the human body, if the work is done during a hand operation such as sawing, chiselling, filing, etc. If, however, the operation is performed by a machine, the energy is usually from *electrical* sources which is then converted to *mechanical* form by an electric motor. It is *energy*, mechanical form, that we are directly interested in, because it is energy in this form that we use for performing most of our metal cutting and other workshop and factory operations.

Energy, whatever its form, is measured in the same units in which work is measured, joule.

Figure 122 shows how most of the energy used by machine tools and other machines is obtained.

Note: A few generators are driven by either water turbines or oil engines, but in Britain by far the most are driven by steam engines or turbines.

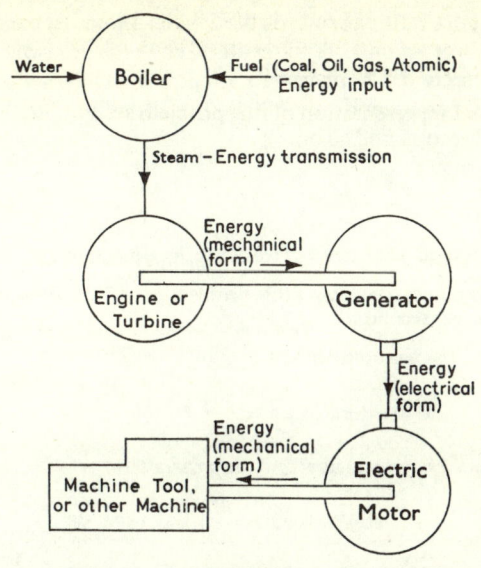

Fig. 122

Power

In machining operations such as turning, milling, shaping, boring, etc., it is the general practice to measure the speed at which the metal is removed in metre per second, i.e., the cutting speed is measured in m/s. Sometimes we talk of the rate at which the metal is cut, i.e. the number of metre length removed in one second. Thus, cutting speed and rate of cutting are really the same thing and we measure both in m/s.

We already know that work is done when metal is cut, so if we know the rate at which the metal is cut we can find the rate at which work is done during the cutting operation. This is very useful. If we know the rate at which work is being done we can find the *power* being used during the process, because

POWER is the RATE of doing WORK,

i.e., POWER is WORK DONE *per second*

This applies to all operations in which work is done; e.g. generating electricity, compressing air or pumping liquid.

Unit of power

Power is measured in watt, kilowatt or megawatt, generally and

$$1 \text{ watt} = 1 \text{ joule/s},$$
$$1 \text{ kilowatt} = 10^3 \text{ watt},$$
$$1 \text{ megawatt} = 10^6 \text{ watt}.$$

This means that 1 watt of power is used when 1 joule of energy is expended over a period of one second. A kilowatt of power is used when 10^3 joule are spent over one second. A megawatt of power involves the using of 10^6, a million, joule per second; or 500 000, half-million, in half-second; or 2×10^6 joule in 2 second and so on.

Examples

1. What power is used when 500 J of work are done in ½ minute?

    The first step towards solving this problem is to find out how many joules of *work* are done per second.

    $$\text{WORK DONE in 30 s} = 500 \text{ J}$$

    $$\text{Work done in 1 s} = \frac{500}{30} \text{ J}$$

    Having done this, we can now find the *power* used.

    $$\text{POWER used} = \frac{50}{3} \text{ J/s} = \underline{16 \cdot 67 \text{ W}}$$

2. What power is used when 500 J of work are done in 2 minutes?

    As before, we first want to know the work being done per second.

    $$\text{WORK DONE in 2 min} = 500 \text{ J}$$

    $$\text{Work done in 1 s} = \frac{500}{120} \text{ J}$$

    $$\text{POWER used} = \frac{50}{12} \text{ J/s} = \underline{4 \cdot 167 \text{ W}}$$

    Comparing this with the answer to the previous problem we see that a quarter of the power is required when the time allowed is quadrupled.

3. What power is used when 10^3 J of work are done in 5 seconds?

    In this problem we will first find the *work* done in 1 second.

    $$\text{WORK DONE in 5 s} = 10^3 \text{ J}$$

    $$\text{Work done in 1 s} = \frac{100}{5} = 200 \text{ J}$$

    $$\text{POWER used} = 200 \text{ J/s} = 200 \text{ W}$$

We must remember that before calculating *power* we have to obtain the *work* done per second.

Quite often the power used on a machine tool has to be carried to the tool by means of a belt, and it is carried from an electric motor, or a line shaft. Because of this, it is necessary for us to consider the power transmitted, or carried, by a belt.

Horse-power transmitted by a belt drive

Figure 123 shows a belt drive with the information needed in our calculation marked on it.

The *driver* pulley causes a tensile force in the tight side of the belt equal to T_1 newton but, because the belt uses some of this force in driving the *driven*

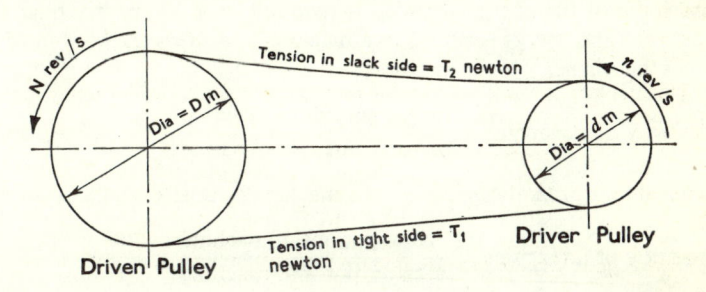

Fig. 123

pulley, the tensile force in the slack side of the belt is reduced to T_2 newton. Therefore, the effective pull in the belt is:

EFFECTIVE PULL in belt

$= (T_1 - T_2)$ newton

EFFECTIVE DRIVING FORCE on DRIVEN pulley

$= (T_1 - T_2)$ newton

WORK DONE in 1 rev of DRIVEN pulley

$= (T_1 - T_2) \times$ circumference of DRIVEN pulley

$= (T_1 - T_2) \times \pi D$ joule

The DRIVEN pulley makes N rev/s,

∴ WORK DONE per s

$= (T_1 - T_2) \times \pi D \times N$ joule

POWER transmitted by belt

$= (T_1 - T_2)\pi DN$ watt

or, using circumference and the speed of the *driver* pulley:

POWER transmitted by belt

$= (T_1 - T_2)\pi dn$ watt.

It does not matter whether we use the diameter of the large pulley or the smaller pulley so long as we remember to use the speed of the pulley whose diameter we take, i.e. use N with D with *n* with *d*.

Efficiency

We already know that to get *work* from a machine we have to supply *energy* to it. What we now have to learn is that all the energy that we supply to a machine is not given back to us as work, i.e., we do not get as many joules of work from a machine as joules of energy we supply to it. This is because some of the energy supplied is required to drive the machine. The ratio between the energy supplied to a machine and the work got out of it is known as the *efficiency* of the machine.

$$\text{EFFICIENCY of a machine} = \frac{\text{WORK got out of machine}}{\text{ENERGY put into machine}}.$$

It is usual to express the *efficiency* of a machine as a percentage, so we have

$$\text{EFFICIENCY of a machine} = \frac{\text{WORK got out of machine}}{\text{ENERGY put into machine}} \times \frac{100}{1}$$

If it is not expressed as a percentage, then it is just a ratio, i.e., it has no units.

Because the work got out of a machine is always less than the energy put into the machine, its *efficiency* is always less than 100%, i.e., less than 1. If an efficiency is not given in a problem we assume that it is 100%.

Examples

1. During a shaping operation the cutting force at the tool point is 400 N and the average cutting speed 0·25 m/s. Find the work done per second and the power used in cutting the metal.

    WORK DONE per s in cutting metal = cutting force × cutting speed

    $$= 400 \times 0 \cdot 25 \text{ J}$$

    Work done cutting metal = 100 J/s

    POWER used cutting metal = 100 W

2. To machine a particular casting 0·3 m long on a shaping machine, the machine is set to make 1 cutting stroke per s. The cutting force at the tool point is 500 N. Calculate the power used in cutting the metal.

    The solution of a problem of this type should be set out as follows, so that it can be followed more easily:

    WORK DONE in 1 stroke = CUTTING FORCE × LENGTH being cut

    $$= 0 \cdot 3 \times 500 \text{ J}$$

    It is not necessary to work this out, since we are not particularly interested in knowing the value of work done per stroke.

    Next find an expression for the work done per second.

WORK DONE per s

= WORK DONE per stroke × no. of CUTTING STROKES per s

= 0·3 × 500 × 1 = 150 J

POWER used in cutting metal = 150 J/s = <u>150 W</u>

3. Holes are being punched in steel plate 5 mm thick. The average force on the punch during the operation is 20 kN. How many joules of work are done during the punching of 1 hole?

FORCE on punch = 20×10^3 N

DISTANCE moved = 5×10^{-3} m

WORK DONE punching 1 hole = $20 \times 10^3 \times 5 \times 10^{-3}$

= <u>100 J</u>

4. A man, when sawing a piece of metal, makes an average stroke of 250 mm and saws at the rate of 30 cutting strokes per min. If he applies an average force of 160 N to the saw, how much work does the man do, in kilojoule per min?

DISTANCE moved by cutting force per stroke = 250×10^{-3} m

DISTANCE moved by cutting force per min = 0·250 × 30 m

WORK DONE by cutting force = 7·5 × 160 = 1200 J

WORK DONE by man in sawing metal = <u>1·2 kJ</u>.

5. In a slotting machine operation 0·75 kilowatt is used on cutting metal at a speed of 25 m/min. What is the force acting on the tool point?

In this problem we are asked to calculate a force. Let this force be F newton, and then proceed as before to find the work done by F and from this the power in terms of F.

WORK DONE by the force acting on tool point per s

$$= \frac{F \times 25}{60}$$

POWER used in cutting metal = $\frac{5F}{12}$ watt

but the power used in cutting metal = 0·75 kW = 750 W

hence, $\frac{5F}{12} = 750$

$$F = \frac{750 \times 12}{5}$$

$$F = 1800 \text{ N}$$

The cutting force on the tool point = <u>1·8 kN</u>

146   MATERIALS AND MACHINES IN THE WORKSHOP

6. A 75 mm dia. bar is turned in a lathe at a speed of 2 rev/s. The force on the tool point due to the cutting of metal is 450 N. What power is used?

In solving a problem of this type we shall find that the following arrangement is very convenient to follow.

WORK DONE in 1 rev of the bar = 450 N × circumference of bar in m

WORK DONE in 2 rev of the bar = $450 \times 75 \times 10^{-3} \times \pi \times 2$ joule/s

since a joule per second is equivalent to a watt

POWER used = $0.450 \times 150\pi$ = $\underline{67.5\pi \text{ W}}$

Note that it is not necessary to do any working out until we get to the last line.

We could also have solved this problem in the following way:

DISTANCE MOVED per s (i.e., cutting speed) = $75 \times 10^{-3} \times \pi \times 2 = 0.15\pi$ m

WORK DONE per s in cutting metal = $450 \times 0.15\pi$ joule

POWER used in cutting metal = $\underline{67.5\pi \text{ W}}$ as before

7. A cast iron pulley is turned at 28 m/min. The force on the tool point is 700 N. Calculate the power used in cutting metal.

This problem is a little more straightforward than the previous one, because we are given the cutting speed.

WORK DONE in cutting metal = $\dfrac{28}{60} \times 700$ J/s

POWER used in cutting metal = $\underline{327 \text{ W}}$

8. An electric motor which has a speed of 25 rev/s, and a pulley of 280 mm dia. drives a line shaft by means of a belt 125 mm wide. The safe load for the belt must not exceed 15 N per mm width of belt. Calculate the maximum safe power that can be transmitted by the belt.

First we will determine the belt speed, i.e., the distance travelled by the belt in 1 second.

In 1 rev of the pulley the belt travels $280 \times 10^{-3} \times \pi$ metre

and in 1 s the belt travels $0.28\pi \times 25$ m

∴ Belt speed is $0.28\pi \times 25$ m/s

There is no need to work this out, because we want to use it in a further calculation.

Next find the force in the belt.

Force in belt 1 mm wide = 15 N

Force in belt 125 mm wide = 15×125

Again there is no need to work this out because we have to use it in a further calculation.

WORK DONE by belt force in 1 s = $15 \times 125 \times 0.28\pi \times 25$ J

POWER transmitted by belt = $15 \times 125 \times 0.28\pi \times 25 = 41\,210$ W

= $\underline{41.2 \text{ kW}}$

WORK AND POWER   147

9. During a turning operation a 2 mm deep cut is used with a feed of 0·5 mm per rev of the lathe spindle. The pressure on the tool point is 700 N/mm², and the cutting speed is 0·5 m/s. What power is used in cutting metal?

In this problem we are given the pressure acting on the tool point due to cutting metal. We must use this to find the force on the tool point.

First draw the tool in its cutting position (see Fig. 124).

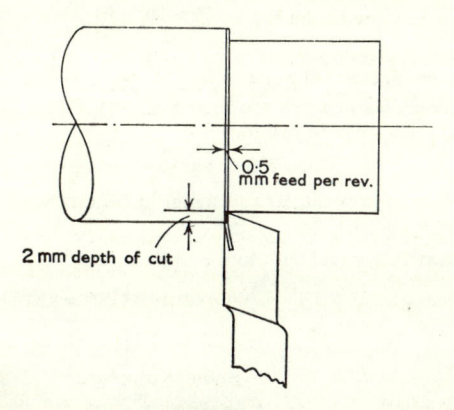

Fig. 124

We can see from Fig. 124 that the area of metal in contact with the tool point can be found as follows:

AREA OF METAL in contact with tool point = 2 × 0·5 = 1 mm²

Since the area of contact is 1 mm², the force on the tool point is 700 N.

WORK DONE per min by force on tool point
$$= 700 \times 0·5 = 350 \text{ joule/s}$$
POWER used in cutting metal = $\underline{350 \text{ watt}}$

Note: We have just found out that FORCE = PRESSURE × AREA in solving part of this problem. This formula is worth remembering.

10. When turning a 75 mm dia. bar with a depth of cut of 2·5 mm and a feed of 0·25 mm per rev of the lathe spindle, 1·2 kW are used. The spindle speed is 400 rev/min. What is the pressure on the tool point?

Begin by letting the force on the tool point be F newton, and then proceed to set out the problem in the manner we have used previously. The mean dia. is 72·5 mm.

DISTANCE moved in 1 rev = $72·5\pi \times 10^{-3}$ metre

Distance moved in 1 min = $72·5\pi \times 10^{-3} \times 400$ m

CUTTING SPEED = $\dfrac{29\pi}{60}$ m/s

WORK DONE per s in cutting metal = $\dfrac{F \times 29\pi}{60}$ joule

POWER used in cutting metal $= \dfrac{29\pi\, F}{60}$ watt,

but the power used during the cutting operation is 1·2 kW,

$$\therefore \dfrac{29\pi\, F}{60} = 1\cdot 2 \times 10^3$$

$$F = \dfrac{72 \times 10^3}{29\pi}$$

The force on the tool point = 790 N

We have already learned that FORCE = PRESSURE × AREA, and in this problem it is the pressure that we have to find.

Thus, 790 = PRESSURE × AREA

The area we require is the area of metal in contact with the tool point and we can find this as follows:

AREA OF METAL in contact with tool point

$$= \text{DEPTH OF CUT} \times \text{RATE OF FEED}$$
$$= 2\cdot 5 \times 0\cdot 25$$
$$= 0\cdot 625 \text{ mm}^2$$

Therefore, we have

PRESSURE × 0·625 = 790

$$\text{PRESSURE} = \dfrac{790}{0\cdot 625}$$

Hence, the pressure on the tool point during the operation

$$= \underline{1265 \text{ N/mm}^2}$$

11. (a) The table of a milling machine is raised 60 mm by means of the traverse screw. If the mass of the table is 300 kg, how much work is done?

(b) If 700 joules of energy are used in doing the operation, what is the efficiency of the lifting mechanism?

(a) The 'weight' of the table is $300 \times 9\cdot 81$ newton since a 1 kg mass has a 'weight' of 9·81 newton, i.e. the gravitational force on a 1 kg mass is 9·81 newton. Then it will require a vertical force of 2943 N to lift table.

WORK DONE in lifting table $= 2943 \times 60 \times 10^{-3}$ joule

WORK DONE in lifting table $= \underline{176\cdot 6 \text{ J}}$

(b) $\qquad\qquad\qquad \text{EFFICIENCY} = \dfrac{\text{WORK got out}}{\text{ENERGY put in}} \times 100\%$

EFFICIENCY of lifting mechanism $= \dfrac{176\cdot 6}{700} \times 100 = \underline{25\cdot 2\%}$

12. In a belt drive to a countershaft the tension in the tight side of the belt is 500 N and in the slack side 180 N. The driving pulley, 250 mm diameter, is keyed to the armature of an electric motor that develops 6 kW at 20 rev/s.

WORK AND POWER

Calculate the power transmitted by the belt, and the efficiency of the drive.

Remember that to solve a problem of this type, we have the following equation which we worked out on page 143.

POWER transmitted by belt $= (T_1 - T_2)\pi d n$

Substituting the figures that we are given,

$$\text{POWER transmitted by belt} = \underset{\text{(newton)}}{(500-180)\pi} \times \underset{\text{(metre)}}{250 \times 10^{-3}} \times 20 \text{ joule}$$

$$= 320\pi \times 0.25 \times 20$$

$$= \underline{5 \text{ kW}}$$

The power supplied to the belt is 6 kW, so we have a loss of power between the input power to the belt and the output. This, is a loss of energy per unit time due to belt slip chiefly.

$$\text{EFFICIENCY of the drive} = \frac{5}{6} \times 100 = \underline{83\%}$$

Exercise 8

Section A

1. A milling machine table 'weighs' 1500 N and carries a vice and component which together 'weigh' 100 N. How much work is done when the table is raised 80 mm?
2. The two tool boxes of a planing machine are in use during the machining of a casting. The cutting force on each tool is 1500 N and the cutting speed 0·8 m/s. How many joules of work are done per second in cutting metal?
3. The tup of a drop hammer together with the half die which is fastened to it have a mass of 300 kg. Calculate the amount of work done when the tup and half die are lifted through a height of 1·2 metre.
4. During a shaping machine operation, the cutting force on the tool point is 600 N and the length of the surface machined 400 mm. If the ram makes 1 cutting stroke per second, how many joules of work are done in 1 second?
5. A fitter, filing a block of metal, makes on average 23 forward strokes every minute and the average cutting force per stroke is 100 N. Determine the work done, in joule/s, cutting metal if the mean stroke length is 250 mm.
6. The cutting force on the point of a lathe tool is 250 N. If the diameter of the surface being cut is 90 mm and the spindle speed 2 rev/s, calculate the work done, per second.
7. A strip of steel 50 mm wide, 2 mm thick is to be cut on the guillotine shears. The maximum shear strength of the strip is 320 N/mm². Calculate the work done in cutting the strip if the mean blade force is 90% of the maximum value.
8. Whilst part of a shaft is being machined in a lathe the force on the tool point due to cutting is 600 N and the feed force is 200 N. The rate of feed is 1·25 mm/s and the cutting speed is 0·5 m/s. Find the work done during the operation.
9. A broach is pulled through a component with a force of 750 N at a speed of 0·5 m/s. What power must be supplied to the broach?
10. An electric motor with a 260 mm dia. pulley drives a line shaft through a belt 150 mm wide. The safe working load for the belt is 15×10^3 N per metre width. If the motor speed is 20 rev/s, what is the power transmitted by the belt?

Section B

1. The traverse rate in a particular milling operation is 2 mm per second, and the feed force is 1 kN. Calculate the power being used to traverse the table.
2. The depth of cut during a turning operation is 6 mm and the rate of feed 0·3 mm per rev of the spindle. Calculate the work done per second in cutting metal if the cutting speed is 2 m/s and tool point pressure 900 N/mm^2.
3. A pump on a milling machine delivers a maximum of $2·5 \times 10^{-3}$ m^3 of cutting solution per second. It lifts the solution through a vertical height of 1·2 m. Calculate the power required by the pump which has an efficiency of 80%. Assume 1 m^3 of solution has a mass of 900 kg.
4. An electric motor drives a countershaft by means of a belt which has a speed of 10 m/s. If the effective pull in the belt is 1·750 kN. Find the power supplied by the motor assuming the drive is 90% efficient.
5. The cutting forces on a two fluted drill are each 250 N. The drill cuts at the rate of 1 m/s. If the drill spindle is driven by an electric motor through a train of gears which have an efficiency of 85%, find the power supplied by the motor during the operation.
6. A pinion fitted to a shearing machine has to transmit 65 kW. The speed of the pinion is 7·5 rev/s, and its pitch circle diameter is 260 mm. If the tooth load is not to exceed 90 N per mm width of tooth, what is the minimum thickness that the wheel may be?
7. On a wheel lathe a locomotive tyre has its bore machined to 1·8 m dia. The cutting force acting on the tool point is 0·9 kN and the spindle speed 1·5 rev/s. A feed of 0·15 mm per rev is used, the feed force being 600 N. Determine the power required.
8. The 'weight' of a milling machine table together with its job is 3·2 kN. It is lifted 6 mm when the screw that lifts it makes 1 rev. Calculate the energy that must be supplied to the lifting mechanism, for 1 rev of screw, if its efficiency is 24%.
9. In a belt drive the tension in the tight side of the belt is 700 N and that in the slack side 120 N. The speed of the driving pulley is 4 rev/s and its diameter 350 mm. Determine the power transmitted by the belt.
10. The tool head of a side planing machine is driven at a speed of 0·2 m/s by a force of 18 kN. The efficiency of the drive mechanism is 32%. Calculate the power supplied by the driving motor.
11. A lift and its passengers have a mass of 800 kg and it is to be lifted through a vertical height of 40 m by means of wire ropes 'weighing' 40 N per metre length, at a uniform velocity. Find the greatest force required at the periphery of the drum, the maximum torque on the drum, assuming drum diameter is 0·9 m and the work done raising the lift 40 m.
12. A helical spring has a load gradually applied to it up to a maximum value of 70 N, and the extension at this load is 45 mm. How much work is done in stretching the spring? It is found that by adding a further 23 N the spring extension is increased to 60 mm. What percentage of the total work done is accomplished by the final 20 N? Find the total strain energy in the spring.
13. The buffer spring on a draw-bench must be capable of absorbing 40 joule of energy without compressing more than 16 mm. Calculate the average and maximum compressing forces. Assume the load is gradually applied.
14. A cam follower spring on an automatic lathe is compressed 18 mm in 0·2 s. The spring stiffness is 1 N/mm, i.e. 1 N compresses the spring 1 mm. How much energy is used per second in compressing the spring?
15. In a wire drawing operation, aluminium alloy wire 5 mm diameter is being drawn. Its proof stress is 200 N/mm^2 and modulus of elasticity 74×10^3 N/mm^2.

What amount of energy, per metre length of wire, will be used in extending the wire assuming it is stressed to the proof stress limit during drawing?

16. The pressure on an engine piston 100 mm diameter varies from 1·6 N/mm² maximum at the beginning of the stroke to 0·1 N/mm² at the end of the stroke. The stroke length is 150 mm. What power is developed by the piston if it makes 20 working strokes per second? Assume direct variation between piston force and distance travelled.

17. A 2 kW electric motor drives a ventilating fan delivering air through a duct, vertical height 6 m. How many cubic metre of air will the fan deliver per second if 1 m³ of air has mass, 3·7 kg? Efficiency of equipment 62%.

18. At a water storage station, a pump is delivering 0·3 m³ of water per second through a vertical height of 120 m. What power motor is required to drive the pump assuming an overall efficiency of 75%, 1 m³ of water has a mass of 10³ kg?

19. At a hydro-electric station, 300 m³ of water are delivered to a water turbine per hour, the water falling through a vertical height of 40 m. Assuming 90% efficiency, what is the output power of the turbine?

20. A train of 10 trucks, each having a 'weight' of 100 kN is hauled along an inclined track at a speed of 1 m/s. The track is inclined at 30° to the horizontal. If the tractive resistance is 6 N per 1 kN 'weight,' what power electric motor is required assuming the overall efficiency is 85% and the pulling force parallel to the track?

Chapter 9. Belt and Pulley Speeds

BELT DRIVES are used extensively in industry for transmitting power from electric motors to machine tools of various types, e.g. lathes, milling machines, punching and shearing machines. The belts are of the flat or vee type generally though belts having a circular cross-section are used.

Calculations

Since belt and pulley drives are used, we need to learn how to calculate belt and pulley speeds. To help us we will use Fig. 125. Here we have a pulley whose diameter is measured in metre and its speed in rev/s.

When the pulley makes 1 revolution, a line marked on the belt moves through a length equal to the circumference of the pulley,

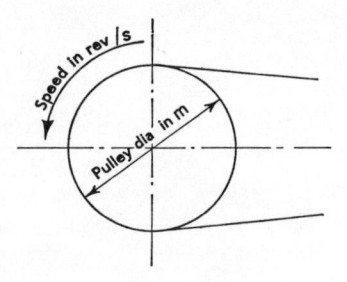

Fig. 125

DISTANCE moved by the belt for 1 rev of pulley
   = circumference of pulley.

DISTANCE moved by belt in 1 second
   = circumference of pulley × pulley speed,

BELT SPEED (metre/s)
   = π × PULLEY DIA. (m) × PULLEY SPEED (rev/s).

Strictly speaking, the diameter that we use should be measured from the centre of the belt, on either side of the pulley. The 'diameter' should be:

DIAMETER OF PULLEY + BELT THICKNESS.

If, however, the belt thickness is not given, then the pulley diameter is used. In fact, the belt speed calculated from the pulley diameter is that most commonly used, and when calculating a pulley speed from a belt speed it is usual practice to ignore belt thickness. It is the inner surface of the belt that contacts the pulley.

Examples

1. An electric motor drives a line shaft by means of a flat open belt. The motor pulley is 400 mm dia. and has a speed of 25 rev/s. Calculate the speed of the belt.

$$\text{BELT SPEED} = \pi \times \text{PULLEY DIA.} \times \text{PULLEY SPEED}$$

$$= \pi \times 400 \times 25 \text{ mm/s}$$

$$= \frac{\pi \times 400 \times 25}{10^3} \text{ m/s}$$

$$= \underline{31\cdot 4 \text{ m/s}}$$

Let us suppose that the thickness of the belt used in Example 1 is 5 mm. We will now calculate the average belt speed using a diameter of 400+5 = 405 mm = 0·405 m.

$$\text{BELT SPEED} = \pi \times 0\cdot 405 \times 25 \text{ m/s}$$

$$= \underline{31\cdot 8 \text{ m/s}}$$

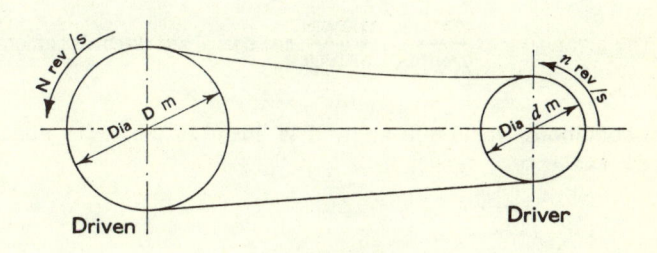

Fig. 126

We can see that, on comparing the two answers, there is a difference. The belt speed 31·4 m/s is really the speed of the inner surface of the belt which is in contact with the pulley. The value 31·8 m/s is the speed of the centre line of the belt, i.e., the mean belt speed. Because the belt has thickness, its outer surface is travelling faster than its inner surface, since the outer surface is further away from the centre of the pulley about which the belt is rotating. The two surfaces have different speeds because they are at different *radii* from the centre of rotation.

We will now turn to the calculation of pulley speed, using Fig. 126.

We have just learned that:

$$\text{BELT SPEED} = \pi \times \text{PULLEY DIA.} \times \text{PULLEY SPEED}$$

Since we are now concerned with finding pulley speed, we use the speed of the *inner surface* of the belt, which is the surface that contacts the pulley.

We are not here concerned with belt thickness. As there are two pulleys, one driving the other, we shall find it easier to call the pulley diameters D and d, and their speeds N and n respectively (see Fig. 126). The belt speed will be given as follows:

$$\text{BELT SPEED} = \pi \times D \times N$$

or

$$\text{BELT SPEED} = \pi \times d \times n$$

Note: It does not matter which pulley we work on, as long as we take the speed and diameter of the same pulley. Hence we have:

$$\pi d n = \pi D N$$

or

$$\frac{\pi d}{\pi D} = \frac{N}{n}$$

i.e., $\quad \dfrac{\text{Dia. of DRIVER (metre)}}{\text{Dia. of DRIVEN (metre)}} = \dfrac{\text{Speed of DRIVEN (rev/s)}}{\text{Speed of DRIVER (rev/s)}}$

or $\quad \dfrac{\text{Dia. of DRIVEN (metre)}}{\text{Dia. of DRIVER (metre)}} = \dfrac{\text{Speed of DRIVER (rev/s)}}{\text{Speed of DRIVEN (rev/s)}}$

Note: The arrangement $\dfrac{\text{DRIVER}}{\text{DRIVEN}}, \dfrac{\text{DRIVEN}}{\text{DRIVER}}$ is a useful aid towards remembering the formula.

These conditions would still apply if in Fig. 126 the larger pulley were driving the smaller one.

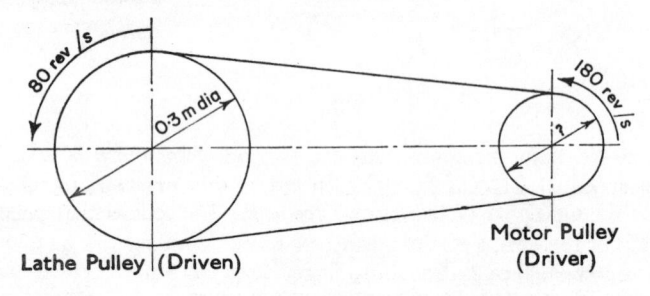

Fig. 127

Examples

1. The driving pulley on the all geared headstock of a centre lathe must have a speed of 8 rev/s. It is to be driven by an electric motor which has a speed of 18 rev/s. If the lathe pulley has a diameter of 0·3 metre, calculate the diameter of the motor pulley in order to obtain the required lathe pulley speed.

   First make a sketch of the drive as shown in Fig. 127.

BELT AND PULLEY SPEEDS 155

It is the diameter of the motor pulley that we want to find, i.e., the diameter of the *driver* pulley. We will begin the calculation by writing the term we have to calculate and then build the remainder of the formula round it.

$$\frac{\text{Dia. of DRIVER}}{\text{Dia. of driven}} = \frac{\text{Speed of driven}}{\text{Speed of driver}}$$

Next put the given values into the formula:

$$\frac{\text{Dia. of DRIVER}}{0.3 \text{ m}} = \frac{8}{18}$$

Now rearrange the formula:

$$\text{Dia. of DRIVER} = \frac{8}{18} \times 0.3$$

$$= 0.133 \text{ m } (133 \text{ mm})$$

2. A countershaft, which has a pulley 950 mm dia. keyed to it, is to be driven by an electric motor which has a speed of 25 rev/s. The motor has a pulley 230 mm dia. What will be the speed of the countershaft?

Again we make a sketch, as in Fig. 128.

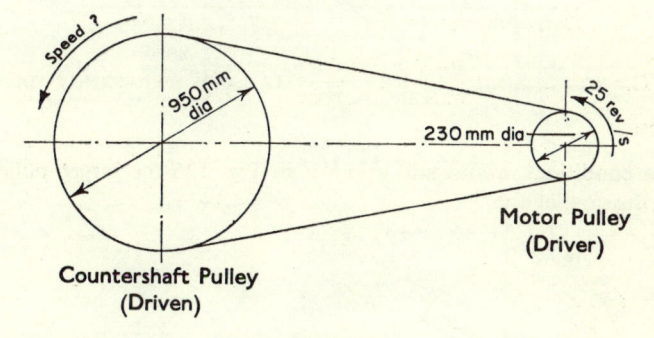

Fig. 128

Next, we write down the unknown quantity, i.e., the speed of *driven*, and build the remainder of the formula round it. The countershaft pulley is the *driven*.

$$\frac{\text{Speed of DRIVEN}}{\text{Speed of driver}} = \frac{\text{Dia. of driver}}{\text{Dia. of driven}}$$

Now we substitute the given values:

$$\frac{\text{Speed of DRIVEN}}{25} = \frac{230}{950}$$

Rearranging this,

$$\text{Speed of DRIVEN} = \frac{23}{95} \times 25$$

Speed of COUNTERSHAFT = 6 rev/s

3. A 250 mm dia. grinding wheel must have a surface speed of 40 m/s. Keyed to the same spindle as the grinding wheel is a 90 mm dia. pulley. This pulley is driven through vee belts by an electric motor whose speed is 25 rev/s. Calculate the speed of the grinding wheel in rev/s, the diameter of the pulley to be fitted to the motor, in order to obtain the correct grinding wheel speed, and the belt speed.

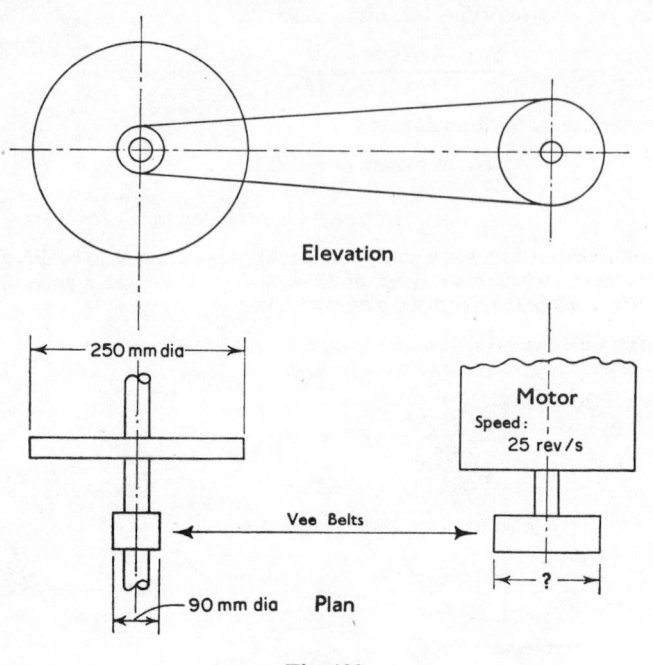

Fig. 129

First make a sketch of the drive, as in Fig. 129.

We are told that the surface speed of the grinding wheel is 40 m/s. If we calculate the circumference of the wheel and then find how many times this circumference divides into 40, we shall know how many revolutions the wheel makes in 1 second.

Circumference of grinding wheel $= \pi \times$ dia. of wheel

$$= \pi \times \frac{250}{10^3} = 0 \cdot 25\pi \text{ metre}$$

SPEED of grinding wheel $= \dfrac{40}{0 \cdot 25\pi} = 50 \cdot 95$

$$= 51 \text{ rev/s}$$

We begin the solution to the second part of the problem by writing down the **unknown quantity** and then building the remainder of the equation round it.

BELT AND PULLEY SPEEDS

We are asked to find the diameter of the motor pulley, i.e., the diameter of the *driver*.

$$\frac{\text{Dia. of DRIVER}}{\text{Dia. of driven}} = \frac{\text{Speed of driven}}{\text{Speed of driver}}$$

$$\text{Dia. of DRIVER} = \frac{50 \cdot 95}{25}$$

Remember that the rev/s of the 90 mm dia. pulley is the same as the rev/s of the grinding wheel, since both are fixed to the same spindle.

$$\text{Dia. of DRIVER} = \frac{50 \cdot 95}{25} \times 90 = 183 \cdot 4$$

$$= \underline{183 \text{ mm}}$$

$$\text{BELT SPEED} = \pi \times \text{PULLEY DIA.} \times \text{PULLEY SPEED}$$

$$= \pi \times \frac{90}{10^3} \times 50 \cdot 95$$

$$= \underline{14 \cdot 4 \text{ m/s}}$$

4. A belt which has a speed of 12 m/s drives a pulley of 280 mm dia. If there is a 6% belt slip, what will be the speed of the pulley?

We know that:

$$\text{BELT SPEED} = \pi \times \text{PULLEY DIA.} \times \text{PULLEY SPEED}$$

If the belt slip is 6%, then its effective speed is 94% of the given speed,

$$\text{Effective belt speed} = 12 \times 0 \cdot 94 \text{ m/s}$$

$$\frac{280\pi}{10^3} \times \text{pulley speed} = 12 \times 0 \cdot 94$$

$$\text{Pulley speed} = \frac{12 \times 0 \cdot 94}{280\pi} \times 10^3$$

$$= \underline{12 \cdot 8 \text{ rev/s}}$$

If there had been no belt slip, the pulley speed would have been higher.

$$\text{Pulley speed} = \frac{12 \times 10^3}{280\pi}$$

$$= \underline{13 \cdot 6 \text{ rev/s}}$$

Because of the 6% belt slip the pulley speed is reduced from 13·6 to 12·8 rev/s.

5. The drive to a shaping machine consists of an electric motor with a pulley, 110 mm dia. which drives a 370 mm dia. pulley fixed to an intermediate shaft. Also keyed to this intermediate shaft is a 330 mm dia. pulley that drives the machine pulley, 220 mm dia. The motor speed is 25 rev/s. What is the speed of the machine pulley?

We will begin by making a diagram of the drive (see Fig. 130).

Next we will find the speed of the 370 mm dia. pulley. It is a *driven* pulley. The motor pulley is the *driver*.

$$\frac{\text{Speed of DRIVEN}}{\text{Speed of driver}} = \frac{\text{Dia. of driver}}{\text{Dia. of driven}}$$

Substituting the figures that we are given:

$$\frac{\text{Speed of DRIVEN}}{25} = \frac{110}{370}$$

$$\text{Speed of DRIVEN} = \frac{110}{370} \times 25 = 7\cdot43 \text{ rev/s}$$

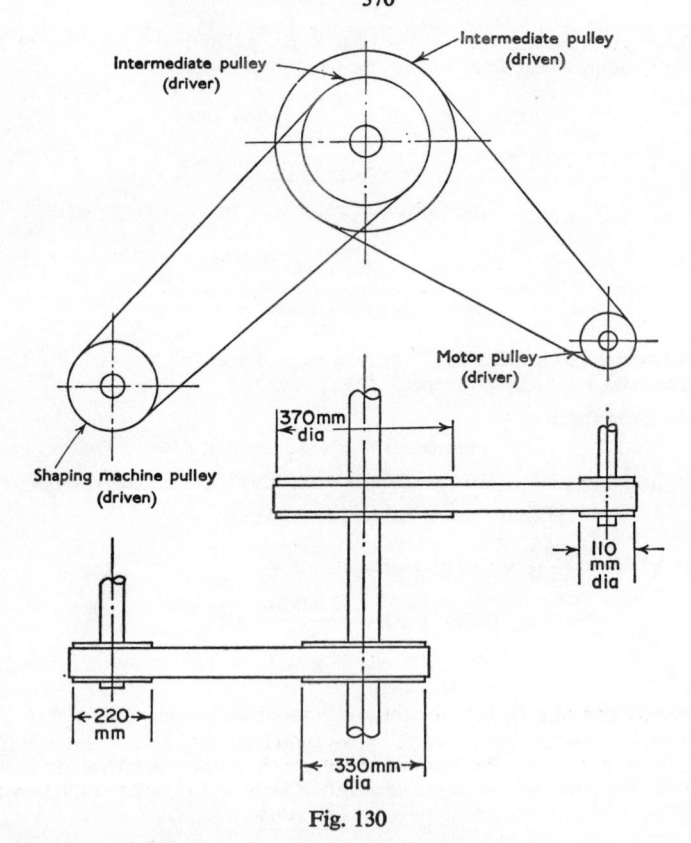

Fig. 130

Since the 370 mm dia. pulley is fastened to the intermediate shaft, the speed of the shaft must be 7·43 rev/s. This must also be the speed of the 330 mm dia. driver pulley. The machine pulley is the *driven*.

$$\frac{\text{Speed of DRIVEN}}{\text{Speed of driver}} = \frac{\text{Dia. of driver}}{\text{Dia. of driven}}$$

$$\frac{\text{Speed of DRIVEN}}{7\cdot43} = \frac{330}{220}$$

$$\text{Speed of DRIVEN} = \frac{330}{220} \times 7\cdot43 = \underline{11 \text{ rev/s}}$$

BELT AND PULLEY SPEEDS   159

6. A grinding wheel has to make 30 rev/s in order to obtain the correct speed at the grinding surface. The grinding wheel spindle has a 120 mm dia. pulley keyed to it, which is driven by a pulley fastened to an intermediate shaft having a speed of 8 rev/s. This in turn is driven by an electric motor with a speed of 25 rev/s. Determine which of the following pulleys could be used in order to make up a suitable drive: 450 mm, 390 mm, 300 mm, 175 mm, 120 mm. diameters.

First we will draw a side elevation of the drive and include the given details (see Fig. 131).

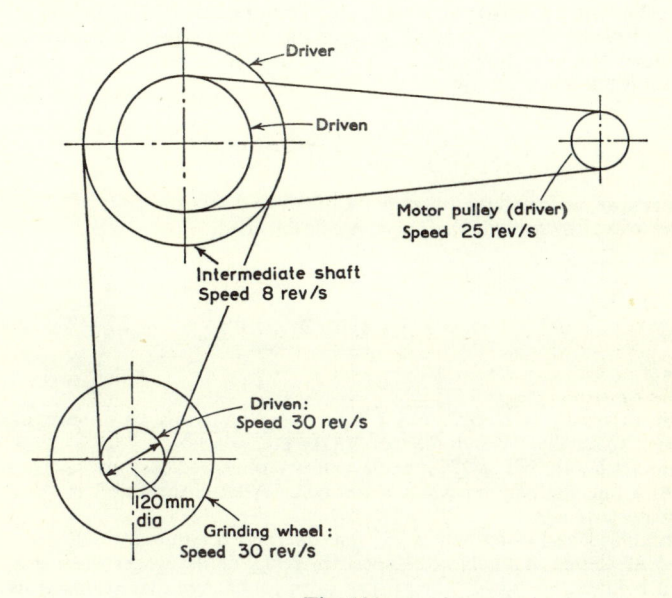

Fig. 131

We will now find what diameter pulley we require to fix to the intermediate shaft in order to drive the grinding wheel spindle at the correct speed. In this instance, the pulley on the intermediate shaft is the *driver*, and the 120 mm dia. pulley on the grinding wheel spindle is the *driven*, so we have:

$$\frac{\text{Dia. of DRIVER}}{\text{Dia. of driven}} = \frac{\text{Speed of driven}}{\text{Speed of driver}}$$

$$\frac{\text{Dia. of DRIVER}}{120} = \frac{30}{8}$$

$$\text{Dia. of DRIVER} = \frac{30}{8} \times 120 = \underline{450 \text{ mm}}$$

We must have the 450 mm pulley keyed to the intermediate shaft.

Next we have to determine what diameter pulleys are to form the drive between the motor and intermediate shafts. In other words we have to find two unknown

MATERIALS AND MACHINES IN THE WORKSHOP

diameters. We cannot do this with the formula that we have but we can obtain the ratio of the two diameters. The motor pulley is the *driver*.

Hence,
$$\frac{\text{Dia. of DRIVER}}{\text{Dia. of DRIVEN}} = \frac{\text{Speed of driven}}{\text{Speed of driver}}$$

$$= \frac{8}{25}$$

This ratio tells us that we need an 8 mm dia. pulley for the *driver* and a 25 mm dia. pulley for the *driven*. Or they must be multiples of 8 and 25. We certainly cannot have 8 mm and 25 mm because pulleys having these diameters are not available. We have, however, a 120 mm pulley and a 390 mm pulley and these will almost give us the required ratio,

$$\frac{\text{Dia of DRIVER}}{\text{Dia. of DRIVEN}} = \frac{120}{390} \text{ (1·5 cancels)} \simeq \frac{8}{26}$$

Therefore, we can fit the 120 mm dia. pulley to the motor, and the 390 mm dia. pulley to the intermediate shaft to complete the drive.

Exercise 9

Section A

1. A belt driven by an electric motor has a speed of 10 m/s. It drives a machine pulley 230 mm dia. What is the speed of the pulley in rev/s?
2. An electric motor with a 270 mm dia. pulley has a speed of 25 rev/s. The motor drives a line shaft by means of a flat belt. What is the speed of the belt in metre per second?
3. A grinding wheel when new is 250 mm dia. and is required to have a surface speed of 45 m/s. Calculate the speed in rev/s, of the wheel when it is new. What is its surface speed when it has worn to 200 mm dia. assuming its rev/s remains the same?
4. The driving pulley of an electric motor is 210 mm dia. and it drives a flat leather belt that is 6 mm thick. What is the average speed of the belt in m/s? The motor speed is 25 rev/s.
5. A lathe pulley 220 mm dia. is driven by a 105 mm dia. motor pulley having a speed of 20 rev/s. What is the speed of the lathe pulley?
6. A 105 mm dia. pulley keyed to the spindle of a grinding machine is driven by a belt which has a speed of 15 m/s. What is the speed, in rev/s, of the grinding wheel?
7. The vertical spindle of a simple bench drill has two pulleys keyed to it, one 220 mm dia. and the other 160 mm dia. Both in turn can be driven by a single vee belt that has a speed of 10 m/s. Calculate the two possible spindle speeds.
8. A 0·8 m dia. pulley, fixed to a countershaft which makes 3 rev/s drives a 300 mm dia. shaping machine pulley. What is the speed of the machine pulley?
9. A belt which has a speed of 16 m/s drives a 130 mm dia. machine pulley. Calculate the speed of the pulley, if there is no belt slip, if the belt slip is 7%.
10. An electric motor with a speed of 25 rev/s and a driving pulley 280 mm dia. has to drive a line shaft at a speed of 6 rev/s. Find what diameter pulley must be fitted to the line shaft.

BELT AND PULLEY SPEEDS

Section B

1. A chain making machine is driven from a countershaft which revolves at 4 rev/s by a 375 mm dia. pulley which drives a 410 mm dia. pulley keyed to the machine shaft. What is the speed of the machine pulley?
2. An electric motor speed 25 rev/s has a 160 mm dia. pulley which drives a lathe pulley at a speed of 8 rev/s. Calculate the diameter of the lathe pulley.

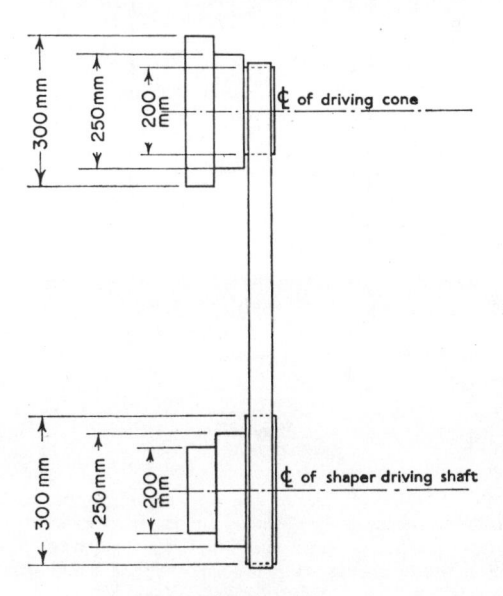

Fig. 132

3. The sketch in Fig. 132 shows a cone pulley drive to a shaping machine. Calculate the three possible speeds for the shaper driving shaft if the speed of the driving cone is 7 rev/s.
4. Figure 133 shows a drive to a lathe from an electric motor, speed 30 rev/s, via an intermediate shaft. Calculate the diameter of the pulley that must be fitted to the lathe if an input speed of 7 rev/s is required.
5. An overhead countershaft is to be driven by an electric motor, speed 20 rev/s. Fitted to the motor is a driving pulley 380 mm dia. If the speed of the countershaft is to be 4 rev/s, what diameter pulley must be fitted to it?
6. The spindle of a sensitive drilling machine can be driven at two speeds depending upon which of the two pulleys that are keyed to the spindle is driven by the belt from the electric motor. The motor pulley is 130 mm dia. and the machine pulleys are 260 mm and 160 mm dia. respectively. If the motor speed is 20 rev/s, find the two possible spindle speeds.
7. A 250 mm dia. grinding wheel is to have a surface speed of 50 m/s. Keyed to the same spindle as the wheel is a 100 mm dia. pulley driven from an overhead line shaft by a pulley 600 mm dia. Find the line shaft speed.
8. A 0·33 m dia. lathe pulley is driven by a belt which has a speed of 3 m/s. Calculate the speed of the pulley, if there is no belt slip, if there is a belt slip of 6%.

162 MATERIALS AND MACHINES IN THE WORKSHOP

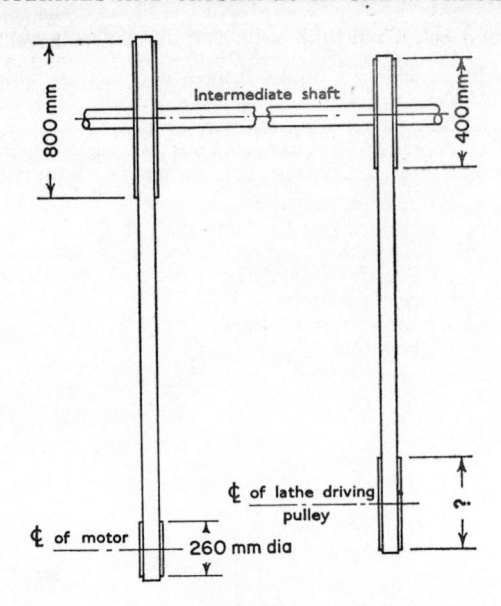

Fig. 133

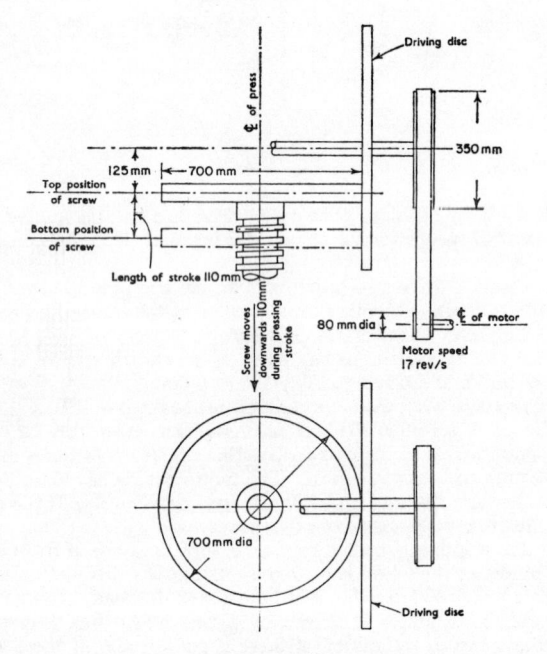

Fig. 134

9. By means of a belt 6 mm thick a 190 mm dia. pulley, speed 10 rev/s, drives a 260 mm dia. pulley. Calculate the belt speed in m/s, neglecting the belt thickness, taking into account the belt thickness, and the speed of the driven pulley.
10. The driving mechanism for the downward stroke of a Screw Press is shown in Fig. 134. The press screw has a stroke length of 110 mm and it screws through a nut in the frame of the press which is not shown. An 80 mm dia. pulley is keyed to the armature of an electric motor, speed 17 rev/s, and this pulley drives a 350 mm dia. pulley keyed to the same shaft as the driving disc. This disc drives, by friction, the pulley fixed to the end of the screw. The efficiency of the drive is 90%. Calculate to the nearest rev/s the speed of the screw:
    (a) at the beginning of the stroke;
    (b) at the point mid-way through the stroke;
    (c) at the end of the stroke.

Chapter 10. Gear Wheels

There are very few, if any, machine tools without a gear wheel on them somewhere. Even the old belt driven lathes with the cone pulleys have a back gear, which incorporates the gear wheels. They also have 'change wheels', which are the gear wheels used to make up the drive to the leadscrew. The formulae for determining change wheels, and for the gear wheels to be used on the dividing head in a 'spiral' milling operation, are probably the most used of all machine shop equations.

Gear wheel speeds and gear trains

It is obvious that we must be able to calculate gear speeds and sizes correctly if we are to be good technicians.

In order to learn something of the relationship between two gear wheels in mesh, we will first study the two gear wheels blanks (gear wheels before the teeth are cut) shown in Fig. 135.

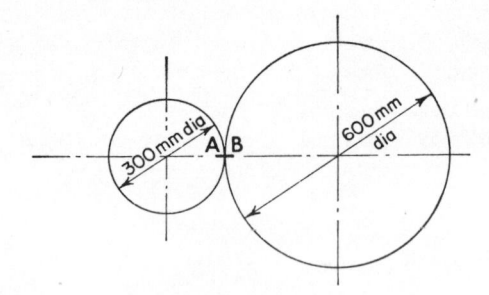

Fig. 135

These are so arranged that one can drive the other because of the friction between them. Assume that one of these blanks is 300 mm, and the other one 600 mm diameter. They could be any diameter, we have used 300 mm and 600 mm for convenience. Suppose we now scribe a line on each wheel, A and B, so that the lines coincide. Let the smaller blank make 1 complete revolution, driving the larger blank without any slip. If we now look at the line A we see that it is back where it started after moving through a distance equal to the circumference of the small gear blank, i.e., after moving through a distance of $\pi \times 300$ mm. The mark B has moved a similar distance round the circumference of the larger blank, i.e., it has moved $\pi \times 300$ mm, which is half the circumference of the larger wheel, see Fig. 136.

GEAR WHEELS

Therefore, when the small blank makes 1 revolution, the large blank makes ½ revolution. Or, when the small blank revolves twice, the large blank revolves once, and so on. We could state this in the following manner:

$$\frac{\text{DISTANCE moved by a line on the periphery of small gear blank}}{\text{DISTANCE moved by a line on the periphery of large gear blank}} =$$

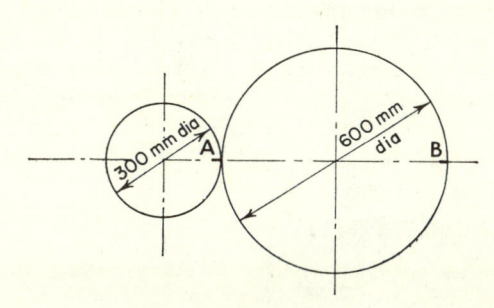

Fig. 136

Suppose we think of *rev/s* now rather than just *revs*, i.e., we think of speed, and let the speed of the small wheel be represented by n rev/s and that of the large wheel by N rev/s. We will represent the diameter of the small blank by d, and that of the large blank by D. Then,

DISTANCE moved per s by a line on the periphery of small gear blank $\Big\} = \pi \times d \times n$

and

DISTANCE moved per s by a line on the periphery of large gear blank $\Big\} = \pi \times D \times N$

As we already know that the two lines move through equal distances in the same interval of time, we have,

$$\pi \times d \times n = \pi \times D \times N$$

or

$$\frac{\pi \times d}{\pi \times D} = \frac{N}{n}$$

π cancels and the equation becomes

$$\frac{d}{D} = \frac{N}{n}$$

which can be written:

$$\frac{\text{Dia. of DRIVER (metre)}}{\text{Dia. of DRIVEN (metre)}} = \frac{\text{Speed of DRIVEN (rev/s)}}{\text{Speed of DRIVER (rev/s)}}$$

We took the small blank to be the DRIVER and the large blank to be the DRIVEN. The large blank could have been taken as the driver.

Because greater forces can be transmitted, without the possibility of slip occurring, in practice we have a toothed gear wheel driving a toothed gear wheel, rather than a disc driving a disc by means of friction. In the case of two gear wheels we have their pitch circles rolling together in the same way as the discs revolved, see Fig. 137.

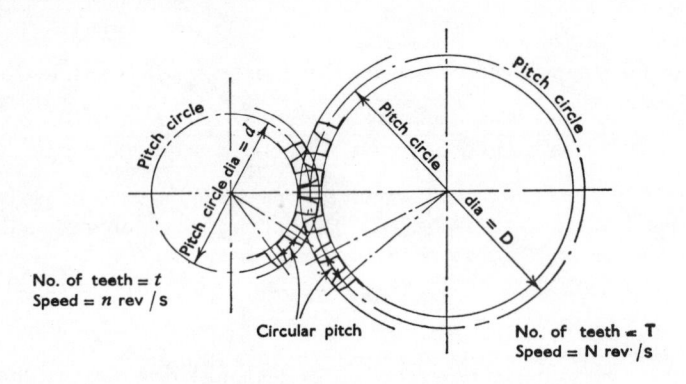

Fig. 137

The distance from the centre of one tooth to the centre of the next, measured on the pitch circle, is known as the *circular pitch*. The circular pitch is the same for both gear wheels. Indeed, it is the same for any number of gear wheels that mesh with each other. If we study Fig. 137 carefully, we can see that in every circular pitch there are two half teeth, i.e., to every circular pitch there is a tooth. There are then, in the complete gear wheels, as many *teeth* as there are *circular pitches*. This applies to all gear wheels. On further inspection of Fig. 137, it becomes obvious that if we multiply the circular pitch by the number of teeth (i.e., number of circular pitches) we shall obtain the circumference of the *pitch circle*,

PITCH CIRCLE circumference of large gear wheel

$$= \text{CIRCULAR PITCH} \times T$$

and

PITCH CIRCLE circumference of small gear wheel

$$= \text{CIRCULAR PITCH} \times t$$

Going back to the original formula we had:

$$\pi d n = \pi D N$$

Replacing πd by circular pitch $\times t$, for this gives pitch circle circumference of the small gear wheel, and πD by circular pitch $\times T$, for this gives pitch circle circumference of the large gear wheel,

GEAR WHEELS

we now have:

$$\text{CIRCULAR PITCH} \times t \times n = \text{CIRCULAR PITCH} \times T \times N$$

$$\frac{\text{CIRCULAR PITCH} \times t}{\text{CIRCULAR PITCH} \times T} = \frac{N}{n}$$

but CIRCULAR PITCH cancels,

$$\therefore \frac{t}{T} = \frac{N}{n}$$

or, if we take the small gear wheel as driving the large gear wheel, we have:

$$\frac{\text{No. of teeth on DRIVER}}{\text{No. of teeth on DRIVEN}} = \frac{\text{Speed of DRIVEN (rev/s)}}{\text{Speed of DRIVER (rev/s)}}$$

This equation is more suitable than the one which includes the pitch circle diameter, because it is far easier for us to obtain the number of teeth on a gear wheel than it is to measure the pitch circle diameter.

We must remember that, when using this formula for calculating the change wheels for screw-cutting on the centre lathe.

Rev/s of DRIVEN becomes threads per inch on LEADSCREW,

Rev/s of DRIVER becomes threads per inch to be CUT,

$$\frac{\text{Teeth on DRIVER (spindle wheel)}}{\text{Teeth on DRIVEN (leadscrew wheel)}} = \frac{\text{Threads per inch LEADSCREW}}{\text{Threads per inch to be CUT}}$$

Velocity ratio

With a gear wheel drive there may be an increase, or decrease in speed, or no change in speed. The *driver* may make the *driven* go faster or slower than itself, or it may drive it at the same speed. Thus, we have high gears and low gears, a high gear being one in which there is a relatively high output rev/s, and a low gear giving a low output speed. We often find it an advantage in the workshop to express a change in speed as a ratio. For instance, we say there is a 2 to 1 reduction driver to driven, meaning of course, a 2 to 1 reduction in speed. We might state that the speed ratio is 1 to 2, driver to driven, meaning that there is an increase in speed. If we express ourselves completely, we say that the

SPEED (or VELOCITY) RATIO is 2 to 1, or in the second instance 1 to 2.

Suppose a driver wheel drives the driven wheel $3\frac{1}{2}$ times faster than itself, then we say there is a speed or gear 'step-up' of 1 to $3\frac{1}{2}$, i.e., 2 to 7. For this drive

$$\text{VELOCITY RATIO} = \frac{\text{Speed of DRIVER}}{\text{Speed of DRIVEN}} = \frac{1}{3\frac{1}{2}} = \frac{2}{7}$$

It is usual practice to express such ratios in whole numbers. This *velocity ratio* tells us that when the driver makes 2 revolutions the driven makes 7 revolutions.

Let us now consider the case in which the driver drives the driven at a speed $3\frac{1}{2}$ times slower than itself. In this instance we would say that there is a speed, or gear reduction of $3\frac{1}{2}$ to 1,

$$\text{VELOCITY RATIO} = \frac{\text{Speed of DRIVER}}{\text{Speed of DRIVEN}} = \frac{3\frac{1}{2}}{1} = \frac{7}{2}$$

From this we understand that when the *driver* makes 7 revolutions the *driven* makes 2 revolutions.

If a gear wheel with a speed of 270 rev/s drives another gear wheel at 36 rev/s, then the *velocity ratio* of the drive is:

$$\text{VELOCITY RATIO} = \frac{\text{Speed of DRIVER}}{\text{Speed of DRIVEN}} = \frac{270}{36} = \frac{15}{2}$$

If a gear wheel with a speed of 25 rev/s drives another gear wheel at 215 rev/s, we have:

$$\text{VELOCITY RATIO} = \frac{\text{Speed of DRIVER}}{\text{Speed of DRIVEN}} = \frac{25}{215} = \frac{5}{43}$$

Examples

1. In a particular gear on a sensitive drilling machine, a 30 tooth pinion, speed 20 rev/s drives a gear wheel which has 130 teeth. What is the speed of the driven gear wheel?

    Begin by writing down the required formula, remembering to write first the ' unknown ', which is the speed of the driven wheel.

$$\frac{\text{Speed of DRIVEN}}{\text{Speed of driver}} = \frac{\text{Teeth on driver}}{\text{Teeth on driven}}$$

$$\frac{\text{Speed of DRIVEN}}{20} = \frac{30}{130}$$

$$\text{Speed of DRIVEN} = \frac{30}{130} \times 20$$

$$= \underline{4\cdot 6 \text{ rev/s}}$$

Rough check

We know that a small gear wheel is driving a larger gear wheel. Such an arrangement will give a reduction in speed. Therefore, we expect the speed of the *driven* wheel to be some value less than 20 rev/s.

As the reduction in speed is related to the number of teeth, we expect the speed of the 130 tooth gear wheel to be approximately $\frac{1}{4}$ of 20, since 30 is approximately $\frac{1}{4}$ of 130. Therefore, the speed of the *driven* wheel should be a little less than 5 rev/s, which it is.

GEAR WHEELS

2. Figure 138 shows a plan view of the gear wheels and hand wheel by which the carriage of a centre lathe can be traversed. The hand wheel is keyed to the same spindle as the 20 tooth pinion. This pinion meshes with a 60 tooth gear wheel, which is keyed to the same spindle as the 18 tooth pinion, which in turn meshes with a rack fixed to the underside of the long slide. The pitch of the rack teeth is 6 mm. Find the distance traversed by the carriage along the slide, when the hand wheel makes 1 revolution.

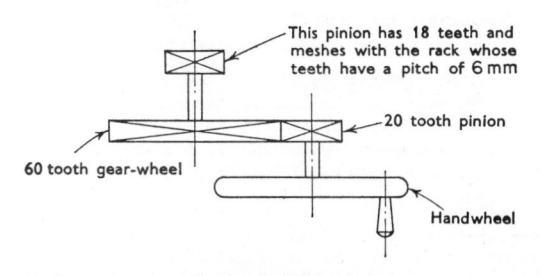

Fig. 138

In this problem we have to consider revs instead of rev/s. When the handwheel makes 1 revolution, the 20 tooth pinion makes 1 revolution, both are keyed to the same spindle. We next want to know what part of a revolution the 18 tooth pinion makes. As this pinion is keyed to the same spindle as the 60 tooth gear wheel, its movement will be the same.

The driver is the 20 tooth pinion, and the driven the 60 tooth gear wheel.

Begin by writing the 'unknown', which is the revs of the driven, and then proceed to build the remainder of the formula.

$$\frac{\text{Revs of DRIVEN}}{\text{Revs of driver}} = \frac{\text{Teeth on driver}}{\text{Teeth on driven}}$$

$$\frac{\text{Revs of DRIVEN}}{1} = \frac{20}{60}$$

$$\text{Revs of DRIVEN} = \frac{1}{3}$$

i.e., when the hand wheel makes 1 rev the 18 tooth pinion makes $\frac{1}{3}$ rev.

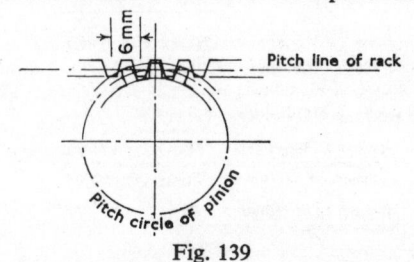

Fig. 139

As the rack teeth have a pitch of 6 mm, the circular pitch of the 18 tooth pinion is 6 mm. Since we have already learned, page 166, that on any gear wheel there are as many circular pitches as teeth. We know that the pitch circle circumference of the 18 tooth pinion is 18×6;

PITCH CIRCLE circumference of 18 tooth pinion = $18 \times 6 = 108$ mm

If we look, in Fig. 139, at the pinion meshing with the rack, we can see that the pitch circle of the pinion rolls on the *pitch line* of the rack.

When the 18 tooth pinion makes $\frac{1}{3}$ of a revolution, it rolls a distance along the rack equal to $\frac{1}{3}$ of 108 mm, i.e., $\frac{1}{3}$ of its pitch circle circumference.

Therefore, when the 18 tooth pinion makes $\frac{1}{3}$ rev, it rolls along the rack a distance = $\frac{1}{3}$ of 108 = 36 mm.

As the 18 tooth pinion is fixed to the carriage of the lathe, it will cause the carriage to be traversed 36 mm along the long slide.

For 1 rev of the hand wheel, carriage is traversed 36 mm.

3. A lathe input pulley revolves at 14 rev/s. In a particular gear an 18 tooth pinion fixed to the same shaft as the pulley meshes with a 130 tooth gear wheel which is keyed to an intermediate shaft. Also keyed to the intermediate shaft is an 85 tooth gear wheel which meshes with a gear wheel having 100 teeth and which is splined to the lathe spindle. Calculate the speed of the lathe spindle in this gear.

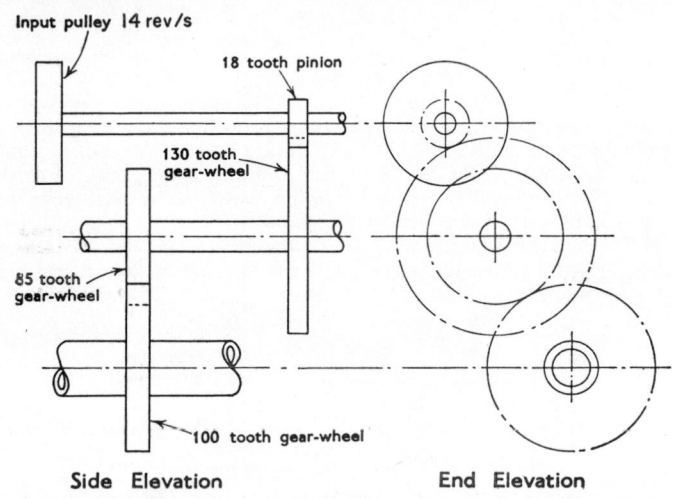

Fig. 140

First make a diagram of the gear train (see Fig. 140).

We have to deal with this problem in two parts, beginning by finding the speed of the 130 tooth gear wheel, which is the driven. The 18 tooth pinion, with a speed of 14 rev/s, is the driver.

$$\frac{\text{Speed of DRIVEN}}{\text{Speed of driver}} = \frac{\text{Teeth on driver}}{\text{Teeth on driven}}$$

$$\frac{\text{Speed of DRIVEN}}{14} = \frac{18}{130}$$

$$\text{Speed of DRIVEN} = \frac{18}{130} \times \frac{14}{1}$$

$$= \frac{252}{130} \text{ rev/s}$$

GEAR WHEELS

This is also the speed of both the intermediate shaft and the 85 tooth gear wheel. There is no need to work this out, because we are not asked for the speed of the intermediate shaft. We will use it as it is to find the speed of the lathe spindle.

In the second part of the problem the 85 tooth gear wheel is the driver and the 100 tooth gear wheel is the driven.

$$\frac{\text{Speed of DRIVEN}}{\text{Speed of driver}} = \frac{\text{Teeth on driver}}{\text{Teeth on driven}}$$

$$\frac{\text{Speed of DRIVEN}}{\frac{252}{130}} = \frac{85}{100}$$

$$\text{Speed of DRIVEN} = \frac{85}{100} \times \frac{252}{130} = \underline{1 \cdot 6}$$

Hence the spindle speed of the lathe is 1·6 rev/s.

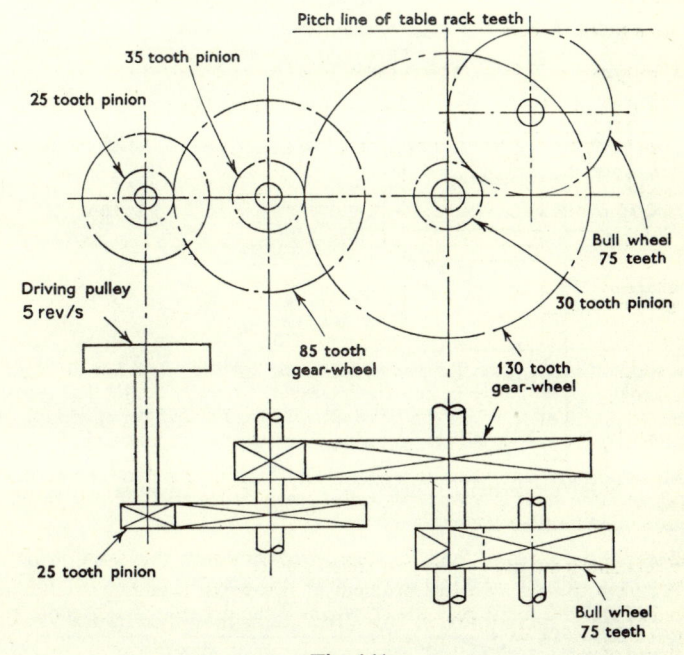

Fig. 141

4. In Fig. 141 the drive to a planing machine table is shown. The driving pulley is driven at 5 rev/s by an electric motor. Keyed to the same shaft as this pulley is a 25 tooth pinion driving an 85 tooth gear wheel which is fixed to the same shaft as a 35 tooth pinion. This pinion drives a gear wheel that has 130 teeth. Fixed to the same shaft as this gear wheel is a 30 tooth pinion which drives the bull wheel. This wheel has 75 teeth and meshes with a rack that is bolted to the underside of the table. The pitch of the rack teeth is 25 mm. Calculate the bull wheel speed and the table speed.

172   MATERIALS AND MACHINES IN THE WORKSHOP

This time we will use the usual method for solving a problem on gear trains, but in shorter form. There are three speed reductions in our problem and the ratios are:

$$\frac{25}{85}, \frac{35}{130} \text{ and } \frac{30}{75}$$

If we multiply the speed of the driver, 5 rev/s, by these three ratios in turn, we shall obtain the speed of the driven.

$$\text{Speed of BULL WHEEL} = \frac{5}{1} \times \frac{25}{85} \times \frac{35}{130} \times \frac{30}{75} = 0.16 \text{ rev/s}$$

The next part of the problem is similar to the second part of Example 2, except that in this case the rack rolls over the bull wheel instead of the wheel over the rack.

CIRCULAR PITCH of RACK TEETH = 25 mm

CIRCULAR PITCH OF BULL WHEEL TEETH = 25 mm

CIRCUMFERENCE of BULL WHEEL PITCH CIRCLE = 25 × 75 mm

In 1 rev of BULL WHEEL, RACK movement = 1875 mm

i.e., the pitch line of the rack moves a distance equal to the circumference of the bull wheel pitch circle.

In 0·16 rev of BULL WHEEL, RACK movement = 1875 × 0·16 mm

$$\text{TABLE SPEED} = \frac{1875 \times 0.16}{10^3}$$

$$= \underline{0.3 \text{ m/s}}$$

5. In a single train of gears for screwcutting the spindle wheel has 20 teeth, the intermediate wheel 55 teeth, and the leadscrew wheel 70 teeth. The leadscrew has a pitch of 6 mm. When the lathe spindle speed is 0·75 rev/s, what distance will the leadscrew move the carriage along the long slide in 1 second?

One important fact to note in a gear train of this type is that the intermediate wheel, or *idler*, as it is sometimes called, may be ignored. If it is taken into consideration its affect is cancelled.

First find the speed of the idler (the driven) and then the speed of the leadscrew wheel. We will do it this way to show that the idler may be ignored.

$$\frac{\text{Speed of IDLER (DRIVEN)}}{\text{Speed of spindle wheel (DRIVER)}} = \frac{\text{Teeth on spindle wheel (DRIVER)}}{\text{Teeth on idler (DRIVEN)}}$$

$$\frac{\text{Speed of IDLER}}{0.75} = \frac{20}{55}$$

$$\text{Speed of IDLER} = \frac{20}{55} \times \frac{0.75}{1}$$

Since we do not want to know the speed of the idler, there is no point in working out these figures. We will use them as they are, to find the speed of the leadscrew wheel. The idler is the driver.

GEAR WHEELS

$$\frac{\text{Speed of LEADSCREW WHEEL (DRIVEN)}}{\text{Speed of idler (DRIVER)}} = \frac{\text{Teeth on idler (DRIVER)}}{\text{Teeth on leadscrew wheel (DRIVEN)}}$$

$$\frac{\text{Speed of LEADSCREW WHEEL}}{\frac{20}{55} \times \frac{0.75}{1}} = \frac{55}{70}$$

$$\text{Speed of LEADSCREW WHEEL} = \frac{55}{70} \times \frac{20}{55} \times \frac{0.75}{1}$$

The effect on the speed of the 'idler' teeth cancels. Hence in the first place we could have stated:

$$\text{Speed of LEADSCREW WHEEL} = \frac{20 \times 0.75}{70} = \frac{1.5}{7} \text{ rev/s}$$

In 1 revolution of the leadscrew wheel, the lathe carriage moves a distance equal to the pitch of the leadscrew.

For 1 rev of leadscrew wheel, carriage movement

$$= 6 \text{ mm}$$

In 1 s, CARRIAGE movement $= 6 \times \dfrac{1.5}{7} = \dfrac{9}{7} = \underline{1.28 \text{ mm}}$

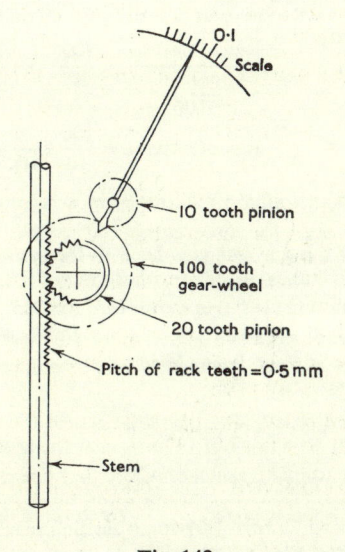

Fig. 142

6. Figure 142 shows the mechanism of a dial test indicator and gives details of the stem rack and gears. A scale which can measure to 0·01 mm is marked on the face of the instrument. Calculate the angle through which the finger of the gauge moves when the stem is lifted 0·01 mm.

We are told that the *pitch* of the stem rack teeth is 0·5 mm.

The *circular pitch* of the teeth on the 20 tooth pinion is 0·5 mm.

Since there are as many *circular pitches* on a gear wheel as teeth (see page 166), the circumference of the pitch circle of the 20 tooth pinion is:

PITCH CIRCLE circumference

$$= 20 \times 0.5 = 10 \text{ mm}$$

This means that if the stem were lifted 10 mm, the 20 tooth pinion would make 1 revolution, but the stem is lifted only 0·01 mm, therefore the pinion makes part of a revolution,

when stem lifts 0·01 mm, rotation of 20 tooth pinion

$$= \frac{0.01}{10} \text{ of a revolution}$$

$$= \frac{1}{1000} \text{ of a revolution.}$$

This means that the 100 tooth gear wheel also makes $\frac{1}{1000}$ of a revolution.

Next we must find what part of a revolution is made by the 10 tooth pinion. This pinion is driven by the 100 tooth gear wheel. Hence we have:

$$\frac{\text{Rev of DRIVEN (10 tooth pinion)}}{\text{Rev of DRIVER (100 tooth gear wheel)}} = \frac{\text{Teeth on DRIVER}}{\text{Teeth on DRIVEN}}$$

$$\frac{\text{Rev of DRIVEN}}{\frac{1}{1000}} = \frac{100}{10}$$

$$\text{Rev of DRIVEN} = \frac{100}{10} \times \frac{1}{1000} = \frac{1}{100}$$

Therefore the 10 tooth pinion makes $\frac{1}{100}$ of a revolution when the stem is lifted 0·01 mm.

Since the finger of the dial gauge is fixed to the same spindle as that of the 10 tooth pinion, it too makes $\frac{1}{100}$ of a revolution.

In 1 complete revolution the finger moves through 360°,

In $\frac{1}{100}$ of a revolution the finger moves through $\frac{1}{100}$ of 360° = 3°·6.

When the stem lifts 0·01 mm the finger moves through an angle of 3°·6.

Exercise 10

(Remember to make sketches where they are not given)

Section A

1. An electric motor speed 25 rev/s, driving a slotting machine has a pinion with 25 teeth which drives a gear wheel, which is keyed to the input shaft, having 175 teeth. What is the speed of the gear wheel?
2. On the hand feed of a drilling machine an 18 tooth pinion meshes with a rack whose teeth have a pitch of 6 mm. Through what angle, in radian, would the feed force, which is applied to the handle, move when a hole 50 mm deep is being drilled?

(*Note:* A similar type of mechanism is shown in Fig. 112.)

GEAR WHEELS

3. A planing machine table is driven by a gear wheel having 80 teeth, pitch 28 mm, which meshes with a rack bolted to the underside of the table. What will be the table speed in m/s, when the speed of the gear wheel is 0·5 rev/s?
4. (a) A bar of steel 75 mm diameter is being cut in a lathe at a speed of 0·6 m/s. What is the speed of the lathe spindle?
   (b) If the drive to the spindle consists of a driven gear wheel, having 75 teeth, keyed to the spindle, and a driver having 65 teeth, which is fixed to the intermediate shaft, what is the speed of the intermediate shaft?
5. When cutting a particular screw thread on a centre lathe having a leadscrew with a 6 mm pitch, a 20 tooth gear wheel drives a 90 tooth gear wheel through an intermediate wheel. What will be the pitch of the thread being cut?
6. The stroke wheel (or bull gear wheel) of a shaping machine has 100 teeth and is driven by a 27 tooth pinion making 12 rev/s. How many cutting strokes are made per second by the ram?

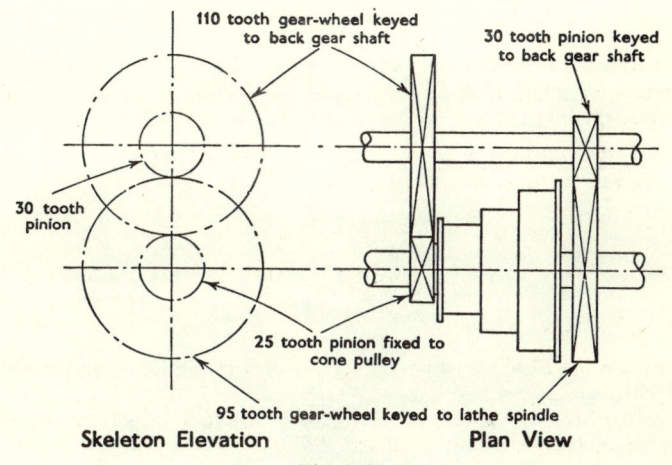

Fig. 143

7. The back gear of a centre lathe is shown in Fig. 143. A 25 tooth pinion fixed to the cone pulley drives a 110 tooth gear wheel, keyed to one end of the back gear shaft. Keyed to the other end of this shaft is a 30 tooth pinion which drives a 95 tooth gear wheel fixed to the lathe spindle. Calculate the speed of the lathe spindle when the cone pulley revolves at 4 rev/s.
8. During a milling operation a 100 mm dia. slab cutter is being used at a cutting speed of 0·34 m/s. What is the arbor speed?
   The arbor has a 60 tooth gear wheel fixed to it and this is driven by a 54 tooth gear wheel which is keyed to the intermediate shaft. Calculate the speed of the intermediate shaft.
9. In a certain gear, a drilling machine spindle makes 22 rev/s. It has keyed to it a 27 tooth gear wheel which is driven by a 31 tooth gear wheel fixed to the intermediate shaft. Also keyed to the intermediate shaft is a gear wheel with 21 teeth, driven by a 24 tooth gear wheel keyed to the driving shaft. Calculate the speed of the driving shaft.
10. A pinion with 27 teeth has to drive a gear wheel having 97 teeth at a speed of 6 rev/s. What must be the speed of the pinion?

Section B

1. A lathe driving pulley revolves at 8 rev/s. Keyed to the same shaft as the pulley is a gear wheel with 51 teeth, which meshes with a 68 tooth gear wheel fixed to the intermediate shaft. Also fixed to the intermediate shaft is a gear wheel with 44 teeth which drives a 55 tooth wheel keyed to the lathe spindle. What is the spindle speed?

2. The bull wheel, or stroke wheel, of a shaping machine has 112 teeth and it is driven by a 24 tooth pinion which has a speed of 3 rev/s. Keyed to the same shaft as the bull wheel, and on the outside of the bull wheel casing, is a 52 tooth gear wheel which drives another 52 tooth wheel. The second 52 tooth gear wheel drives the feed, or traverse rod. The traverse ratchet is taking two teeth at a time on a 22 tooth ratchet wheel which is keyed to the table traverse screw, pitch 6 mm. Through what distance is the table traversed per second? See Fig. 144 for diagram of the mechanism.

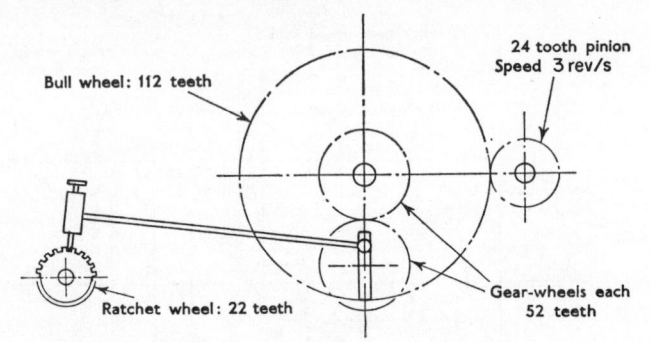

Fig. 144

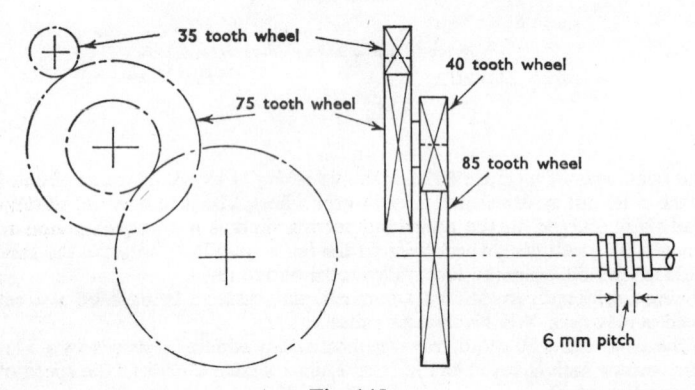

Fig. 145

3. In a thread cutting operation on a centre lathe, the following compound train of gear wheels is used: a 35 tooth spindle wheel drives a 75 tooth first stud wheel and a 40 tooth second stud wheel drives an 85 tooth leadscrew wheel. The leadscrew has a 6 mm pitch. Calculate how many revolutions the lathe spindle must make in order to move the carriage 25 mm. See Fig. 145 for the arrangement of this compound train of gear wheels.

GEAR WHEELS

4. A drive to a drilling machine spindle consists of an electric motor, speed 25 rev/s which drives an intermediate shaft, the speed reduction between them being in the ratio of $2\frac{1}{2}$ to 1. Keyed to the intermediate shaft is a gear wheel having 32 teeth, which meshes with a 57 tooth gear wheel fixed to the machine spindle. Find the cutting speed in m/s of a 20 mm dia. drill.

5. The drive to a centre lathe is as follows: an electric motor, speed 22 rev/s and pulley 250 mm dia., drives a countershaft through a flat belt and a 1 metre dia. pulley. Also on the counter-shaft is a 375 mm dia. pulley which drives the 230 mm dia. input pulley to the all-geared headstock of the lathe. In a particular gear there is a velocity ratio of 3 to 1 between the input pulley and the lathe spindle. Calculate the speed of the lathe spindle.

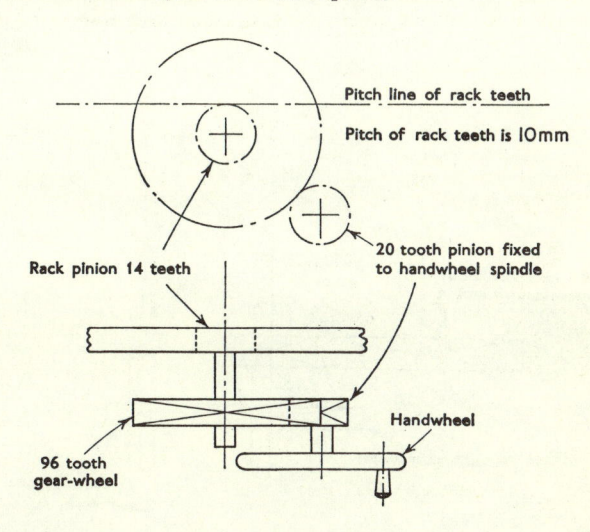

Fig. 146

6. A machine tool gear box has an input shaft with a speed of 9 rev/s. The output shaft must not have a speed of more than 6·6 rev/s. There is an intermediate shaft between the output and the input shafts, and the following gear wheels are available for making up a suitable drive:
    30 tooth; 50 tooth; 60 tooth; 90 tooth; 110 tooth.
    Calculate the speed ratio between the input and the output shafts and select 4 of the above gear wheels to make a suitable drive.

7. Figure 146 shows a hand traverse mechanism which is fitted to the carriage of a centre lathe, with details of the gear wheels and rack. Calculate the movement of the carriage along the slide when the handwheel makes 1 revolution.

8. In a drilling operation, 15 mm dia. holes are drilled. The cutting speed is 1 m/s. The gear train between the drill spindle and the driving shaft is as follows: a 41 tooth gear wheel on the spindle is driven by a 31 tooth pinion on the intermediate shaft. Also on the intermediate shaft is a 33 tooth gear wheel, which is driven by a 35 tooth pinion on the driving shaft. What is the speed of the driving shaft in rev/s?

9. Figure 147 shows the drive to a planing machine table. The speed of the driving shaft is 4 rev/s, and the pitch of the rack teeth is 20 mm. Calculate the table speed in metre per second.
10. In top gear a lathe spindle has a speed of 16 rev/s. The drive from the driving shaft to the spindle consists of a 40 tooth pinion fixed to the driving shaft, which drives a 52 tooth gear on the intermediate shaft. Also fixed to the intermediate shaft is a 36 tooth gear wheel which drives a 54 tooth gear wheel on the spindle. Calculate the speed of the driving shaft.

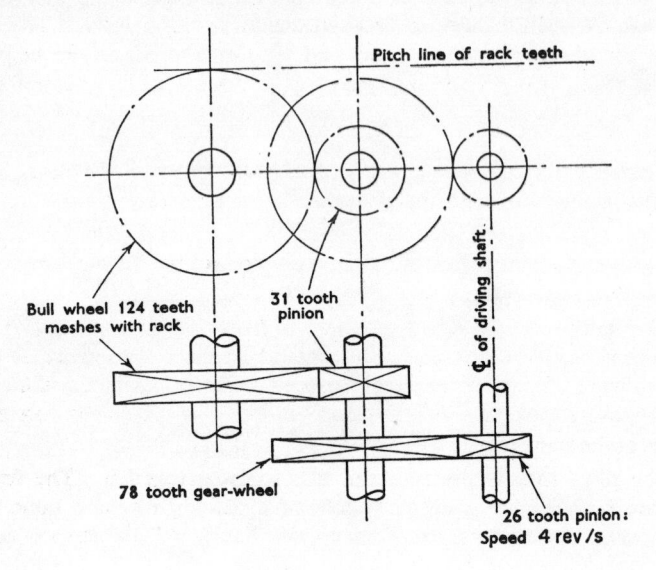

Fig. 147

Chapter 11. Friction

What is *friction*? It is the force which resists the sliding of one surface over another. Because of the *friction* between surfaces in contact, a force has to be applied to the table of a planing machine to cause it to slide on its guideways. When the force is removed, the table stops sliding. The same is true of the shaping machine ram and the lathe carriage, or the milling machine table. It is *friction* that prevents a component from slipping when gripped in the jaws of a vice, or prevents a job from sliding off a magnetic chuck.

The amount of friction, or *frictional resistance*, between two clean, dry (not lubricated) surfaces in contact is governed by the following:

1. The *roughness* (or *smoothness*) of the surfaces. The smoother the surfaces the lower the frictional resistance because the interlocking between the surfaces is lessened.

2. The *metals* in contact, e.g., there is likely to be a greater *frictional resistance* when steel surfaces slide on steel surfaces than there is when a brass surface slides on a cast iron surface, or if steel slides on white metal. This is because molecular bonding is more likely to take place between two surfaces of the same metal.

3. The *force* that is pressing the two surfaces together. The frictional resistance between the sliding surfaces of a planing machine table will be greater when the table carries a heavy job than it will be when it carries a lighter one.

The area of contact between the two surfaces does not affect the frictional resistance, i.e., the frictional resistance remains the same even if the area of contact is halved. If the area is halved the pressure between the surfaces will be doubled, providing the force pressing the surfaces together remains the same.

Disadvantages of friction

Knowing what friction is, we can conclude that it is undesirable to have it in cases where we need one surface to slide over another, because we have to waste energy in overcoming the frictional resistance. It is, then, a disadvantage to have friction between:

1. The shaping machine ram and its slides,
2. The planing machine table and its slides,
3. Shaft journals and bearings,
4. Any parts of machine tools, or engines, that have to slide on other parts.

In order to reduce *frictional resistance* between sliding surfaces to a minimum, we find that it is a good policy to obey the following rules:

1. Make a proper selection of metals that are to slide together, e.g., a steel shaft should slide on white metal, bronze or brass bearings, brass or bronze sliders should slide on a cast iron slide. Generally, surfaces made of the same metal are not put in sliding contact, although this does not apply strictly in the case of cast iron. The chief reason for this is that cast iron has self-lubricating properties since it contains free graphite. Another reason is that the sliding speed may be relatively slow.

2. Surfaces in contact should be made as smooth as possible. A bearing surface on a shaft is ground, whilst the bearing itself is scraped. Slideways on good machine tools are scraped. Some surfaces are lapped, or honed.

3. Lubricate the surfaces, the lubricant used being the correct one for the job and not just any type.

4. Sliding friction can be greatly reduced by the use of roller bearings and ball races wherever it is possible and appropriate to use these.

Advantages of friction

There are instances where friction is an essential part of a machine tool or a workshop process, and the following examples give some indication of this:

1. Friction is used on clutches, e.g., plate clutch, cone clutch, etc.

2. Friction transmits the drive between belt and pulley, e.g. vee and flat belt drives.

3. It is a friction force which prevents components gripped in chucks, between vice jaws, and in many clamping devices, from slipping.

4. Friction caused by clamping holds the shaper, slotter and lathe tools in position in their tool posts. It is friction that prevents the grinding wheel from rotating on its spindle and the slitting saw on the arbor.

5. Friction is used on brakes, motor vehicle tyres depend upon it for successful driving and braking.

6. Locking devices for bolts, etc., rely on friction for their successful operation, e.g., lock nuts, spring washer, nylon insert in a nut, etc.

7. Components held on magnetic chucks, or plates, rely on friction to prevent them from slipping. The clamping force intensifies the frictional force between the contact surfaces.

In cases where we use friction to advantage, we obviously make great efforts to keep it at a high value. The following list gives some of the methods used for doing this:

1. Use of the correct materials. ' Ferodo ' is used for clutch and brake linings, rubber and leather for belt drives.

2. Belt paste, or resin powder is used on belts, to reduce slip.

FRICTION

3. The gripping surfaces of vice jaws, chuck jaws and other clamping devices are made rough. If the surfaces being gripped are finished machined, they are given some protection by covering with a softer material (e.g., copper or lead sheet). There is a resultant reduction in the holding force.

The magnitude (size) of the *friction force* between two surfaces can be determined as follows:

The following can be done using a shaping machine ram, lathe carriage, or tailstock; first with the slides dry, and then with them lubricated. The effect too of different types of lubricants on friction can be observed.

Determination of friction force between two surfaces in contact

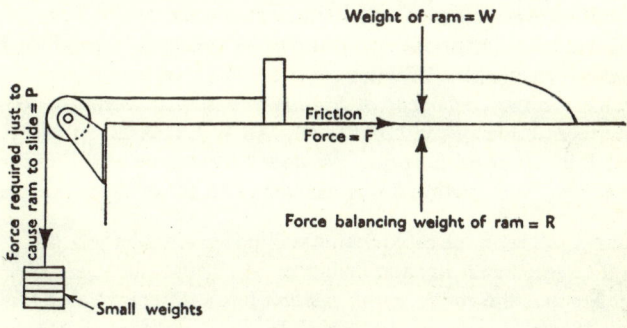

Fig. 148

Note: The forces F, P, R, and W are measured in *newton* or multiples of this unit.

In Fig. 148 a shaping machine ram, without tool post, is shown resting on its slide. The 'weight' of the ram is W and the reaction to this is R, i.e., R is the force in the slide balancing W. Force F is the *frictional resistance* between the sliding surfaces. It is the *friction force* which will resist the sliding of the ram over the slide.

Disconnect the ram from its driving mechanism so that it is free to be pushed, or pulled, along its slide. Fix a pulley and bracket on the machine frame in the position shown in such a way that the groove in the pulley rim is in line with the centre line of the ram. Then fix one end of a piece of cord to the ram as near to the slide as possible and let the cord pass over the pulley; it being horizontal and hang vertically downwards. By adding masses to the lower end of the cord, we can find how much force is required to cause the ram to move slowly, at uniform speed, over the slide. What we are doing really is using the gravitational pull on the masses to produce the force P and balance this against the *frictional force* F. The instant the ram begins to move with uniform speed we know that P is balancing F, i.e., we know that P is equal to F. Thus we have measured the *friction force* between the surfaces in contact, and we can use it as follows:

At the instant the ram begins to move the system is in equilibrium and we have:
$$F = P \text{ and } R = W$$
and $\dfrac{F \text{ (newton)}}{R \text{ (newton)}} = \dfrac{P \text{ (newton)}}{W \text{ (newton)}} =$ COEFFICIENT OF FRICTION (a number).

In deciding which is P and which is W in a particular problem, it is useful to remember that the force P always acts in the *direction* of movement, or possible movement, and W at *right angles* to it pressing the contact surfaces together.

The *coefficient of friction* is a ratio. It has not units, for the units of $\dfrac{P}{W}$ or $\dfrac{F}{R}$ cancel. This ratio is a measure of frictional resistance.

Take values of P for about six values of W and plot a graph of P against W, P on the vertical axis. Of what use is the slope of this graph?

The value for the *coefficient of friction* will remain constant for any one pair of surfaces under given conditions. As W increases, so will R, and so will F and P in direct proportion. We have an example of this in the planing machine table. The friction force between the table slides will be greater when a heavy job is on the table than it will be for a lighter job, but $\dfrac{P}{W}$ and $\dfrac{F}{R}$, will give the same result in each instance.

Remember that the *coefficient of friction* has a different value for different materials. It can, of course, be varied by the condition of the surfaces in contact, depending on whether they are rough or smooth, or whether or not they are lubricated. In our experiment, the surfaces were clean and dry initially.

How does lubrication affect the coefficient of friction?

Lubricated surfaces and Friction

Two important properties of a lubricant are *oiliness* and *viscosity*. A simple definition of oiliness is to say that it is a measure of the ease with which lubricated surfaces slip over each other; e.g., the lubricant which has the greatest oiliness is the one which gives the lowest frictional resistance.

The vegetable oil, castor oil, has a high degree of oiliness, this is generally true of vegetable oils, whilst mineral oils are not nearly so well endowed with this property. Vegetable oil more readily wets the surface of a bearing, it has a greater affinity for the metallic surface. In order to improve the oiliness of some lubricants, fine particles of chemically pure graphite are mixed with the lubricants. The effect is that a thin layer of graphite adheres to the sliding surface and encourages the adsorption (holding) of the oil to the surface. The graphite itself has certain lubricating properties. The metallic chemical molybdenum–disulphide is added to lubricants to improve their lubricating qualities. This too adheres to the metallic surfaces of bearings, gear-wheels, etc., it reduces the dry, and wet, frictional resistance.

FRICTION

Viscosity is a measure of the readiness with which adjacent layers of a lubricant slip over each other. A lubricant with a high viscosity has a greater drag on the motion of a journal in its bearing than one of low viscosity. The tendency in motor vehicle engine lubrication is to use oil of constant low viscosity in order to reduce frictional losses at all temperatures.

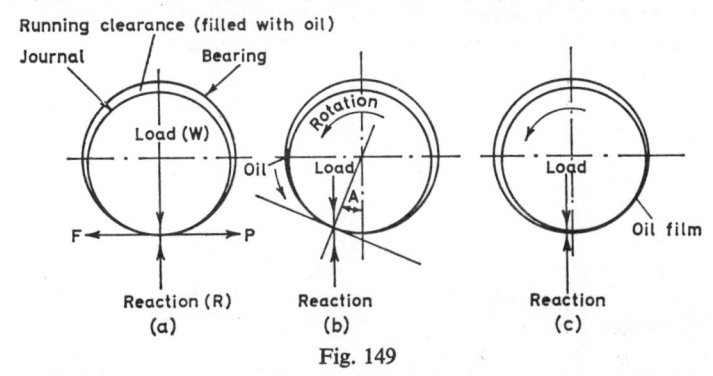

Fig. 149

The ideal form of lubrication is to have the sliding surfaces separated by a film of lubricant, so that such friction that is present arises from the sliding of one layer of oil over another. This condition is not easily obtained. It may be obtained in the ordinary journal bearing. In this type of bearing the white metal shell is bored to a diameter slightly larger than the journal diameter, e.g., it has a running clearance, see Fig. 149(a). When the shaft is stationary the journal contacts the bearing at its lowest point, on the vertical centre line. The crescent shaped clearance is filled with oil. When the shaft begins to rotate the journal climbs the bearing wall, see Fig. 149(b). The line of contact between the two is shifted to a position higher up the bearing surface. As the journal climbs the bearing wall its path becomes increasingly steeper. It reaches a point beyond which it cannot climb, because the friction force between the contact surfaces would not be great enough to support it, i.e., angle, A, has a limiting value. The rotating shaft pulls some of the lubricant round with it and this causes an increase in the pressure of the lubricant in the relatively narrow clearance space before the line of contact. As the speed of the shaft increases, the pressure in this converging gap is progressively increased and the line of contact between the bearing surfaces is moved down the bearing due to the force created by the increasing oil pressure. The oil pressure becomes so great that the lubricant eventually squeezes between the journal and bearing. The shaft is now running on an oil film and the journal takes up a position indicated by Fig. 149(c). The film of oil can only exist if its pressure is great enough to balance that caused by the downward force of the journal load.

The instant rotation begins the force to overcome friction is given by:

$$P = \mu W, \text{ or } F = \mu R$$

If there were no lubricant present, i.e., there was dry friction, this force would remain virtually constant as the speed started to increase. Since a lubricant is present the force tends to decrease and become least at the instant the oil film completely separates the journal and the bearing surfaces. There will be some slight force required, e.g. a wet friction force, to cause the journal surface to shear the oil film. When the speed is increased beyond the point at which the oil film is formed, the force to overcome friction gradually increases with speed, see Fig. 150. Thus there is an apparent

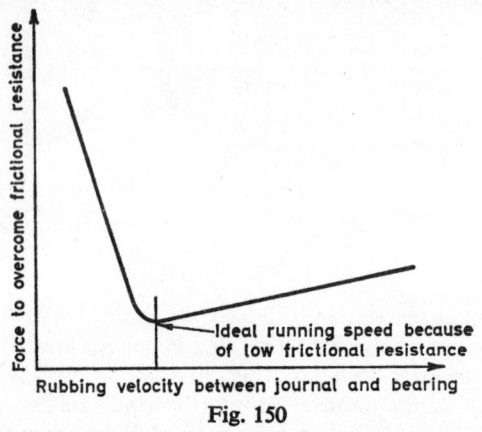

Fig. 150

ideal speed of operation. In practice it is usual to run at a speed in excess of this so that an instant overload, causing a drop in speed, may be partly offset by the drop in frictional resistance. This condition occurs on high speed machinery such as steam turbines with speeds of 50 rev/s, or more. There are many instances when the state does not exist.

Film lubrication can exist on plane surfaces too, but the speeds on many, if not all, machine tool slides will be far too low for a film of lubricant to separate, completely, the sliding surfaces. Such a film would occur on certain types of thrust bearings fitted to high speed machinery.

Examples
1. The table of a small planing machine has a mass of 500 kg and has clamped to it a casting of mass 100 kg. If the coefficient of friction for the table slides is 0·06, calculate the force that must be applied to the table in order to overcome the frictional resistance of the slides.

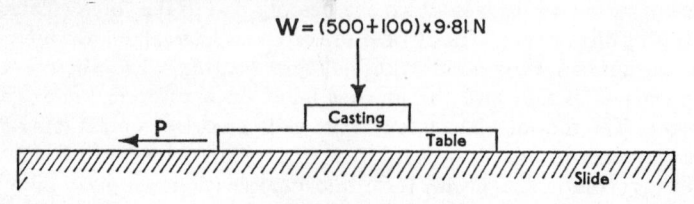

$W = (500+100) \times 9\cdot81 \, N$

Fig. 151

FRICTION

Begin by making a simple diagram of the machine table with the job on it, as in Fig. 151.

We must now determine the gravitational effect on a mass of 600 kg, remembering that the gravitational pull on a 1 kg mass is 9·81 N. Gravitational pull on combined mass of table and job = 600 × 9·81 newton and this is the W value.

$$\text{We know that } \frac{P}{W} = \text{COEFFICIENT OF FRICTION}$$

Therefore, by substituting the values for W and the coefficient of friction, we get:

$$\frac{P}{600 \times 9 \cdot 81} = 0 \cdot 06$$

From this we can find the value of P,

$$\frac{P}{5 \cdot 886 \times 10^3} = 0 \cdot 06$$

$$P = 0 \cdot 06 \times 5 \cdot 886 \times 10^3 = \underline{353 \cdot 2 \text{ N}}$$

A force of 353 N must be applied to the table to cause it to slide. Extra force will have to be supplied for the cutting of metal and the traversing of tools, etc.

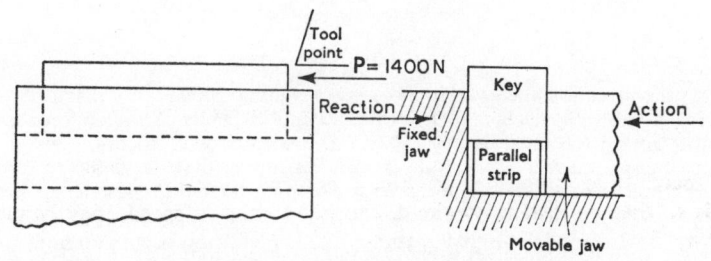

Fig. 152

2. A rectangular key is gripped between the jaws of a machine vice for the purpose of shaping its faces. The maximum cutting force that will act on the key during the operation is 1400 N. Find the smallest gripping force that must be applied to the jaws of the vice by the screw if the key is not to slip through the jaws. Coefficient of friction for contact surfaces can be taken as 0·21.

Again we begin by making a sketch of the ' set-up ' (see Fig. 152).

If the key is to slip it must do so on two pairs of surfaces. The ' action ' (force in vice screw) is pressing one pair together, whilst the ' reaction ' to this force presses the other pair of surfaces together.

$$\text{We know that, } \frac{P}{W} = \text{COEFFICIENT OF FRICTION}$$

In this case we have: W = Action + Reaction

$$\frac{1400}{\text{Action} + \text{Reaction}} = 0 \cdot 21$$

$$\frac{1400}{0 \cdot 21} = \text{Action} + \text{Reaction}$$

Action + Reaction = 6·67 × 10³ N

Since the action force applied by the vice screw is equal to the reaction force of the fixed jaw,

$$\text{Action force} = \frac{6 \cdot 67 \times 10^3}{2} = 3 \cdot 33 \times 10^3 \text{ N}$$

Smallest gripping force = <u>3·33 kN</u>

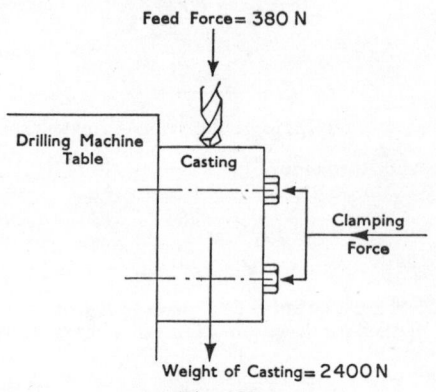

Fig. 153

3. In order to drill some holes in a casting 'weighing' 2400 N, it is clamped to the side of a drilling machine table by two 20 mm dia. bolts. The feed force on the drill during the operation is 380 N. The coefficient of friction between the casting and the face of the table is 0·28. Assuming that the clamping force in each bolt is equal, calculate the least clamping force that must be applied by each bolt if the casting is not to slip.

We shall help ourselves to understand the problem if we first make a sketch (see Fig. 153).

Now we must determine which of the forces is W and which is the P value. Remember the W force is the one which presses the surfaces together. In this problem, the W force is the clamping force. The force which tends to cause movement, or sliding, is the P force. We can see from Fig. 153 that it is the 'weight' (gravitational effect) of the casting plus the feed force that are tending to cause a downward movement of the casting. Therefore the P value is (2400 + 380) N and we have to find W.

$$\text{Now, } \frac{P}{W} = \text{COEFFICIENT OF FRICTION}$$

In this case:

$$\frac{2400 + 380}{W} = 0 \cdot 28$$

$$\frac{2780}{W} = 0 \cdot 28$$

$$2780 = 0 \cdot 28 \times W$$

$$\frac{2780}{0 \cdot 28} = W$$

$$\underline{9926 = W.}$$

FRICTION

The minimum clamping force that must be applied by the bolts is 9926 newton.
We must remember that the clamping force in each bolt is the same; we are told so in the question. Therefore, as there are two clamping bolts, the force in each is half of the total force,

i.e., minimum clamping force in each bolt

$$= \frac{9 \cdot 926 \times 10^3}{2}$$

$$= \underline{4 \cdot 963 \text{ kN}}$$

In practice, the clamping force would be made greater than this minimum.

4. In an operation on a surface grinder, a component 'weighing' 15 N is held on a magnetic chuck. The coefficient of friction between the component and the chuck is 0·22 and the cutting force at the face of the grinding wheel is 100 N. Find the least magnetic force with which the chuck must draw the component to it, in order that the component shall not slip.

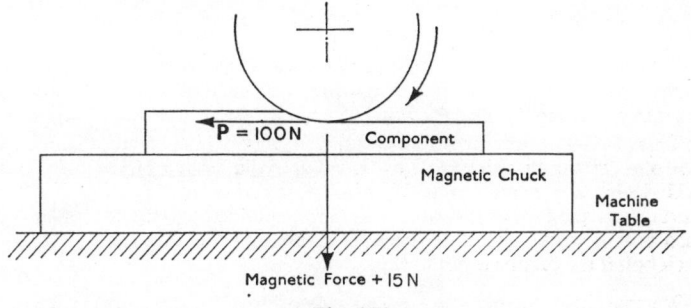

Fig. 154

We will again make a simple sketch of the parts upon which the forces are acting (see Fig. 154).

In this problem we have the 100 N force as the P force and the magnetic force + 'weight' of component as the W force.

$$\frac{P}{W} = \text{COEFFICIENT OF FRICTION}$$

$$\frac{100}{\text{Magnetic force} + 15} = 0 \cdot 22$$

$$100 = 0 \cdot 22 \,(\text{Magnetic force} + 15)$$

$$\text{Magnetic force} + 15 = \frac{100}{0 \cdot 22}$$

and $\text{Magnetic force} = \dfrac{100}{0 \cdot 22} - 15$

$$= 454 \cdot 5 - 15 = \underline{439 \cdot 5 \text{ N}}$$

There must be a magnetic pull of at least 439·5 N to prevent the component from being pushed off the chuck. The magnetic force actually used would be greater than this.

5. The table of a large planing machine 'weighs' 40 kN and clamped to it is a casting 'weighing' 50 kN. The speed of the table on the forward stroke is 0·1 m/s and on the return stroke 0·25 m/s. If the coefficient of friction for the table slide is 0·075, calculate the work done per second in overcoming friction (*a*) on the forward stroke, (*b*) on the return stroke.

We again make a sketch of the arrangement (see Fig. 155).

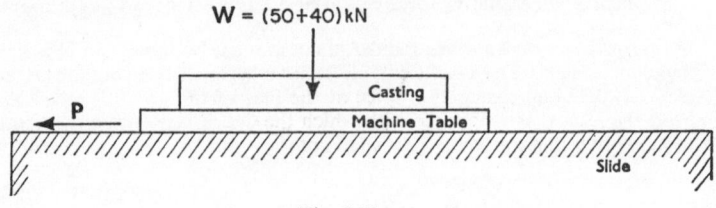

Fig. 155

Next we must find the force, P newton, that must be applied to the table in order to overcome the slide friction.

$$\frac{P}{W} = \text{COEFFICIENT OF FRICTION}$$

$$\frac{P}{(50+40) \times 10^3} = 0·075$$

$$\frac{P}{90 \times 10^3} = 0·075$$

and
$$P = 0·075 \times 90 \times 10^3$$
$$= 6750 \text{ N}$$

Thus a force of 6750 N must be applied to the table to cause it to slide on the forward and backward strokes.

Remembering that on page 135 we learned that the work done by a force is given by:

WORK DONE by a FORCE = FORCE × DISTANCE moved by FORCE

WORK DONE per second against FRICTION on forward stroke

$$= 6750 \times 0·1 = \underline{675 \text{ J}}$$

and

WORK DONE per second against FRICTION on return stroke

$$= 6750 \times 0·25 = \underline{1688 \text{ J}}$$

6. The load carried by one of the spindle bearings on a centre lathe is 280 N, and the bearing diameter is 80 mm. If the coefficient of friction for the bearing is 0·06. Calculate the torque and the power required to overcome friction when spindle speed is 3 rev/s.

FRICTION

The main difference between this problem and the previous ones is that here we have sliding taking place between curved surfaces. However, the procedure for solving the problem is exactly as before. The P value acts tangentially to the curved surface of the bearing (see Fig. 156).

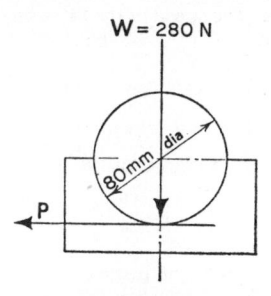

Fig. 156

We begin our calculations by finding the value of P, this being the force that is required to cause the shaft to slide over the bearing.

$$\frac{P}{W} = \text{COEFFICIENT OF FRICTION}$$

$$\frac{P}{280} = 0 \cdot 06$$

$$P = 0 \cdot 06 \times 280 = 16 \cdot 8 \text{ N}$$

This force acts at a radius of $\frac{80}{2} = 40$ mm

The TORQUE required to overcome the bearing friction

$$= 16 \cdot 8 \times \frac{40}{10^3}$$

$$= \underline{0 \cdot 67 \text{ Nm}}$$

To find the power used, we must begin by finding the work done in 1 revolution of the shaft, then the work done per second.

WORK DONE against friction in 1 rev of shaft

$$= P \times \text{circum. of shaft}$$

$$= 16 \cdot 8 \times \frac{80\pi}{10^3} \text{ joule}$$

WORK DONE against friction in 1 second

$$= 1 \cdot 68 \times 0 \cdot 8\pi \times 3 \text{ J}$$

WORK DONE against friction = 12·66 joule/s

POWER used in overcoming friction = 12·66 watt

190   MATERIALS AND MACHINES IN THE WORKSHOP

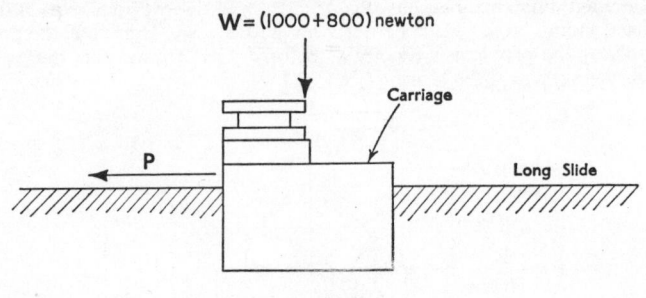

W = (1000+800) newton

Fig. 157

7. A lathe carriage 'weighs' 1000 N and during a roughing operation the tangential cutting force on the tool point is 800 N. The coefficient of friction between the carriage and its slide is 0·11. Find the force that must be applied to the carriage in order to overcome the frictional resistance of the slide. See Fig. 157 for a line diagram of the carriage and slide.

$$\frac{P}{W} = \text{COEFFICIENT OF FRICTION}$$

We must note that the value of W is equal to the force due to the 'weight' of the carriage plus the cutting force,

$$W = (1000+800) \text{ N}$$

$$\frac{P}{1800} = 0\cdot11$$

$$P = 0\cdot11 \times 1800$$

$$= \underline{198 \text{ N}}$$

A force of 198 N must be applied to the carriage, parallel to the long slide, in order to cause it to slide.

8. The sketch in Fig. 158 shows a foot brake fitted to a small winch. Find the force F newton that the operator must apply to the foot pedal in order to stop the brake drum, when the torque on the drum is 16 Nm. The coefficient of friction between the brake drum and the brake block is 0·48.

There are three distinct parts to this problem.
First, we have to calculate the tangential force, P, on the drum, using the torque and the radius.
Secondly, we have to calculate W, using the friction formula.
Thirdly, we have to find F, by taking moments about the pivot of the foot lever.

Remembering that in Chapter 7, page 122, we learned that:

TORQUE = TANGENTIAL FORCE × RADIUS

FRICTION

Hence in this problem we have:

$$16 = P \times \frac{120}{10^3}$$

$$\frac{16}{0\cdot120} = P$$

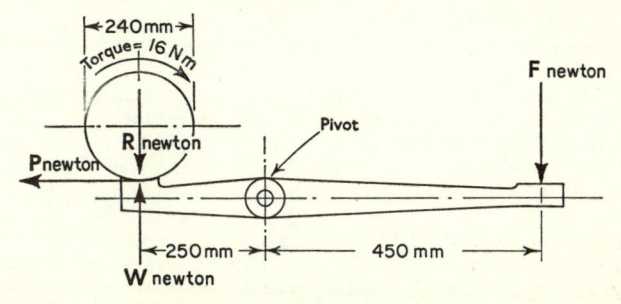

Fig. 158

or the tangential force on the brake drum $= \frac{16}{0\cdot120}$ N

We have learned in this chapter that

$$\frac{P}{W} = \text{COEFFICIENT OF FRICTION}$$

Therefore in this problem we have:

$$\frac{16}{0\cdot120} \times \frac{1}{W} = 0\cdot48$$

$$\frac{16}{0\cdot120} = 0\cdot48\,W$$

$$W = \frac{16}{0\cdot120 \times 0\cdot48} = \frac{16}{0\cdot0576}\,\text{N}$$

As before, there is no need to work this value out any further. We can use it in the next calculation as it is.

Take moments about the pivot and consider R in place of W, since R = W.

$$\text{CLOCKWISE MOMENT} = F \times 0\cdot450\,\text{Nm}$$

$$\text{ANTI-CLOCKWISE MOMENT} = \frac{16}{0\cdot0576} \times 0\cdot250\,\text{Nm}$$

When the *clockwise moment* balances the *anti-clockwise moment* we have:

$$F \times 0\cdot450 = \frac{16}{0\cdot0576} \times 0\cdot250$$

$$F = \frac{16}{0\cdot0576} \times \frac{0\cdot250}{0\cdot450}$$

$$= 154\cdot4\,\text{N}$$

Exercise 11

Section A

1. Show, by making sketches, four parts of a lathe where friction is used to advantage.
2. Make sketches of two parts of a milling machine in which friction is a disadvantage.
3. The table of a horizontal milling machine has a mass of 300 kg and the vice and job, 14 kg. If the coefficient of friction for the slides is 0·09, calculate the force that must be applied to the table to cause it to slide.
4. Figure 159 shows a set-up for a milling operation. The cutter applies a 450 N cutting force to the component, which is held between the jaws of a machine vice. The coefficient of friction between the vice jaws and the job is 0·27. Calculate the minimum gripping force that must be applied to the vice jaws by the screw in order that the component shall not slip through the jaws.

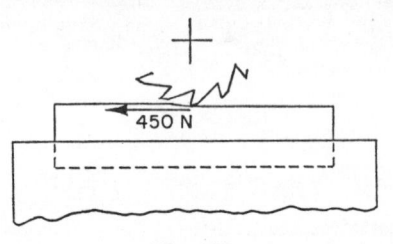

Fig. 159

5. During an operation on a surface grinder the torque on the 240 mm dia. wheel is 2·5 Nm. Calculate the minimum magnetic force required to hold the component rigidly to the chuck if the coefficient of friction between the chuck and the component is 0·21. Neglect the effect of the mass of the component.
6. The ram of a shaping machine has a mass of 80 kg, has a stroke 220 mm long, speed 2 strokes per s (one forward one backward) and the cutting force, 500 N. If the coefficient of friction for ram slide is 0·07. Find the total work done per second.
7. A 350 mm dia. brake drum fitted to an electric crane has a turning moment of 24 Nm on it. The coefficient of friction between the brake drum and the brake shoe is 0·45. Calculate the normal force that must be applied to the brake shoe in order to stop the drum.
8. An angle plate 'weighing' 60 N is clamped to the side of a shaping machine table, to support a job 'weighing' 50 N by two 12 mm dia. bolts, equally stressed. If the coefficient of friction between plate and machine table is 0·17, find the stress in the shank of each bolt due to the minimum clamping force.
9. The table of a small planing machine has a 'weight' of 10 kN and carries a job 'weighing' 1·4 kN. The cutting force on the tool point is 1·4 kN, and the coefficient of friction for the table slides is 0·08. If the table speed on the forward stroke is 0·1 m/s, calculate the work done cutting metal as a percentage of the total work done on the forward stroke.
10. When setting a shaft between centres in a lathe, a force of 200 N is applied to the tailstock spindle along its axis. If the coefficient of friction between the tailstock base and the long slide is 0·16. Find the minimum clamping force that must be applied to the 25 mm dia. clamping bolt and the stress due to this force in the bolt shank.

Section B

1. A shaping machine tool is gripped in a vertical position in the tool post, by a clamping force of 9 kN. If the coefficient of friction between the tool and the tool post is 0·31, calculate the greatest force that can be applied in a downward, or upward, direction before the tool would slip.
2. A casting of 'weight' 3·7 kN is clamped to the side of a drilling machine table by two tee-headed bolts, 20 mm dia., that pass through clearance holes in the casting. The coefficient of friction between the table and the casting is 0·29. Calculate the shear stress in the shank of each bolt if the clamping force per bolt is 4·2 kN.
3. The chuck and spindle of a vertical boring machine 'weigh' 4·4 kN and they are supported on a phosphor bronze thrust bearing fixed to the bed of the machine, as shown in Fig. 160. The bore of the thrust bearing is 240 mm dia., outside dia. 340 mm and coefficient of friction 0·09. During a certain operation, a component 'weighing' 260 N is held in the chuck and it is revolved at a speed of 2 rev/s. Calculate the work done per second in overcoming frictional resistance of the bearing.

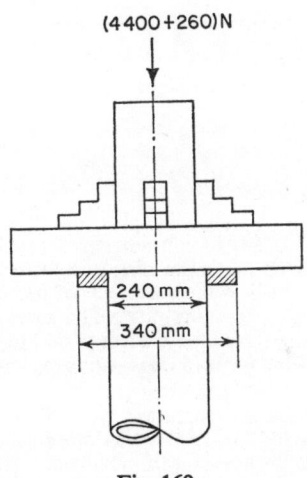

Fig. 160

4. The table of a milling machine, and job have a mass of 300 kg. It can be traversed by hand, by applying a force at the end of a handle 225 mm long, which operates a square threaded screw pitch, 6 mm. This screw engages a nut fixed to the underside of the table and lies parallel to its longitudinal centre line. The coefficient of friction for the table slide is 0·11. Calculate the minimum force that must be applied to the end of the handle in order to cause the table to slide. Assume the mechanism is 25% efficient.
5. A slitting saw of 110 mm outside diameter is held fast on a 30 mm dia. milling machine arbor by friction, the cutter being gripped between distance pieces which have an outside diameter, 44 mm. The coefficient of friction for the contact faces is 0·29, whilst the tangential cutting force on the saw is 210 N. Calculate the minimum clamping force that must be applied by the arbot nut to prevent the slitting saw from slipping, and (b) the size and type of stress in the arbor due to the nut force.

194 MATERIALS AND MACHINES IN THE WORKSHOP

6. The contact surfaces of a cone clutch fitted to a centre lathe have a mean radius of 85 mm. During a particular gear there is a torque increase of 1 to 3 between the clutch and the lathe spindle. The diameter of the work being turned is 82 mm and the cutting force on the tool point is 580 N. Find the smallest force, normal to the contact surfaces, that is required to press the friction surfaces of the clutch together if the coefficient of friction for these surfaces is 0·46.

7. A grinding wheel, 300 mm dia. when new, is prevented from turning on its spindle by gripping it between two discs, as shown in Fig. 161. During normal use the cutting force at the periphery of the wheel is 72 N. Calculate the smallest clamping force required between the two plates if the coefficient of friction is 0·35.

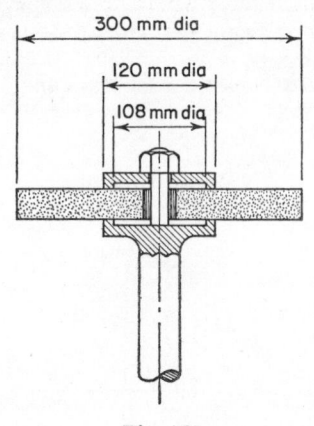

Fig. 161

8. The ram of a shaping machine 'weighs' 810 N and the coefficient of friction for its slide is 0·065. During a particular operation the ram is cutting on a 400 mm stroke making 1 cutting and 1 return stroke per s. The average cutting force on the tool point during the operation is 510 N. Calculate the work done per second in cutting metal, the total work done against friction.

9. A casting 'weighing' 65 kN is clamped to a planing machine table that 'weighs' 30 kN. The table speed on the forward stroke is 0·07 m/s and on the return stroke 0·15 m/s. The coefficient of friction for the table slides is 0·085. Two tools are in use, the cutting force on each being 1·6 kN. Calculate the average power being used during the operation.

10. A machine vice 'weighing' 100 N is bolted to the top surface of a shaping machine table by two 10 mm dia. bolts. The maximum cutting thrust that the vice will have to withstand is 1·5 kN. Determine the maximum shear stress in the shank of each bolt due to the thrust if the clamping force in each bolt is 2·8 kN. The coefficient of friction between the vice and the table is 0·16. The 'weight' of the heaviest job that will be held in the vice is 45 N.

11. A flywheel and its shaft have a mass of 1500 kg and are supported in two bearings, 80 mm diameter, which have a coefficient of friction of 0·06. Calculate the energy lost, per second, in bearing friction when the speed is 25 rev/s.

Chapter 12. Torque and Power

Relationship between torque and power

In Fig. 162, a circle is shown which represents the rim of a belt pulley. We could also take it as representing the pitch circle of a gear wheel. In the first instance, the tangential force would be the effective pull in the belt. In the second instance, it would be the tooth force.

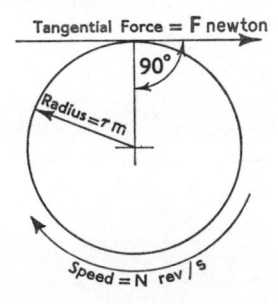

Fig. 162

We will begin by determining the work done by F in 1 revolution of the pulley. Remember what we learned in Chapter 8, page 135, namely that,

WORK DONE by a force = FORCE × DISTANCE moved by force

WORK DONE by F in 1 rev of pulley = F × CIRCUMFERENCE of pulley
(joule)   (newton) × (metre)

$$= F \times 2r\pi \text{ joule}$$

WORK DONE by F in 1 second = $F \times 2r\pi \times N$ joule

$$= F \times r \times 2\pi N \text{ joule}$$

but FORCE × RADIUS = TORQUE
(newton) (metre)   (N m)

(we learned this in Chapter 7, page 122).

Therefore we have:

$$F \times r = T$$

WORK DONE by F = $2\pi NT$ joule/s

POWER developed = $2\pi NT$ watt

Since, a *joule per second* = a *watt*

$$P = 2\pi NT$$

or $$\frac{P}{2\pi N} = T$$

We must remember that these equations are equally true for gear wheels, electric motor armatures, lathe spindles, in fact all shafts that transmit a torque.

Examples

1. When turning a bar of metal 110 mm dia., the cutting force on the tool point is 1·2 kN and the spindle speed 7 rev/s. Calculate the torque on the lathe spindle and the power used in cutting metal.

   Begin by making a sketch, see Fig. 163.

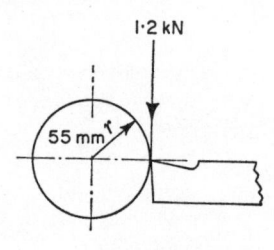

   Fig. 163

   TORQUE on lathe spindle = CUTTING FORCE × RADIUS

   $$= 1{\cdot}2 \times 10^3 \times \frac{55}{10^3}$$

   $$= 66 \text{ Nm}$$

   POWER used in cutting metal = $2\pi NT$ (N rev/s, T newton metre)

   $$= 2\pi \times 7 \times 66 = 2900 \text{ watt}$$

   $$= 2{\cdot}9 \text{ kW}$$

2. In a drilling operation the power being used at the drill spindle is 0·4 kW and the speed of the spindle is 6 rev/s. What is the torque on the drill?

   Remembering that POWER = $2\pi NT$, we have

   $$0{\cdot}4 \times 10^3 = 2\pi \times 6 \times T$$

   $$\frac{400}{12\pi} = T$$

   TORQUE on drill spindle = <u>10·62 Nm</u>

TORQUE AND POWER

3. To drive a milling machine a 5 kW motor with a 230 mm dia. driving pulley is used. The motor speed is 17 rev/s. Calculate the torque on the motor pulley and the effective pull in the belt which the motor pulley drives when maximum is used.

We know that
$$\text{POWER} = 2\pi NT \text{ watt}$$
and in this problem we have,
$$5 \times 10^3 = 2\pi \times 17 \times T$$

$$\frac{5000}{34\pi} = T$$

TORQUE on motor pulley = <u>46·84 Nm</u>

We know that
$$\text{TORQUE} = \text{FORCE} \times \text{RADIUS}$$
which, in this example, will be
$$\text{TORQUE} = \text{EFFECTIVE PULL in belt} \times \text{RADIUS}$$

$$46 \cdot 84 = \text{effective pull in belt} \times \frac{115}{10^3}$$

$$\frac{46 \cdot 84 \times 10^3}{115} = \text{effective pull in belt}$$

EFFECTIVE PULL in belt = <u>407·2 N</u>

4. A centre lathe is driven by a 1 kW electric motor having a speed of 25 rev/s, the motor pulley is 110 mm dia. and drives a 210 mm dia. pulley keyed to the lathe input shaft. Calculate the speed of the lathe pulley and the torque on the lathe input shaft.

The lathe pulley is a *driven* pulley and we want to know its speed.

Hence we have,

$$\frac{\text{Speed of DRIVEN}}{\text{Speed of driver}} = \frac{\text{Dia. of driver}}{\text{Dia. of driven}}$$

$$\frac{\text{Speed of DRIVEN}}{25} = \frac{110}{210}$$

$$\text{Speed of DRIVEN} = \frac{11}{21} \times 25$$

Speed of headstock pulley = 13 rev/s.

An important fact we must note is that the power available at the lathe pulley is equal to the power at the motor assuming 100% efficiency for the drive. This is because the power available does not change with a variation in speed. It is the torque that varies as the speed varies (see page 201) for a given power.

$$\text{POWER} = 2\pi NT \text{ watt}$$

where T is the torque on the lathe pulley, in newton-metre and N is in rev/s.
Hence, $\quad\quad\quad\quad\quad 1 \times 10^3 = 2\pi \times 13 \times T$

$$T = \frac{1000}{26\pi} = 12 \cdot 26$$

The torque on the lathe input shaft is 12·26 Nm

5. A milling machine is using 1·5 kW at the motor. On a particular job a 150 mm dia. cutter is used, at a speed of 1 rev/s. Find the torque on the arbor and the force at the periphery of the cutter.

Remember that if the motor is using 1·5 kW, there are 1·5 kW of power available at the arbor, assuming 100 % efficiency in transmission.
We know that,
$$\text{Power} = 2\pi NT$$
In this example, $\quad 1 \cdot 5 \times 10^3 = 2\pi \times 1 \times T$

$$T = \frac{1 \cdot 5 \times 10^3}{2\pi} = 238 \cdot 9 \text{ Nm}$$

The torque on the arbor, and therefore on the cutter, is 238·9 Nm

The radius of the cutter = 75×10^{-3} metre

We have learned that TORQUE = TANGENTIAL FORCE × RADIUS

Therefore, in this problem we have:

$$238 \cdot 9 = \text{Tangential force} \times 75 \times 10^{-3}$$
$$\text{Tangential force} = \frac{238 \cdot 9 \times 10^3}{75} = 3 \cdot 185 \times 10^3 \text{ N}$$

Hence, the force at the periphery of the cutter, i.e., the cutting force on the cutter teeth, is 3·185 kN at 75 mm radius.

6. In a belt drive to a lathe the motor pulley is 130 mm dia. and it drives the headstock pulley by means of a flat open belt 50 mm wide. If the effective force in the belt is 7 kN per metre width of belt, calculate the torque on the motor shaft and the power being developed by the motor when its speed is 18 rev/s.

Total EFFECTIVE FORCE in belt = Width of belt $\times 7 \times 10^3$

$$= \frac{50}{10^3} \times 7 \times 10^3$$
$$= 350 \text{ N}$$
$$\text{TORQUE on motor shaft} = 350 \times \frac{130}{2 \times 10^3}$$
$$= 22 \cdot 75 \text{ Nm}$$

It is necessary to have this *torque* in newton-metre, if we are to use it in the formula connecting *torque* and *power*,

POWER developed by motor = $2\pi NT$ watt
$$= 2\pi \times 18 \times 22 \cdot 75 = 2571 \text{ watt}$$
$$= \underline{2 \cdot 57 \text{ kW}}$$

Relationship between torque and gear wheel speed

In Chapter 10 we saw how the speed of gear wheels could be calculated. Now we want to know the relationship between this speed and the torque transmitted by gear wheels.

We have all had the experience of changing gear on a lathe, when we have wanted a slower speed. The reason for wanting a slower speed was probably that we had put on a heavier cut, or had changed from machining a metal, like aluminium, which cuts easily, to a harder to cut metal such as high carbon steel. Of course, what we were really doing was changing *torques*, e.g., changing from a low *torque* to a higher *torque*. The higher *torque* gave us a greater cutting force at the tool point in order to cut the harder—cut resisting—metal. We sometimes change from a low speed to a higher speed, thus obtaining a lower torque. This would be done when a lower torque would suffice and we were more concerned with cutting at a faster speed, e.g., on a finishing cut, where the general rule is ' light cut (low torque) high speed '. Why do we change to a low gear when climbing a steep hill in a motor car?

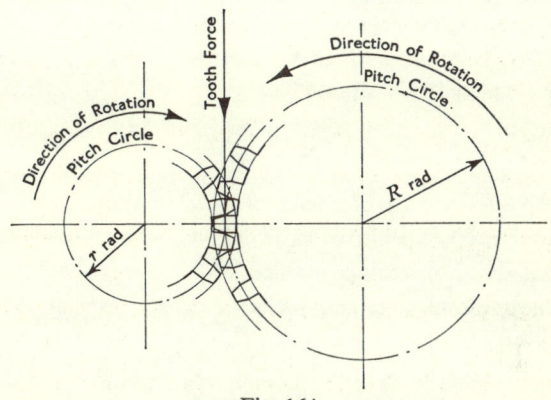

Fig. 164

To help us learn more about this relationship, let us study the two different size gear wheels shown in mesh in Fig. 164.

Suppose the smaller of the two gear wheels drives the other, the driving force at the pitch circle being known as the tooth force. We learned in Chapter 7, page 122, that,

$$\text{TORQUE} = \text{FORCE} \times \text{RADIUS}$$

TORQUE on the smaller gear wheel = TOOTH FORCE × r

and TORQUE on the larger gear wheel = TOOTH FORCE × R

The most important thing that we should note from these two statements is that:

TOOTH FORCE × R is greater than TOOTH FORCE × r

because R is greater than r.

This means that the torque on the larger gear wheel is greater than the torque on the smaller gear wheel, but the speed of the larger wheel will be lower than the speed of the smaller wheel (because the smaller wheel is driving the larger wheel).

Thus, in this arrangement of the gear wheels, we have an increase in *torque* and a decrease in *speed*.

Let us now suppose that the larger gear wheel is driving the smaller gear wheel with the tooth force exactly the same value as before. The *torque* on each wheel will be as in the previous case, but instead of an increase in *torque* we get a reduction, and instead of a decrease in speed we get an increase (because a larger wheel is driving a smaller wheel).

What really happens in a lathe gear box when we change from top gear (high output speed) to low gear (low output speed) is this. We change from the smallest spindle torque to the largest spindle torque.

In fact all gear boxes are really systems of *torques* or *turning moments*, and we select the one most suitable for the job being done.

We must remember that from any gear box,
    high speed is accompanied by low torque.
    low speed is accompanied by high torque.

In Chapter 10, page 165, we found that

$$\frac{\pi d}{\pi D} = \frac{N}{n}$$

where *d* is the pitch circle dia. of small gear wheel (pinion),
    D is the pitch circle dia. of large gear wheel,
    n is the speed of small gear wheel,
    N is the speed of large gear wheel.

We can see that π cancels. Therefore we have:

$$\frac{d}{D} = \frac{N}{n}$$

It would be helpful here if we could use radii instead of diameters, and we can do this by halving the diameter,

$$\frac{\frac{d}{2}}{\frac{D}{2}} = \frac{N}{n}$$

$$\text{or } \frac{d}{2} \div \frac{D}{2} = \frac{N}{n}$$

$$\frac{d}{2} \times \frac{2}{D} = \frac{N}{n}$$

TORQUE AND POWER

It makes no difference to the value of the equation. Instead of expressing radius in terms of diameter, we will replace $\frac{d}{2}$ and $\frac{D}{2}$ by r and R respectively. Hence we get:

$$\frac{r}{R} = \frac{N}{n}$$

Now we will multiply r and R by the tooth force thus:

$$\frac{\text{TOOTH FORCE} \times r}{\text{TOOTH FORCE} \times R} = \frac{N}{n}$$

We must note that the value of the equation is not changed by doing this, for the *tooth force* would cancel if we so desired. Now we have:

$$\frac{\text{TORQUE on smaller gear wheel}}{\text{TORQUE on larger gear wheel}} = \frac{N}{n}$$

Or, if we take the smaller wheel as driving the larger one we get

$$\frac{\text{TORQUE on DRIVER}}{\text{TORQUE on DRIVEN}} = \frac{\text{SPEED of DRIVEN}}{\text{SPEED of DRIVER}}$$

We have already learned from Chapter 10, page 167, that

$$\frac{\text{SPEED of DRIVEN}}{\text{SPEED of DRIVER}} = \frac{\text{TEETH on DRIVER}}{\text{TEETH on DRIVEN}}$$

Therefore, we now have:

$$\frac{\text{TORQUE on driver}}{\text{TORQUE on driven}} = \frac{\text{SPEED of driven}}{\text{SPEED of driver}} = \frac{\text{TEETH on driver}}{\text{TEETH on driven}}$$

For pulley drives this would become:

$$\frac{\text{TORQUE on driver}}{\text{TORQUE on driven}} = \frac{\text{SPEED of driven}}{\text{SPEED of driver}} = \frac{\text{DIA. of driver}}{\text{DIA. of driven}}$$

Examples

1. On part of a drive to a slotting machine a 15 tooth pinion with a speed of 17 rev/s and a torque of 115 Nm meshes with a gear wheel having 91 teeth. What is the torque on the gear wheel?

   The gear wheel is the *driven* wheel, so we have:

   $$\frac{\text{TORQUE on DRIVEN}}{\text{Torque on driver}} = \frac{\text{Teeth on driven}}{\text{Teeth on driver}}$$

   $$\frac{\text{TORQUE on DRIVEN}}{115} = \frac{91}{15}$$

   $$\text{TORQUE on DRIVEN} = \frac{91}{15} \times \frac{115}{1} = \underline{697 \cdot 7 \text{ Nm}}$$

2. A small centre lathe is driven by a 0·75 kW electric motor which has a speed of 25 rev/s. The motor has a driving pulley of 110 mm dia., which drives a 225 mm dia. pulley keyed to the input shaft of the all-geared headstock. Fixed to the same shaft as this 225 mm pulley, is a pinion having 22 teeth that meshes with a gear wheel having 98 teeth and splined to the intermediate shaft. Also fixed to this shaft is a 42 tooth gear wheel which meshes with a 65 tooth gear wheel keyed to the lathe spindle. Calculate the torque on the lathe spindle at maximum power.

To begin make a sketch of the pulley and gear wheels. See the plan view in Fig. 165.

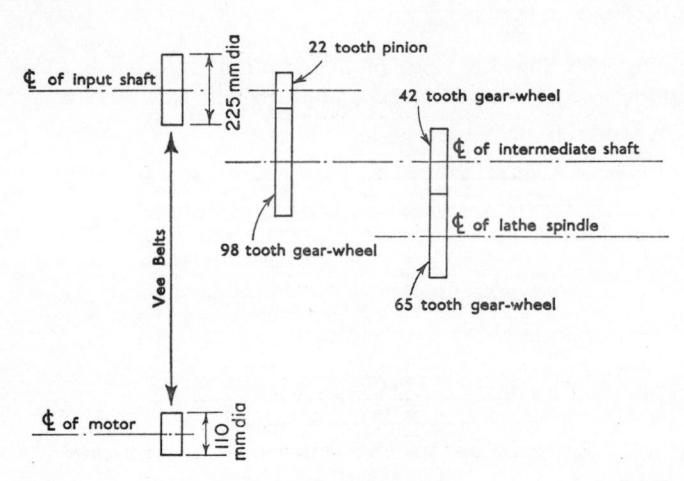

Fig. 165

Next determine the *torque* on the motor shaft, using the formula

$$\text{POWER} = 2\pi NT \text{ watt}$$

In this case we have:

$$0{\cdot}75 \times 10^3 = 2\pi \times 25 \times T$$

$$T = \frac{750}{50\pi}$$

There is no need for us to work this out for our purpose is to use it in a later calculation.

Next we will find the speed ratio between the motor and the lathe spindle.

Velocity Ratio: Driver to Driven

$$\text{V.R.} = \frac{25}{25 \times \dfrac{110}{225} \times \dfrac{22}{98} \times \dfrac{42}{65}}$$

$$\text{V.R.} = \frac{225 \times 98 \times 65}{110 \times 22 \times 42}$$

TORQUE AND POWER 203

Since speed is reduced in this ratio the torque will be increased in the same proportion.

$$\text{Torque on lathe spindle} = \frac{750}{50\pi} \times \frac{225 \times 98 \times 65}{110 \times 22 \times 42}$$

$$= \underline{67\cdot4 \text{ Nm}}$$

3. Figure 166 shows a sketch of a gear train and hand wheel, which is fitted to a lathe carriage for hand traversing. It is a plan view taken from the underside and does not include any part of the apron, or bed. Find what force, F newton the rack pinion applies to the carriage when a force of 20 newton is applied to the handle on the hand wheel.

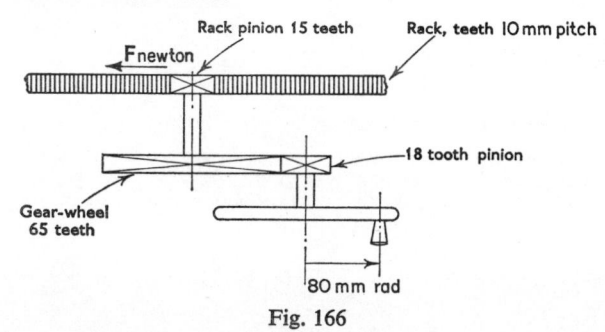

Fig. 166

First we will find the torque on the handwheel.

$$\qquad\qquad\text{(newton metre)} \qquad \text{(newton)} \qquad \text{(metre)}$$
$$\text{Remember that TORQUE} = \text{TANGENTIAL FORCE} \times \text{RADIUS}$$

so we have,

$$\text{TORQUE on handwheel} = 20 \times 0\cdot08$$

$$= 1\cdot6 \text{ Nm}$$

1·6 Nm is also the torque on the 18 tooth pinion, since it is keyed to the same shaft as the hand wheel. We will now work out the *torque* on the 65 tooth gear wheel. The gear wheel is *driven*.

$$\frac{\text{TORQUE ON DRIVEN}}{\text{Torque on driver}} = \frac{\text{Teeth on driven}}{\text{Teeth on driver}}$$

$$\frac{\text{TORQUE on driven}}{1\cdot6} = \frac{65}{18}$$

$$\text{TORQUE on driven} = \frac{65}{18} \times 1\cdot6$$

As this is the *torque* on the 65 tooth gear wheel, it is also the *torque* on the 15 tooth pinion which is keyed to the same spindle.

We must now find the radius of the pitch circle of the pinion, because the force that acts tangential to this circle is the force that acts on the carriage to push it forwards, or backwards.

CIRCUM. of rack pinion pitch circle = 10×15 mm (circular pitch × number of teeth)

We can also write this follows:

$\pi \times$ DIA. of rack pinion pitch circle = 150 mm

$$\text{DIA. of rack pinion pitch circle} = \frac{150}{\pi} \text{ mm}$$

and

$$\text{RADIUS of rack pinion pitch circle} = \frac{150}{2\pi} \text{ mm}$$

Therefore the required force, F newton, is acting at a radius of $\frac{75}{\pi} \times \frac{1}{10^3}$ metre

and TORQUE on rack pinion $= F \times \dfrac{0 \cdot 075}{\pi}$ Nm

Since the *torque* on the rack pinion is $\frac{65}{18} \times 1 \cdot 6$ Nm

$$F \times \frac{0 \cdot 075}{\pi} = \frac{65}{18} \times 1 \cdot 6$$

$$F = \frac{65 \times 1 \cdot 6 \times \pi}{18 \times 0 \cdot 075}$$

$$= \underline{242 \text{ N}}$$

The force acting to push the carriage is approximately 242 newton.

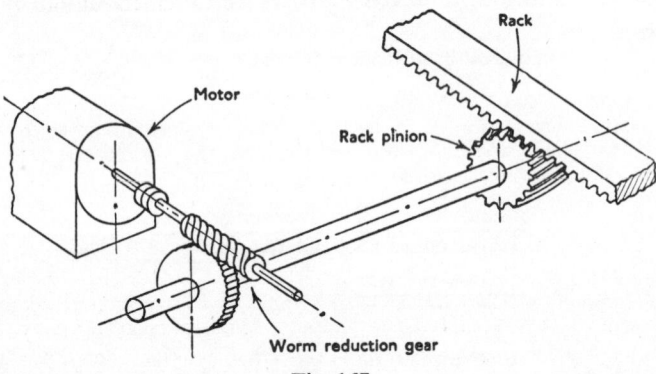

Fig. 167

4. A planing machine table is driven by a 10 kW motor that has a speed of 5 rev/s under certain conditions. Between the motor and the rack pinion there is a 26 to 1 speed reduction. If the rack pinion has a pitch circle diameter of 200 mm, determine the force that the pinion applies to the machine table via the rack which is fixed to the underside of the table. See Fig. 167 for a sketch of this drive.

First we must determine the *torque* on the motor shaft using the formula:

$$\text{POWER} = 2\pi NT \text{ watt}$$

In this case $\quad 10 \times 10^3 = 2\pi \times 5 \times T$

$$T = \frac{10 \times 10^3}{10\pi} = \frac{10^3}{\pi} \text{ Nm}$$

Next we find the *torque* on the worm wheel shaft.

There is a 26 to 1 reduction in speed between the motor shaft and the worm wheel shaft, and the worm wheel shaft is *driven*. Therefore we have

$$\frac{\text{TORQUE on DRIVEN}}{\text{Torque on driver}} = \frac{\text{Speed of driver}}{\text{Speed of driven}}$$

$$\frac{\text{TORQUE on driven}}{\frac{10^3}{\pi}} = \frac{26}{1}$$

$$\text{TORQUE on DRIVEN} = \frac{26}{1} \times \frac{10^3}{\pi} = 8 \cdot 28 \times 10^3 \text{ Nm}$$

The *torque* on the rack pinion is 8·28 kN m

Note: The *torque* on the rack pinion is 26 times greater than the *torque* on the motor shaft, but its *speed* is 26 times less.

Finally, to find the *force* that the rack pinion applies to the rack. We know that:

$$\text{TORQUE} = \text{TANGENTIAL FORCE} \times \text{RADIUS}$$

Therefore we have:

$$\begin{array}{cc} \text{(newton)} & \text{(metre)} \end{array}$$

$$8 \cdot 28 \times 10^3 = \text{TANGENTIAL FORCE} \times \text{PITCH CIRCLE RADIUS}$$

$$8 \cdot 28 \times 10^3 = \text{tangential force} \times \frac{100}{10^3}$$

$$\text{TANGENTIAL FORCE} = 8 \cdot 28 \times 10^3 \times 10$$

$$= 82 \cdot 8 \times 10^3 \text{ N}$$

The force which the rack pinion applies to the table is 82·8 kN.

Exercise 12

Section B

1. The suds pump fitted to a small lathe is driven by a 75 W electric motor having a speed of 20 rev/s. What is the torque on the spindle at maximum power?
2. During a milling operation a slab mill of 250 mm dia. is used at a speed of 5 rev/s. The machine is driven by a 4 kW electric motor. Calculate the force on the cutter at 125 mm radius, at maximum power.
3. A sensitive drilling machine is driven by a 500 W electric motor which drives a pulley, keyed to the drill spindle, at 7 rev/s. Calculate the torque on the spindle, at maximum power.
4. A component 150 mm dia. is being turned. The force on the tool point is 1·2 kN A 1·5 kW motor drives the lathe, and the torque on its armature is $\frac{1}{8}$ of that on the lathe spindle. Calculate the speed of the motor.

5. On a vertical boring machine a steel tyre is bored 1·8 m diameter at a speed of 3·5 rev/s; the cutting force on the tool point being 1·5 kN. Find the torque required on the machine spindle, and the power being used.

6. A drilling machine spindle driven by a 2 kW electric motor has a maximum speed of 16 rev/s and minimum speed 0·5 rev/s. What is the torque on the spindle at each of these speeds at maximum power?

7. When drilling a 16 mm dia. hole it is found that a spindle speed of 3·5 rev/s appears to be satisfactory. Driving the machine is a 1 kW motor. Calculate at maximum power the force on the two cutting edges of the drill at a radius of 8 mm.

8. A blanking press is driven by a 40 kW electric motor which has a speed of 26 rev/s. There is a 40 to 1 speed reduction between the motor shaft and the press crankshaft. Calculate the torque on the crankshaft at maximum power.

9. A shaping machine is driven by a 1 kW motor. During a particular operation the bull gear wheel makes 0·5 rev/s and the crank pin is at 100 mm radius on the bull wheel. Find the tangential force on the crank pin at maximum power.

10. A planing machine table is driven by a 10 kW electric motor, which at maximum torque has a speed of 4 rev/s. There is a 34 to 1 speed reduction between the motor and the rack pinion. This pinion drives a rack that is bolted to the underside of the table (the arrangement of the drive is similar to that shown in Fig. 167). If the pitch circle diameter of the rack pinion is 210 mm, find the maximum driving force that it applies to the machine table.

11. A pinion in the gear box of a large machine tool has a torque of 2300 Nm and speed 17 rev/s. The pinion has 25 teeth and drives a gear wheel having 74 teeth. What is the power at the pinion, and the torque on the shaft to which the 74 tooth gear wheel is keyed?

12. A 12 kW electric motor, speed of 20 rev/s has a 150 mm dia. pulley which drives a 350 mm dia. pulley, keyed to the input shaft of a gear box. Also keyed to this shaft is a pinion having 35 teeth, which meshes with a gear wheel having 130 teeth. Calculate the speed and maximum torque of the shaft to which the gear wheel is keyed.

13. During the milling of a key a 150 mm dia. helical tooth cutter is used, the cutting force, acting at 80 mm radius, being 1 kN. The arbor makes 4 rev/s. Find the torque and power required at the arbor.

14. A trepanning process, being carried out on a drilling machine, makes use of one tool which is at a radius of 110 mm. The spindle to which it is fixed makes 2 rev/s. In the drive to the spindle a 32 tooth pinion drives a 95 tooth gear wheel which is fixed to the spindle. If the force on the trepanning tool is 140 N, calculate the torque on the pinion and the work done per min by the tool.

15. Figure 168 shows the drive to the tool head of a horizontal boring machine. The tip of the tool is at a radius of 320 mm and the cutting force on it is 1·05 kN. The tool makes 2 rev/s. In the drive to the head a 210 mm dia. motor pulley drives a 0·5 m dia. pulley keyed to the same shaft as a 22 tooth pinion. This pinion meshes with a gear wheel having 136 internal teeth, which is fixed to the head. Calculate the torque on the motor shaft and the output power of the motor assuming transmission is 90% efficient.

16. The mean diameter of a bar being turned is 140 mm and the cutting force at the tool point 500 N. The lathe is driven by a 4 kW motor, speed 25 rev/s and the velocity ratio, motor to spindle, is 3 to 1. What percentage of the maximum motor torque is being used in the operation?

17. A 30·2 kN force is required to draw a tube through a die at the rate of 0·1 m/s. This is obtained through a wire rope which is wound on to a drum of 1·1 m mean diameter. The drum is driven by an electric motor through a worm reduction gear of 36 to 1 velocity ratio. Find the torque and power required at the motor shaft if the efficiency of the transmission is 92%?

18. A machine table is elevated by means of a screw at the rate of 0·01 m/s. The mass of the table is 2 Mg and it supports a steel ingot of mass 3·5 Mg. The screw is driven at 0·8 rev/s by an electric motor through a gearbox having a velocity ratio of 19. If the overall efficiency of the drive is 21%, what power and torque are required at the motor?

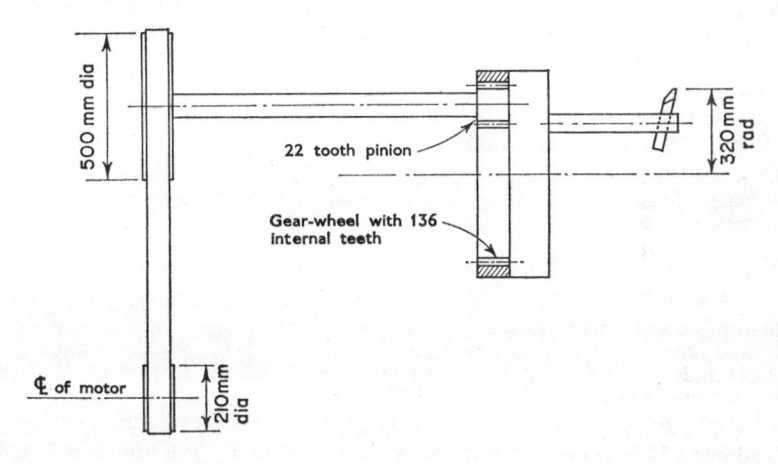

Fig. 168

19. A blanking press exerts a tool force of 20 kN at an average speed of 8 m/s the tool head being driven through a gear box 95% efficient, by an electric motor. What torque is required at the motor shaft if the motor speed is 20 rev/s?

20. The total tooth force on a milling cutter is 2 kN at a radius of 200 mm, the cutter speed being 0·3 rev/s. There is a 49 to 1 velocity ratio between arbor and motor and the efficiency of the transmission is 89%. Calculate the power required at the motor and the motor speed.

Chapter 13. Mechanical Advantage and Velocity Ratio

Lifting machines

It is obvious to all engineering technicians that there is a limit to the load that can be lifted directly by hand. When this limit is reached, we use a lifting machine. This may be hand, or power operated. Examples of these are:

The screw jack (or screw and nut mechanism as used on boring, milling, planing, shaping machines, etc.),
Rope blocks,
Chain blocks,
Winch,
Hydraulic jack, Wedge, Crane.

If we were asked what a lifting machine was, a suitable reply would be 'It is a machine which is used to lift large loads with relatively small efforts'. In order to do this, the effort has to move faster than the load. We only need operate a chain block to appreciate this.

It is worth pointing out at this stage that the following definitions and formulae apply to any mechanisms which embody the lifting machine principle of creating a large force by a smaller force. Examples of such mechanisms are to be found on broaching machines, rivetting machines, feed mechanisms on machine tools, hydraulically and electrically operated presses.

Mechanical advantage

We can say that by using a lifting machine we gain an advantage, namely that of lifting a large load with a relatively small effort. This advantage is known as the *mechanical advantage*, since it is by using a mechanical device that we gain it. We can determine the size of a *mechanical advantage* as follows:

$$\text{MECHANICAL ADVANTAGE} = \frac{\text{LOAD (newton)}}{\text{EFFORT (newton)}}, \text{ (a ratio)}$$

It has no units, for these cancel, as can be seen. Therefore, it is just a number which varies for different loads on any one machine. This is because the frictional resistance of the machine varies with the load. Some part of the effort is used to drive the mechanism.

Velocity ratio

We have gained the mechanical advantage because the load has moved at a speed slower than that of the effort. We could say that when the effort moves through some fixed distance, the load moves through a shorter distance. We shall see this more clearly if we remember that, when raising the table of a milling machine, the lifting handle to which the effort is applied moves faster than the table. For one revolution of the effort handle, the table is lifted only a fraction of the circumference of that revolution. In Chapter 10, page 167, we learned that *velocity ratio* means speed ratio and for a lifting machine can be found as follows:

$$\text{VELOCITY RATIO} = \frac{\text{SPEED OF EFFORT}}{\text{SPEED OF LOAD}}$$

This can be written,

$$\text{VELOCITY RATIO} = \frac{\text{DISTANCE MOVED by EFFORT (mm) in certain time}}{\text{DISTANCE MOVED by LOAD (mm) in same time}}$$

It has no units, i.e., it is just a number. This number remains constant for any one machine. Since it is governed by the construction of the machine.

Note: *Mechanical advantage* and *velocity ratio* are both usually greater than 1.

Efficiency

In Chapter 8, page 144, we learned that no machine is 100 per cent efficient, and this includes lifting machines.

Remembering the equation by which we can calculate efficiency,

$$\text{EFFICIENCY} = \frac{\text{WORK OUTPUT FROM MACHINE}}{\text{ENERGY INPUT TO MACHINE}}$$

we have

$$\text{WORK OUTPUT} = \text{LOAD} \times \text{DISTANCE MOVED by LOAD joule}$$

and

$$\text{ENERGY INPUT} = \text{EFFORT} \times \text{DISTANCE MOVED by EFFORT joule}$$

EFFICIENCY OF A LIFTING MACHINE

$$= \frac{\text{LOAD} \times \text{DISTANCE MOVED by LOAD}}{\text{EFFORT} \times \text{DISTANCE MOVED by EFFORT}}$$

$$= \frac{\text{LOAD}}{\text{EFFORT}} \times \frac{\text{DISTANCE MOVED by LOAD}}{\text{DISTANCE MOVED by EFFORT}}$$

but

$$\frac{\text{LOAD}}{\text{EFFORT}} = \text{MECHANICAL ADVANTAGE}$$

and

$$\frac{\text{DISTANCE MOVED by LOAD}}{\text{DISTANCE MOVED by EFFORT}} = \frac{1}{\text{VELOCITY RATIO}} \text{ (the reciprocal of V.R.)}$$

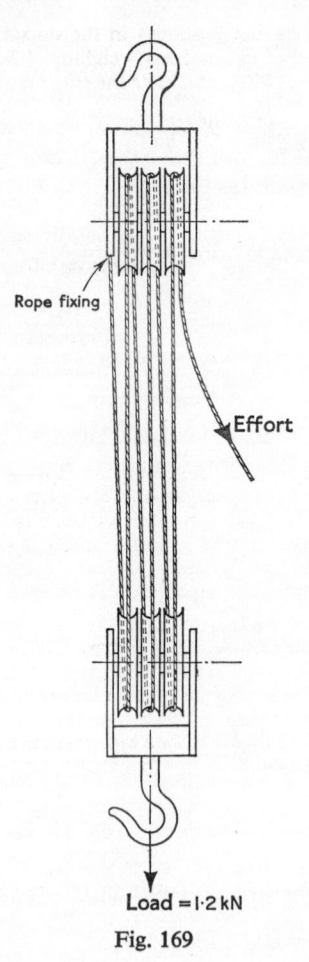

Fig. 169

EFFICIENCY of a LIFTING MACHINE

$$= \text{MECHANICAL ADVANTAGE} \times \frac{1}{\text{VELOCITY RATIO}}$$

$$= \frac{\text{MECHANICAL ADVANTAGE}}{\text{VELOCITY RATIO}}$$

We may express the efficiency as a number which will always be less than 1,

LIFTING MACHINES 211

or as a percentage. In the latter case we would multiply the right-hand side of the equation by 100.

Examples

1. A set of rope blocks having 3 pulleys in the upper block and 3 in the lower is to be used to lift a drilling machine 'weighing' 1·2 kN onto its bench. If the efficiency of the blocks is 90%, calculate the effort required.

First we will make a sketch of the blocks, as in Fig. 169. Next we will write the formulae already learned.

1. MECHANICAL ADVANTAGE $= \dfrac{\text{LOAD}}{\text{EFFORT}}$

2. VELOCITY RATIO $= \dfrac{\text{DISTANCE MOVED by EFFORT}}{\text{DISTANCE MOVED by LOAD}}$

3. EFFICIENCY $= \dfrac{\text{MECHANICAL ADVANTAGE}}{\text{VELOCITY RATIO}}$

We have to calculate an *effort*, and the only formula in which this appears is the first one, but we do not know the *mechanical advantage*. We can find this by using the third formula, but before doing so we shall have to find the *velocity ratio*.

To find the *velocity ratio*, we will suppose that the load is lifted 1 metre and then determine how far the *effort* moves during the same period. Look at it like this: for the load to be lifted 1 m, each of the 6 falls of rope that support the load must be shortened 1 m. This means pulling in 6 m at the loose end of the rope, i.e., when the load moves 1 metre, the effort moves 6 metre. Strictly speaking *displacement ratio* is a better description in cases of this type.

$$\text{VELOCITY RATIO} = \frac{6}{1} = 6.$$

Note: The *velocity ratio* of a set of rope blocks is always equal to the number of pulleys in the two blocks, e.g., in this case there are 6 pulleys. Knowing the velocity ratio we can proceed to find the mechanical advantage, because we are told that the efficiency of the blocks is 90%.

$$\frac{90}{100} = \frac{\text{MECHANICAL ADVANTAGE}}{6}$$

$$\text{MECHANICAL ADVANTAGE} = \frac{90 \times 6}{100} = 5 \cdot 4$$

$$\text{MECHANICAL ADVANTAGE} = \frac{\text{LOAD}}{\text{EFFORT}}$$

$$5 \cdot 4 = \frac{1 \cdot 2 \times 10^3}{\text{EFFORT}}$$

$$5 \cdot 4 \times \text{EFFORT} = 1 \cdot 2 \times 10^3$$

$$\therefore \text{EFFORT} = \frac{1 \cdot 2 \times 10^3}{5 \cdot 4}$$

$$= \underline{\underline{222 \text{ N}}}$$

212 MATERIALS AND MACHINES IN THE WORKSHOP

2. A screw jack used to support one end of a bar of metal being cut in the mechanical saw, has a single start square thread, 6 mm pitch. The effort is applied to a handle, at a radius of 125 mm and an effort of 10 N lifts a load of 300 N. What is the efficiency of the jack?

Begin by making a sketch, as in Fig. 170

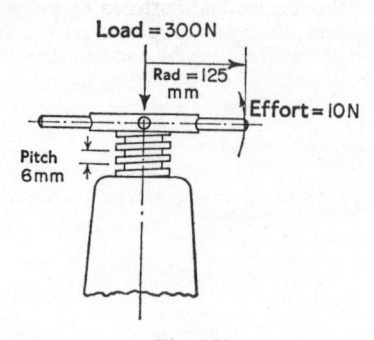

Fig. 170

Let us suppose the screw makes 1 revolution, and then proceed to find the distances moved by the effort and the load.

$$\text{DISTANCE MOVED by EFFORT} = \frac{\text{Circumference of circular path 125 mm}}{\text{radius}}$$

$$= 2\pi \times 125 \text{ mm}$$

$$\text{DISTANCE MOVED by LOAD} = 1 \text{ pitch of the screw}$$

$$= 6 \text{ mm}$$

$$\text{VELOCITY RATIO} = \frac{2\pi \times 125}{6} = \frac{125\pi}{3}$$

$$\text{MECHANICAL ADVANTAGE} = \frac{300}{10} = 30$$

$$\text{EFFICIENCY of screw jack} = 30 \div \frac{125\pi}{3} = \frac{90}{125\pi}$$

$$= 0.23$$

$$= \underline{\underline{23\%}}$$

3. A set of Weston differential chain blocks has one pulley of 150 mm dia. and the other of 200 mm dia. Calculate the velocity ratio of the blocks.

As usual, begin by making a sketch (see Fig. 171).

A differential pulley is a compound pulley in which the two pulleys have different diameters.

Let us now find out how the blocks work. To lift the *load*, the *effort* moves in the direction indicated, but as the 200 mm dia. pulley winds the chain in, the 150 mm dia. pulley unwinds it. However, because the bigger diameter pulley has the larger circumference, the part of the chain supporting the load is shortened,

LIFTING MACHINES

and the amount it is shortened for 1 revolution of the *differential pulley* is found as follows.

The part of the chain supporting the load is shortened by
Circumference of 200 mm dia. pulley − Circumference of 150 mm dia. pulley,
$$\text{Amount chain shortens} = (200\pi - 150\pi) \text{ mm}$$

If the chain supporting the load is shortened by this amount, the load must be lifted half this amount, since it is supported on two falls of chain. It is as if half the amount were taken from one fall and half from the other,

$$\text{Load is lifted } \frac{200\pi - 150\pi}{2} \text{ mm}$$

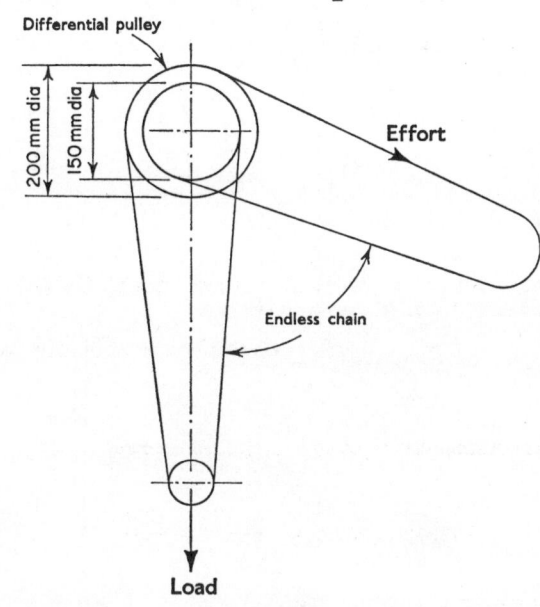

Fig. 171

While the load is being lifted this amount, the effort moves through a distance equal to the circumference of the bigger pulley,

Effort moves 200π mm

We know that,

$$\text{(DISPLACEMENT) or VELOCITY RATIO} = \frac{\text{DISTANCE MOVED by EFFORT}}{\text{DISTANCE MOVED by LOAD}}$$

$$= \frac{200\pi}{\frac{200\pi - 150\pi}{2}} = \frac{200\pi \times 2}{200\pi - 150\pi}$$

$$= \frac{400\pi}{50\pi} = \frac{8}{1}$$

Weston differential chain block pulleys usually have their rims shaped to suit the shape of the chain links, so instead of the pulley rim being curved, it is made up of a series of flats. These flats can be used in determining the *velocity ratio*. For instance, in the last example, instead of being given the two diameters, 150 and 200 mm, we could have been told that the pulleys had 12 and 16 flats respectively. Thus, for 1 revolution of the differential pulley

$$\text{DISTANCE MOVED by LOAD} = \frac{16-12}{2}$$

and DISTANCE MOVED by EFFORT = 16

$$\text{VELOCITY RATIO of blocks} = 16 \div \frac{16-12}{2}$$

$$= \frac{16}{1} \times \frac{2}{16-12} = \frac{16}{1} \times \frac{1}{2} = \underline{8}.$$

4. A set of chain blocks comprises a single start worm and a 40-tooth worm-wheel. Fixed to the worm-wheel is a 125 mm dia. wheel that carries the load chain. Attached to the worm spindle is a 140 mm dia. wheel to which the effort is applied. Calculate the displacement ratio of the blocks. For a diagram of this, see Fig. 172.

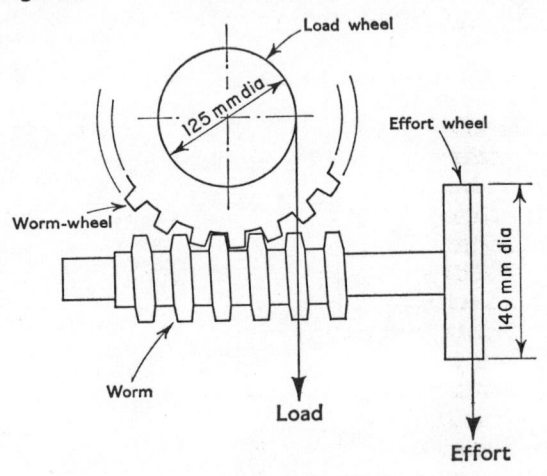

Fig. 172

Suppose the effort wheel makes 1 revolution, i.e., the worm makes 1 rev. The worm-wheel then makes $\frac{1}{40}$ of a rev, it moves 1 tooth the equivalent of the worm pitch.

$$\text{DISTANCE MOVED by EFFORT} = \text{Circumference of effort wheel}$$
$$= 140\pi \text{ mm}$$

$$\text{DISTANCE MOVED by LOAD} = \text{Circumference of load wheel} \times \frac{1}{40},$$

$$= 125\pi \times \frac{1}{40} = \frac{25\pi}{8} \text{ mm}$$

$$\text{DISPLACEMENT RATIO of blocks} = 140\pi \div \frac{25\pi}{8} = \frac{140 \times 8\pi}{25\pi}$$

$$= \underline{45}.$$

LIFTING MACHINES

5. A milling machine table, mass 280 kg, carries a job and vice, combined mass 20 kg. The table is lifted by a 4 mm square thread single start screw, operated through a pair of equal bevel gears by a handle 250 mm long. Calculate the effort that must be applied to the end of the handle in order to lift the table. The efficiency of the lifting mechanism is 25%.

This problem is really a screw jack problem. To solve it we require the equations for mechanical advantage, velocity ratio and efficiency.

First we will find the *velocity ratio*, because we do not have enough information to enable us to use the other formulae.

Let us suppose the effort handle makes 1 revolution.

Distance moved by effort
= Circumference of a circular path of 250 mm
= $2\pi \times 250 = 500\pi$ mm

When the effort handle makes one revolution, the screw lifting the table makes one revolution, and it lifts the table a distance equal to the pitch of the screw. The pitch is 8 mm, as shown in Fig. 173.

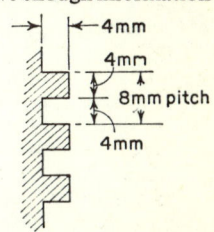

Fig. 173

Hence,

VELOCITY RATIO of lifting mechanism = $\dfrac{500\pi}{8} = 62 \cdot 5\pi$

Having found the *velocity ratio*, we can now use the *efficiency* equation to determine the *mechanical advantage* of the mechanism.

$$\frac{25}{100} = \frac{\text{MECHANICAL ADVANTAGE}}{62 \cdot 5\pi}$$

Mechanical advantage = $\dfrac{25}{100} \times \dfrac{62 \cdot 5\pi}{1} = 15 \cdot 6\pi$

Also, Mechanical advantage = $\dfrac{\text{load}}{\text{effort}}$

$$15 \cdot 6\pi = \frac{(280+20) \times 9 \cdot 81 \text{ newton}}{\text{effort}}$$

Remember that the gravitational pull on 1 kg is 9·81 N.

$15 \cdot 6\pi \times \text{effort} = 300 \times 9 \cdot 81$

EFFORT = $\dfrac{2943}{15 \cdot 6\pi}$

= 60 N

Some mechanisms have screws which have more than one thread, e.g. they may have two, or more, see Fig. 174. In such cases we have to consider the *lead* of the thread instead of the pitch.

Definitions:—the *lead* of a thread is the distance, measured parallel to the axis of the screw, between the centres of the crests of the start and finish of one complete turn of the thread.

The *pitch* is the distance, measured parallel to the axis of the screw, between the centres of two adjacent crests.

On a multi-thread screw the lead is a dimension taken between crests of the same thread, the pitch between crests of different threads. On a single start thread the pitch and lead are equal.

The nut, or screw, will always move, along the screw, a distance equivalent to the lead of the thread in one revolution.

It is recommended that tests be carried out on the mechanisms, illustrated in the foregoing examples, so as to determine the relationship between load and efficiency; also to find the *law of a machine*.

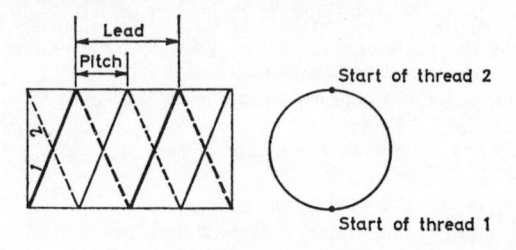

A two-start thread
Fig. 174

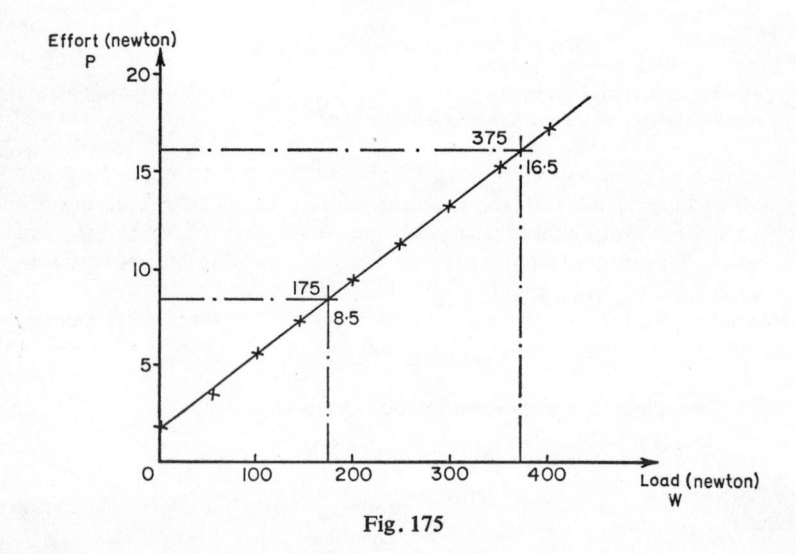

Fig. 175

Law of a machine

In order to obtain the law of a machine a test has to be carried out. This involves applying a series of loads to the machine and determining the efforts required to lift these loads. Having obtained at least six pairs of values a graph of load against effort is then plotted, see Fig. 175. Since the graph is of the straight line form it will follow the law $y = ax + b$ where a and b have constant values, they are known as *constants*. On our graph we have P in place of y and W instead of x; hence the law is written: $P = aW + b$. The problem now is to find the values of a and b. This is done by taking two pairs of co-ordinate values from the graph and substituting them in the law equation:

$$16 \cdot 5 = a \cdot 375 + b \qquad \text{(i)}$$
$$8 \cdot 5 = a \cdot 175 + b \qquad \text{(ii)}$$

Here we have a pair of simultaneous equations which may be solved by subtracting the latter from the former and so eliminating one of the 'unknowns' e.g., b. Because we are subtracting equation (ii) from equation (i) we change the signs of the terms in equation (ii) then proceed by adding the equations.

$$8 = 200a$$
$$\frac{8}{200} = a$$
$$a = 0 \cdot 04$$

To find the value of b, we substitute the value for a in either of the equations (i) or (ii). Substituting 0·04 in equation (ii),

$$8 \cdot 5 = 175(0 \cdot 04) + b$$
$$8 \cdot 5 - 7 = b$$
$$b = 1 \cdot 5$$

Therefore the law of a machine is, $P = 0 \cdot 04W + 1 \cdot 5$. Thus for any load that we may require to lift with the machine under test, we can determine the effort required by substituting the load value in place of W in the 'law equation'. This is the chief reason for wanting to know the law of a machine. The effort required for no load is 1·5 newton. Why is this?

If in such a test the load applied to the machine is in the form of kilogramme masses. We must remember that the gravitational pull on each 1 kg mass is 9·81 newton.

$$\text{Load (newton)} = \text{Load (kg)} \times 9 \cdot 81.$$

Exercise 13

Remember, when calculating *velocity (or displacement) ratio*, to let the effort move through a fixed distance, and then proceed to find how much the load moves.

Section A

1. A set of rope blocks is to be used to lift a 1 kN load. There are two pulleys in the top block and two in the lower block. If an effort of 300 N is required, find the mechanical advantage at this load and velocity ratio of the blocks.
2. In a Weston differential chain block, the larger of the differential pulleys has 12 chain flats on its circumference, whilst the smaller pulley has 10. Calculate the velocity ratio for the block.
3. The table of a shaping machine is lifted by means of a single start, square thread screw pitch 6 mm, and the effort is applied at the end of a handle 240 mm long. If the total mass of the table, vice and job is 74 kg, find the mechanical advantage of the mechanism at this load if its efficiency is 20%.

218 MATERIALS AND MACHINES IN THE WORKSHOP

4. A set of chain blocks has a load wheel 106 mm dia., fixed to a worm wheel with 28 teeth. The single start worm engaging with the worm-wheel has an effort wheel 118 mm dia., keyed to its spindle. Calculate the displacement ratio of the blocks.

5. An electric motor 'weighing' 0·9 kN is to be lifted by means of rope blocks. There are two pulleys in the lower block and three in the upper block. The mechanical advantage is 4·5 at this load. Calculate the actual and ideal efforts at this load and explain the difference between the two.

6. The efficiency of a set of chain blocks is 32%, and the velocity ratio is 37. Calculate the effort required to lift a lathe of mass 500 kg.

7. A set of rope blocks, having 6 pulleys in all, has an efficiency of 77%. The blocks are to be used to lift a shaft of mass 210 kg. What effort is required?

8. A small hand operated winch, velocity ratio 11, is used for lifting billets 'weighing' 700 N each from a furnace. An effort of 120 N has to be applied to the handle of the winch to do the job. What is its efficiency?

9. The pulleys of a Weston chain block have 18 and 15 chain flats respectively and the efficiency of the blocks is 72%. Calculate the effort required to lift a fan-shaft, 'weighing' 705 N.

10. The table of a milling machine 'weighs' 2·4 kN, and carries a fixture which, complete with job, 'weighs' 140 N. To lift the table an effort is applied at the end of a handle 280 mm long. The overall efficiency of the lifting mechanism is 23% and the pitch of the single start lifting screw is 6 mm. Calculate the effort required to lift the table.

Section B

1. Figure 176 shows a hand operated crane that is used on a forging press. Its efficiency is 80%. The billets being lifted have a mass of 80 kg each. Calculate the effort that must be applied to the end of the handle.

2. A shaping machine table, together with job 'weighs' 1500 N. It is raised by a 3 mm square thread, single start screw, operated by a handle 210 mm long. The efficiency of the lifting mechanism is 26%. Find the effort that must be applied to the end of the handle to raise the table.

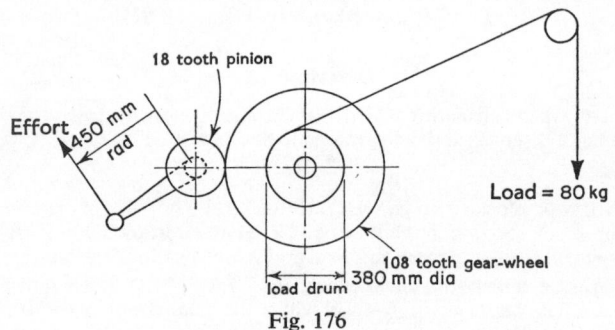

Fig. 176

3. The velocity ratio of the hoist on an overhead electric crane is 22. If its efficiency is 88%, what torque would have to be created by the hoist motor to lift a load of 50 kN? The pitch circle diameter of the motor pinion is 60 mm.

LIFTING MACHINES

4. A set of Weston differential chain blocks, 62% efficient, is to be used to lift a shaping machine ram, of mass 100 kg. What effort has to be applied to the blocks if the differential pulley rims have 16 and 11 chain flats respectively?
5. Suppose a set of chain blocks, efficiency 37%, with a worm and wheel, were used to lift the load given in Question 4, what effort would be required? The worm-wheel has 32 teeth and a 150 mm dia. load wheel keyed to it. The worm is single start and has an effort wheel of 156 mm dia.
6. The tool-head of a side planing machine is driven by a 3-start, 20 mm square thread screw. The total resisting force at the head is 2·8 kN. What effort must be applied to the screw at 0·3 m radius if the efficiency of the mechanism is 37%?
7. The constant torque on an electric motor driving a press is 300 Nm. The velocity ratio of the transmission between the motor and tool head is 19 to 2 and its efficiency is 97%. What is the force at the tool-head at 0·35 m radius?
8. The following results are taken from a test on a worm and worm wheel mechanism:

LOAD (newton)	25	50	75	100	125	150	175	200
EFFORT (newton)	4·0	6·5	9·0	11·5	14·0	16·5	19·0	21·5

The load drum is of 100 mm diameter, and effort drum of 125 mm diameter; the worm is single start and the worm-wheel has 36 teeth. Find the law of the machine and plot a graph of load against efficiency.

9. A shaping machine makes 0·5 stroke per second and the force on the tool point is 950 N. The motor speed is 20 rev/s and the efficiency of the transmission 93%. What is the velocity ratio of the drive and the effort required at the motor?
10. The motor driving a milling machine arbor is developing 0·35 kW at 20 rev/s. The cutter speed is 1 rev/s and the cutter-load 540 N at 100 mm radius. Calculate the velocity ratio and efficiency of the transmission.

Chapter 14. Parallelogram and Triangle of Forces

IN ORDER to learn what *parallelograms* and *triangles of forces* are, we will study the following examples. The first example has no direct connection with the workshop, but it will give us the right idea before moving to workshop examples

Examples

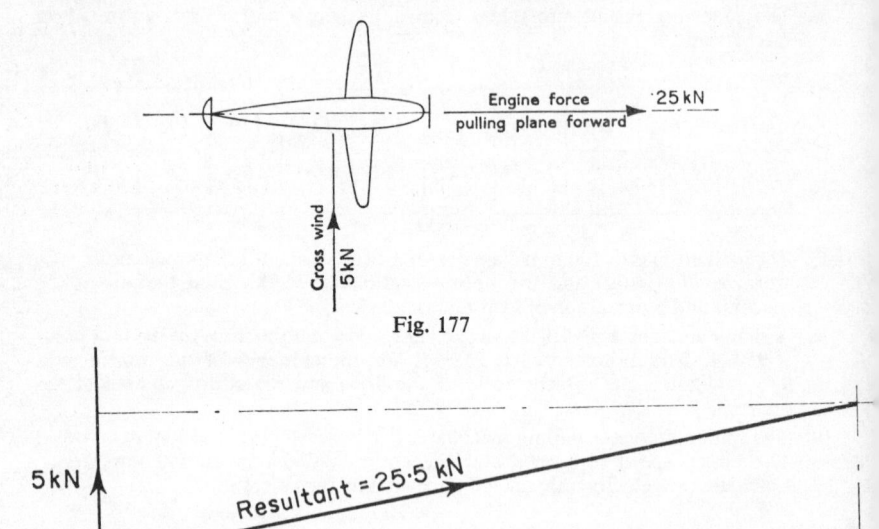

Parallelogram of forces

Fig. 177

Fig. 178

Scale: 10 mm = 2·5 kN

1. Figure 177 shows a small aircraft. The engine force tries to drive the plane forward on its course, while the force of the cross-wind tends to drive the plane in a direction at 90° to the true course. The result is that the aircraft is driven along a course that is somewhere between the lines of the engine force and the force of the cross-wind, i.e., by a force which is the resultant of the 25 kN and the 5 kN forces. This force is known as the *resultant* and is determined by drawing what is known as a *parallelogram* of forces (see Fig. 178).

The method for drawing a *parallelogram* of forces is as follows:

(*a*) We represent the two forces given, by two lines that are drawn parallel to the lines of action of the forces.

PARALLELOGRAM AND TRIANGLE OF FORCES

(b) Commencing at the point where these two lines intersect, we mark off, to some suitable scale, the size of each force on the line representing the force. The scale in our example is: 10 mm = 2·5 kN. A length of 100 mm is stepped off for the 25 kN force, and a length of 20 mm for the 5 kN force.

(c) We then complete the parallelogram by drawing the other two sides, shown in the chain dotted line.

(d) We now draw the diagonal, representing the *resultant*.

(e) Next we scale the diagonal, i.e., we measure it in millimetres and multiply the length by 0·25 kN, in order to obtain the size of the resultant. The value of the resultant force is found to be 25·5 kN, i.e. 102 mm × 0·25.

(f) Finally we require the direction of the resultant with respect to one of the other forces. In this instance it is convenient to measure the angle between the 25 kN engine force and the resultant, with a protractor.

Hence we have the answer to our problem,

RESULTANT FORCE acting on aircraft = 25·5 kN at an angle of 11°·3 to the 25 kN force.

Thus the aircraft is pushed off its course, if correcting action is not taken.

2. In Figure 179 a side elevation of the blade of a parting off tool is shown with the forces acting on its cutting edge. The cutting force is 800 N, whilst the 200 N reaction is caused by the feeding of the tool into the metal. Determine the *resultant* force acting on the tool's cutting edge and the angle at which it acts relative to the 200 N force.

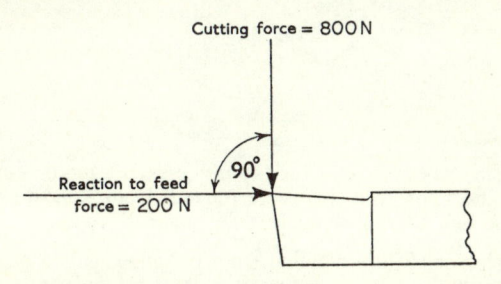

Fig. 179

As in the previous example, we draw lines to represent the cutting force and the reaction to the feed force (Fig. 179a). On these two lines we step off, to scale 1 mm to 10 N, the lengths to represent the two forces, e.g., $\frac{800}{10}$ = 80 mm., and we have a line 80 mm long to represent the 800 N cutting force, $\frac{200}{10}$ = 20 mm., and we have a line 20 mm long to represent the feed force reaction. We then complete the parallelogram and draw the diagonal which represents the magnitude and direction of the *resultant*. Next we measure the length of the diagonal and multiply it by 10 to obtain the magnitude (size) of the resultant. Length of diagonal = 82·5 mm.

We know that 1 mm represents 10 N therefore,

$$82·5 \text{ mm represents } 10 \times 82·5 = 825 \text{ N}$$

Hence, resultant force = 825 N at 75°·5, in a clockwise sense, to the 200 N force. The angle is again measured by protractor.

MATERIALS AND MACHINES IN THE WORKSHOP

Equilibrant. There is the force at the tool point which is equal to the resultant force and directly opposite in action. It is a reaction to it. This force is known as the *equilibrant* because it balances, or equals, the resultant.

A parallelogram of forces may be used to determine the resultant, or equilibrant, of any two forces that are in one plane and act from the same point, e.g. for any two co-planar concurrent forces, irrespective of the value of the angle between the forces.

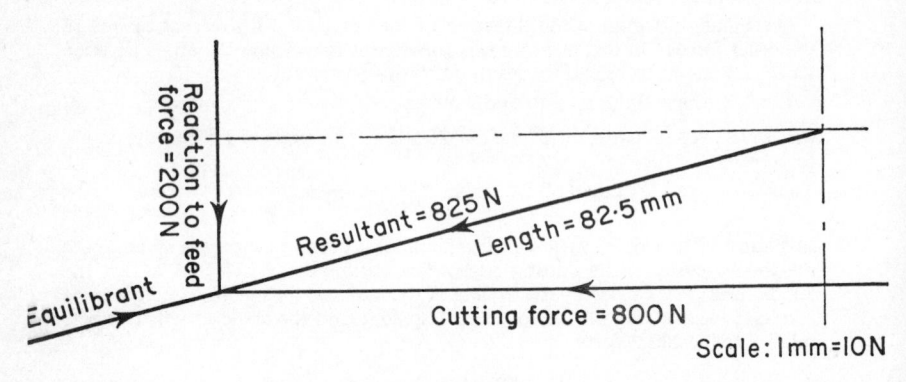

Fig. 179(a)

Triangle of forces

3. A set of chain blocks is used for lifting an electric motor onto its bedplate. The blocks are hung from two wire ropes that are fixed to a roof joist, as shown in Fig. 180. Determine the forces acting in the ropes.

This problem is slightly different from the previous two problems, because instead of having two forces acting at a point we have three, i.e., the forces in the two ropes and the gravitational pull ('weight') on the motor. Such a problem is tackled in the following manner.

(a) We must draw accurately to scale the system, using the details given in the question. This is shown in Fig. 181. The diagram, which is known as the *space diagram*, must be drawn in such a position as to leave space by the side of it for drawing the *triangle of forces* diagram, the force diagram.

(b) Now we must letter (using capital letters) the three spaces between the forces, A, B and C. Commence with the space to the right of the given force and move in a clockwise direction. We can now refer to the 3000 N force as force AB, and to the other forces as BC and CA.

(c) Having drawn the space diagram accurately to scale, we can transfer lines from it, by sliding our set square on the ruler, in order to construct the *triangle of forces*, or *force diagram*. Beginning with the given force AB, we draw a line parallel to it and mark off on this line length *ab* to the scale decided, i.e., 1 mm = 75 N, *ab* would be 40 mm long, i.e., $\frac{3000}{75}$. We must mark an arrowhead on *ab* to indicate the direction of the force.

PARALLELOGRAM AND TRIANGLE OF FORCES 223

(*d*) Next we draw a line from *b* parallel to BC, but we cannot represent this force by a length because we do not know the size of the force BC. We must put an arrowhead on this line indicating the direction in which we are moving, i.e., from *b* to *c*.

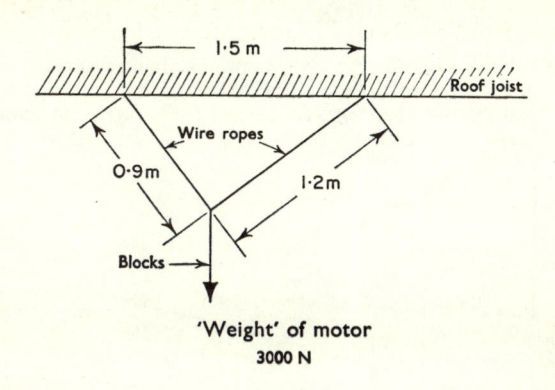

Fig. 180

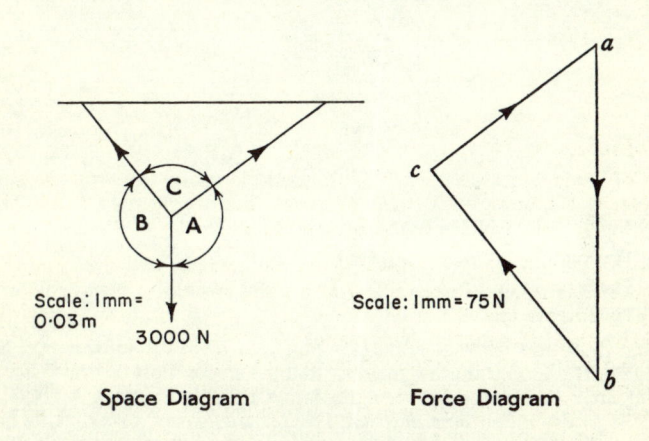

Fig. 181

(*e*) To complete the force diagram, we draw from *a* a line parallel to CA to meet the line drawn from *b*, at *c*. We must put the arrowhead on the line pointing from *c* to *a*. This brings us back to point *a*, and we have moved round in alphabetical order and a clockwise direction. We have taken each force in turn. This procedure must always be adopted when constructing force triangles, or polygons.

(f) Finally, we have to measure the lengths *bc* and *ca* and scale them, in order to obtain the forces in the two wire ropes.

FORCE BC = *bc* = 33 × 75 = 2475
FORCE CA = *ca* = 25 × 75 = 1875

Note: The sum of the forces in the two ropes does not equal the 'weight' of the load placed upon them. It should not, unless the ropes are vertical.

4. Figure 182 shows a small crane which swings about a vertical spindle, fixed to a stanchion, in a forging shop. The maximum safe load that the crane can carry is 20 kN. Determine the forces in the jib and tie of the crane, when it is carrying the maximum load.

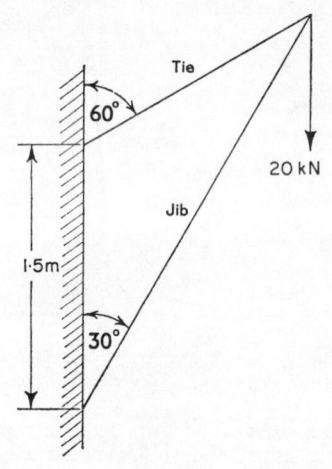

Fig. 182

Here again we have three forces acting from a point. They are

1. The load.
2. The force in the jib.
3. The force in the tie.

First draw the *space diagram* (see Fig. 183) from the information given in Fig. 182. It is important to remember that the angles must be drawn accurately. Next, letter the spaces between the forces on this diagram A, B, C. Now proceed to draw the *force diagram*, i.e., *the triangle of forces*, as in Fig. 184. Begin with the given force AB and use a suitable scale, in this instance 1 mm = 500 N, i.e., *ab* = 40 mm and represents the 20 kN (20 × 10³N) force.

0·5 kN is represented by 1 mm,

1 kN is represented by 1 mm ÷ 0·5 = 2 mm

and 20 kN is represented by 2 × 20 = 40 mm

Draw *bc* parallel to force BC and *ac* parallel to force CA.

Remember to mark the arrowheads on *ab*, *bc* and *ca*, taking note that they follow each other round as before. This is always so.

Finally, measure *bc* and *ca* and scale them. The results are as follows:
Force in jib = *bc* = 70 mm × 0·5 kN = 35·0 kN
Force in tie = *ca* = 40 mm × 0·5 = 20 kN.

Space Diagram
Fig. 183

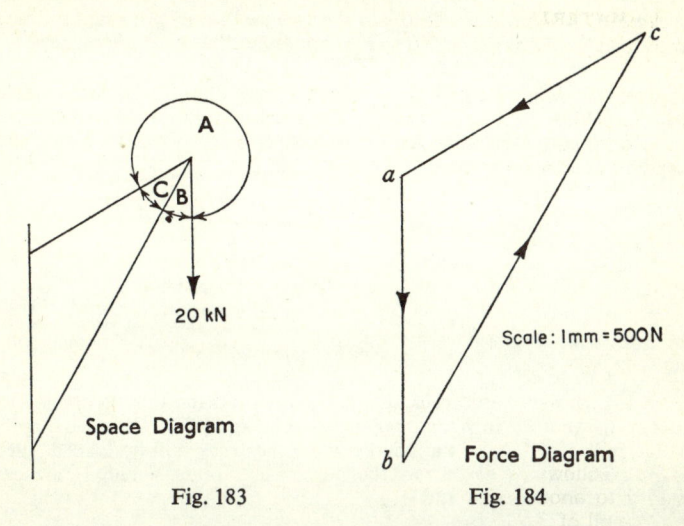

Force Diagram
Scale: 1mm = 500N
Fig. 184

The procedure may be used for any number of co-planar, concurrent forces. If there are more than three forces the force diagram is known as a polygon of forces.

A useful point to note is that if we are given details of two forces and asked to find a third, the resultant of the two, we should use the *parallelogram* method. If we are given details of one force and asked to determine certain information about two others, we should use the *triangle* method.

Exercise 14

Section A

1. The sketch in Fig. 185 shows the supports of a travelling steady with the forces acting on them. Find the *resultant* of these forces and the angle at which it acts in relation to the horizontal support. What is the magnitude of the equilibrant?

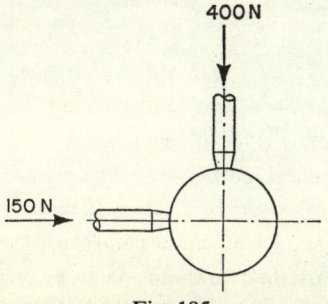

Fig. 185

2. Figure 186 shows a cranked lever which is part of a belt shifting mechanism. Find the resultant force R acting on the hinge pin.

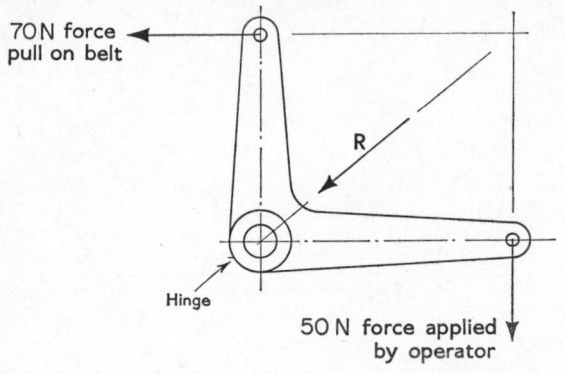

Fig. 186

3. Figure 187 shows a hand truck that is used for transporting components from one bay to another in a maintenance workshop. The workman applies an average pull of 310 N to the handle when the truck is fully loaded. Because this pull is inclined at 25° to the horizontal, forces V and H are caused. Determine the values of V and H.

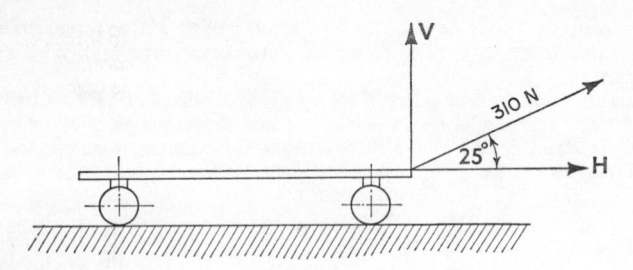

Fig. 187

4. The end elevation of the long slide and part of the carriage of a lathe is shown in Fig. 188 The slide slopes away from the operator at an angle of 30°. The 'weight' of the carriage and tool post is 2·7 kN, and there is a vertical force of 1·5 kN acting on the carriage due to cutting. Determine the magnitudes of the forces F and W.

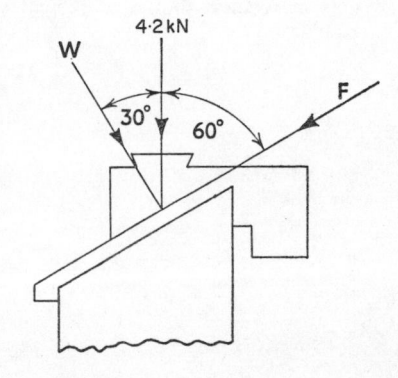

Fig. 188

PARALLELOGRAM AND TRIANGLE OF FORCES 227

5. Figure 189 shows a shaft of 'weight' 1·6 kN, slung in a chain, and supported by a crane hook in a horizontal position. Determine the forces in the two sling chains when the angle between them is 80° and when it is 20°.

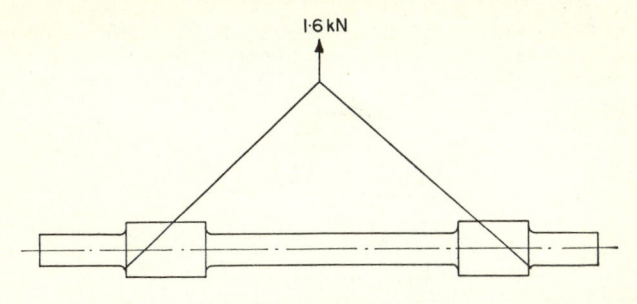

Fig. 189

6. Determine the forces in the jib and tie of the wall crane shown in Fig. 190 using the principle of moments. Check your results graphically.

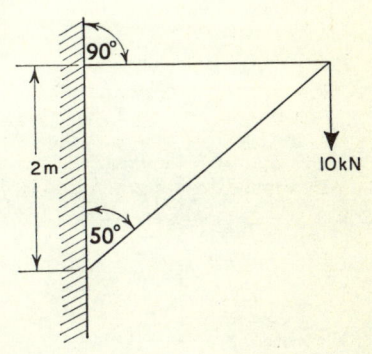

Fig. 190

7. A plan view of a lathe tool cutting a bar of metal is shown in Fig. 191. The feed force is 200 N, and the force maintaining the cut is 40 N. Find the resultant of these two forces.

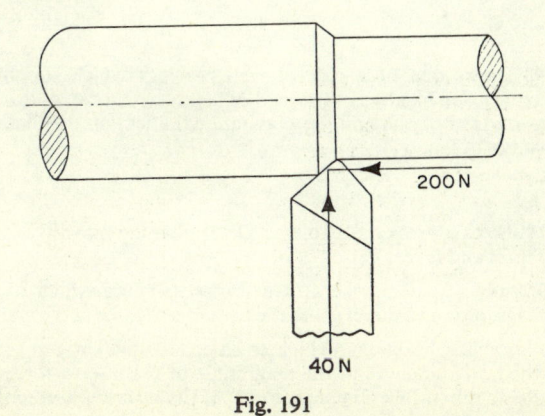

Fig. 191

228   MATERIALS AND MACHINES IN THE WORKSHOP

8. Two wire ropes 1·1 m long and 1·6 m long are suspended from two eye bolts, 1·5 m apart, fixed to a roof joist. The lower ends of the ropes are brought together and a pair of chain blocks hung on them. These blocks are used to lift a load of 6 kN. Determine the force in each rope.

9. In Fig. 192, the swivelling table of a shaping machine is shown set over at an angle of 15°. The 'weight' of the table is 1 kN and there is a downward tool force of 100 N. Find the compressive stress in the leg if its diameter is 25 mm, and the force H.

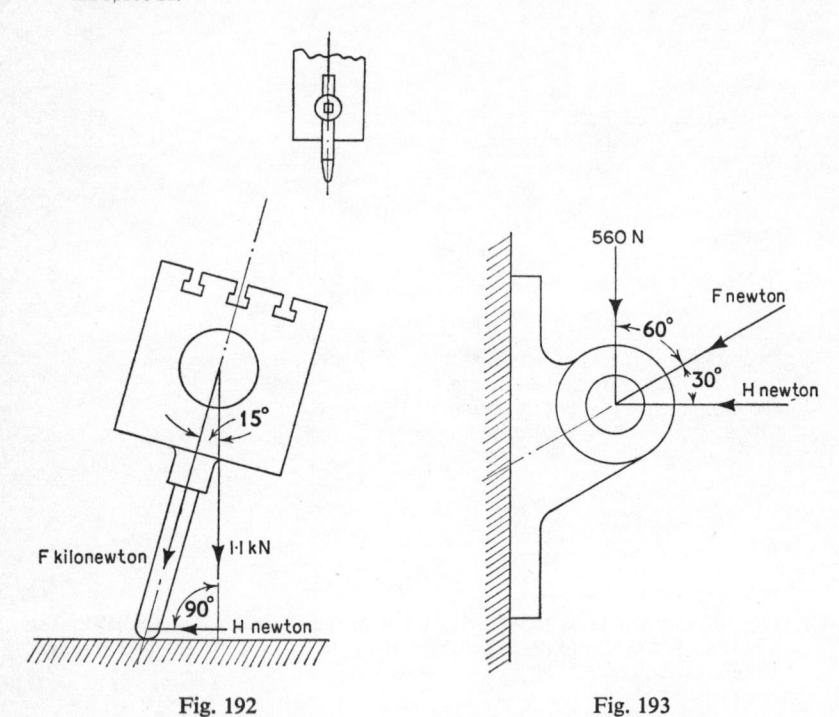

Fig. 192   Fig. 193

10. Figure 193 shows one of a pair of wall bearings which support a countershaft. Each bearing has to carry a load of 560 N as indicated, this being caused by the tension in the belts and the gravitational affect on the mass of the shaft. Find the values of F and H.

Section B

1. Figure 194 shows a press mechanism. Determine the magnitudes of the forces in the two links and force, F.

2. Figure 195 shows two members of a frame structure which meet at a wall. Find the magnitude and direction of the equilibrant force at the wall.

3. A frictionless roller, 'weight' 150 N, is to be pulled up a smooth surface inclined at 33° to the horizontal. Find the magnitude of the pulling force required if it were parallel to the surface; inclined at 12° to the surface; horizontal.

4. A steel ball, 'weight' 60 N, is supported in an unequal vee-groove, one surface of the groove being inclined at 45° to the vertical plane, the other at 30°. What are the magnitudes of the normal reactions at the points where the ball touches the surfaces of the groove?

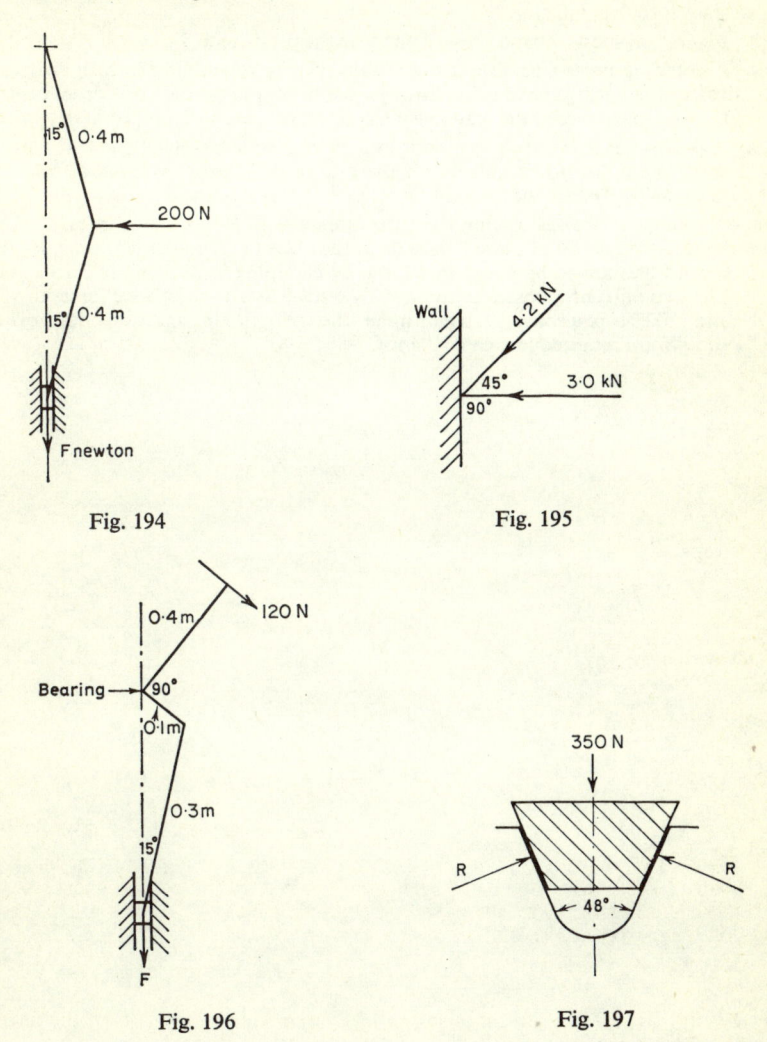

Fig. 194

Fig. 195

Fig. 196

Fig. 197

5. A steel frame made from angle iron is in the form of an isosceles triangle. The base angles are 30°. The frame is in a vertical plane with apex above base, the base member being supported at its ends. The apex carries a vertical load of 30 kN. Find the magnitudes of the forces in each member of the frame.

230   MATERIALS AND MACHINES IN THE WORKSHOP

6. The punch head on a punching machine moves in a vertical plane, and is driven by a connecting rod 0·3 m long which is attached to a crank of radius 90 mm. What is the force on the punch just as it meets the plate to be cut if the crank has moved 60° from its top position in a clockwise direction? The crankshaft torque is 360 Nm.
7. Figure 196 shows a hand press, determine the press force, F.
8. A crane supports a 55 kN load on 2 falls of wire rope each of which is inclined 10° to the vertical centre line through the hook pulley and rope drum centres. By how much would the load in each rope be reduced if the ropes were vertical?
9. A pulley has a 48° vee-groove cut in it to receive a vee-belt. The belt tensions cause the pulley to be pulled into the groove by a force of 350 N as indicated in Fig. 197. Determine the values of the normal reactions, R.
10. A triangular shaped casting has three bosses with a cored hole in each. Lines joining the centres of these bosses form the sides of an equilateral triangle. The cored holes are to be bored in a lathe by clamping the casting to a face plate. The two off-centre bosses cause radially outwards out-of-balance forces of 95 N and 102 N respectively. Determine the magnitude and direction of the equilibrant required to create balance.

Chapter 15. Resolution of Forces

Horizontal and vertical components

We have seen in the previous chapter that when two forces are acting at right-angles to each other, see Figs. 177 and 178, they can be combined into one force known as the *resultant* of the two. Suppose we were given a single force (a resultant) and desired to know the effect of this force in horizontal and vertical directions, e.g., we were given a force F newton acting at an angle θ (theta) to the horizontal plane, see Fig. 198(a), and required to find F_V and F_H, the vertical and horizontal components of F respectively. This could be done by drawing a parallelogram of forces in reverse, see Fig. 198(b).

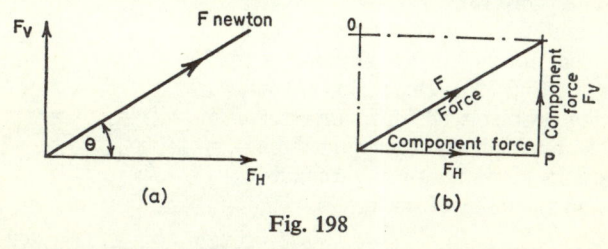

Fig. 198

Having drawn a line to represent the direction of force F, the magnitude of the force would then be stepped off on the line to some suitable scale. This would in fact be the diagonal of the parallelogram. Then by drawing horizontal and vertical lines from either end of this diagonal to meet in points O and P we have obtained the component forces F_V and F_H. It is not really necessary to complete the parallelogram. To construct half of it would be sufficient to give the result required. F_V represents the effect of F in a vertical direction and F_H its effect in a horizontal direction. Their values are obtained by scaling the vertical and horizontal lines meeting in P. We could have found the effect of F in any other directions so long as these were specified. It is usually the vertical and horizontal directions with which we are mostly concerned. Or, to be more exact, the components which act at right-angles to each other. For example, if F itself had been acting in a horizontal direction and we desired to know the values of the component forces, these being at right-angles to each other, we could find them by constructing two lines which are at right-angles to each other, one from each end of the line representing F. We should need to know the direction of one of the components, e.g. angle θ, in Fig. 199.

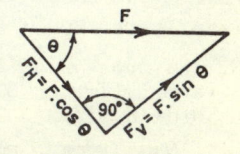

Fig. 199

Actually, when the component forces act at right-angles, it is more convenient to calculate their magnitudes rather than obtain them graphically. This can be done by applying the trigonometry that we have learned in the mathematics lessons. We will find out more about this method. In Fig. 199 we see that F is acting in a horizontal direction. The two lines representing the component forces form the 'opposite' and 'adjacent' sides of a right-angle triangle, F is the hypotenuse. Instead of the sides of the triangle being dimensioned in millimetres they are measured in newtons, i.e. 1 mm represents a specific number of newtons, so we have:

$$\frac{F_H \text{ (Adj.)}}{F \text{ (Hyp.)}} = \cos\theta \quad \text{and} \quad \frac{F_V \text{ (Opp.)}}{F \text{ (Hyp.)}} = \sin\theta$$

$$F_H = F\cos\theta \qquad F_V = F\sin\theta$$

These equations are valid whatever the directions of F and the components, so long as the component forces are at right-angles to each other, see Fig. 200. This theory is useful when solving problems which involve forces acting on inclined planes, e.g. screwthreads, and crank and connecting-rod mechanisms, as is shown by the following examples.

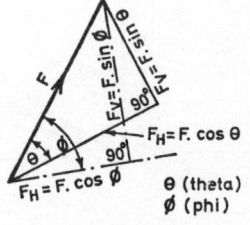

Fig. 200

Examples

1. A force of 100 N acts in a direction which is 42° to the horizontal. Find the magnitudes of its right-angle component forces.

We will first make a sketch of the system, see Fig. 201.

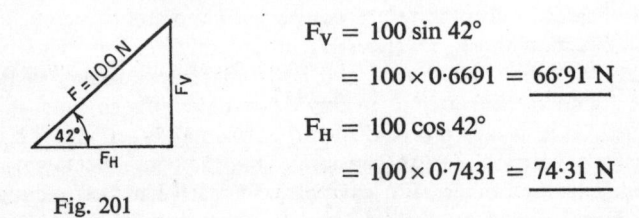

$$F_V = 100 \sin 42°$$
$$= 100 \times 0.6691 = \underline{66.91 \text{ N}}$$

$$F_H = 100 \cos 42°$$
$$= 100 \times 0.7431 = \underline{74.31 \text{ N}}$$

Fig. 201

Thus the effect of the 100 N force in a vertical direction is 66·91 N, and in a horizontal direction 74·31 N.

2. The screw of a small jack is of 'unified' form, i.e. thread angle 60°. It supports a vertical load of 700 N. Find the magnitude of the force which acts normal to the thread face.

Note: 'normal' means at right-angles to the thread face. See Fig. 202.

Let the force acting normal to the thread face be F_N. We are not concerned with the value of the other component force in this problem.

$$\frac{F_N}{700} = \cos 30°$$

$$F_N = 700 \cos 30°$$

$$\log_{10} F_N = \log_{10} 700 + \log_{10} \cos 30°$$

$$= 2·8451 + \bar{1}·9375$$

$$= 2·7826$$

$$F_N = \text{Anti-log } 2·7826$$

$$= \underline{606·1 \text{ N.}}$$

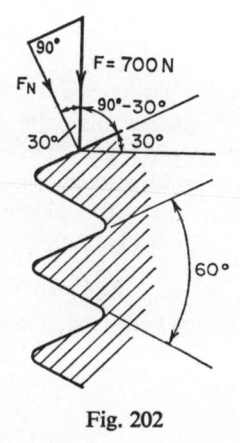

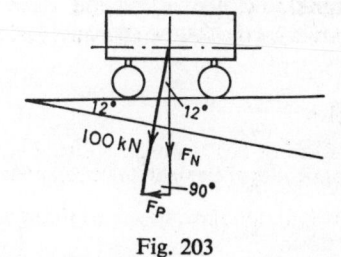

Fig. 202

Fig. 203

Note: For large screw jacks the thread is usually of the *buttress*, or *square* form. Why are vee threads not suitable?

3. A loaded wagon, total 'weight' 100 kN, rests on a track which is inclined at 12° to the horizontal. Calculate the force acting parallel to the track and tending to move the wagon down the track, and also the force acting normal to the track.

We will begin by making a sketch of the system, see Fig. 203.

The total 'weight' of the laden wagon may be taken as acting from its centre of gravity and will be acting vertically downwards (actually towards the earth's centre), e.g. like a plumb-bob. Because the wagon is on an inclined track the load force has component forces, one of which acts normal to the track, F_N, and the other acting parallel to the track, F_P.

To find F_N,

$$\frac{F_N}{100} = \cos 12°$$

$$F_N = 100 \cos 12°$$

$$= 100 \times 0.9781$$

$$F_N = \underline{97.8 \text{ kN}}$$

To find F_P,

$$\frac{F_P}{100} = \sin 12°$$

$$F_P = 100 \sin 12°$$

$$= 100 \times 0.2079$$

$$F_P = \underline{20.8 \text{ N}}$$

Note: The force normal to the plane is much greater than the force parallel to the plane. If the plane were made steeper F_P would increase in magnitude, whilst F_N would decrease. At the two extreme positions $F_P = 100$ kN and $F_N = 0$ kN (track plane vertical) and $F_P = 0$ and $F_N = 100$ kN (track plane horizontal).

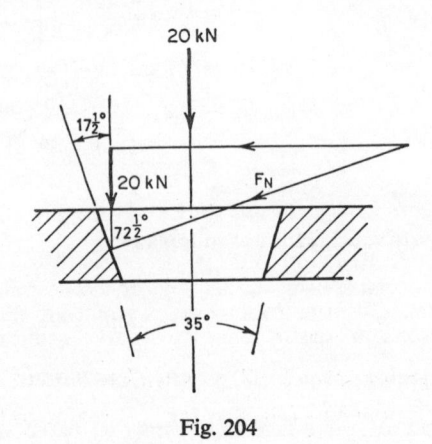

Fig. 204

4. A vertical shaft is supported on a conical thrust bearing of 35° cone angle. The vertical thrust load is 20 kN. What is the bearing load normal to the bearing face? See Fig. 204 for a sketch of the bearing.

RESOLUTION OF FORCES

In this example the given force is in fact a component force, i.e. the vertical component.

$$\frac{20}{F_N} = \cos 72\tfrac{1}{2}°$$

$$20 = F_N \cos 72\tfrac{1}{2}°$$

$$\frac{20}{\cos 72\tfrac{1}{2}°} = F_N$$

$$\log_{10} F_N = \log_{10} 20 - \log_{10} \cos 72\tfrac{1}{2}°$$

$$= 1\cdot 3010 - \bar{1}\cdot 4781$$

$$= 1\cdot 8229$$

$$F_N = \text{Anti-log } 1\cdot 8229$$

$$\underline{F_N = 66\cdot 5 \text{ kN}}$$

5. A loaded mine car 'weighing' 1·2 kN is to be hauled up an incline of 1 in 4 by a wire rope which is horizontal when the pulling force is applied. Calculate the value of the pulling force if the speed of the car is to be constant, see Fig. 205.

The Pulling force P has to counter the tendency of F_P to take the car down the plane. F_P is a component of the force that we require to find, and this force is the reaction to P and therefore equal to P. F_P is also a component of the 1·2 kN force.

$$\frac{F_P}{1\cdot 2} = \sin 14° \, 2'$$

$$F_P = 1\cdot 2 \sin 14° \, 2'$$

$$\log_{10} F_P = \log_{10} 1\cdot 2 + \log_{10} \sin 14° \, 2'$$

$$= 0\cdot 0792 + \bar{1}\cdot 3847$$

$$\log_{10} F_P = \bar{1}\cdot 4639$$

$$\frac{F_P}{\text{Reaction to P}} = \cos 14° \, 2'$$

$$\frac{F_P}{\cos 14° \, 2'} = \text{Reaction to P}$$

$$\log_{10} F_P - \log_{10} \cos 14° \, 2' = \log_{10} \text{Reaction to P}$$

$$\bar{1}\cdot 4639 - \bar{1}\cdot 9868 = \log_{10} \text{Reaction to P}$$

$$\bar{1}\cdot 4771 = \log_{10} \text{Reaction to P}$$

$$P = \text{Anti-log } \bar{1}\cdot 4771$$

$$\underline{P = 0\cdot 3 \text{ kN (300 N)}}$$

Alternative method

If we study Fig. 205 we shall see that the 1·2 kN force and P are 'components' of the force R. Hence,

$$\frac{P}{1 \cdot 2} = \tan 14° 2'$$

$$P = 1 \cdot 2 \tan 14° 2'$$

$$= 1 \cdot 2 \times 10^3 \times 0 \cdot 25 \text{ N}$$

$$\underline{P = 300 \text{ N}}$$

If the pulling force had been parallel to the plane, its value would have been equal to that of F_P.

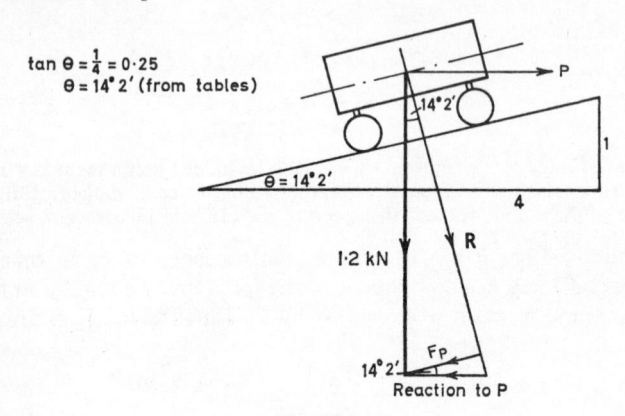

$\tan \theta = \frac{1}{4} = 0 \cdot 25$
$\theta = 14° 2'$ (from tables)

Fig. 205

Exercise 15

Section B

1. A loaded bogey 'weighs' 550 N and is to be hauled up an inclined track, the angle of inclination being 15°. Calculate the pulling force required to move the bogey up the track at constant speed; (i) when it is horizontal, (ii) when it is inclined at 10° to the track.

2. A loaded wagon, mass 5 Mg, is to be hauled up a track, inclined at the rate of 1 in 3½, by a wire rope which is parallel to the incline when pulling takes place. The tractive resistance of the truck is 60 N/Mg mass. What pulling force is required?

3. A gib-head key is tapered on its upper face 10 mm per metre length. It is driven into its keyway on its final fitting by a 2·5 kN force parallel to the key base. What is the magnitude of the force normal to the tapered face? Neglect friction.

4. The tool-head of a small press moves in vertical guides, being actuated by a connecting rod 0·4 m long which is driven by a crank 0·1 m radius. Calculate the vertical and normal forces at the guides when the crank is 30° past its top position on the downward stroke. The torque on the crank shaft is 200 Nm. Neglect friction.

RESOLUTION OF FORCES 237

5. Figure 206 shows a press mechanism. Calculate, for the position shown, the force F on the press head when $P = 3.6$ kN. What are the forces in the two links?

6. Figure 207 shows the driving mechanism for a shaping machine ram. The torque on the crank pin is 70 Nm. Calculate the magnitude of the force F_N which is normal to the rocker arm.

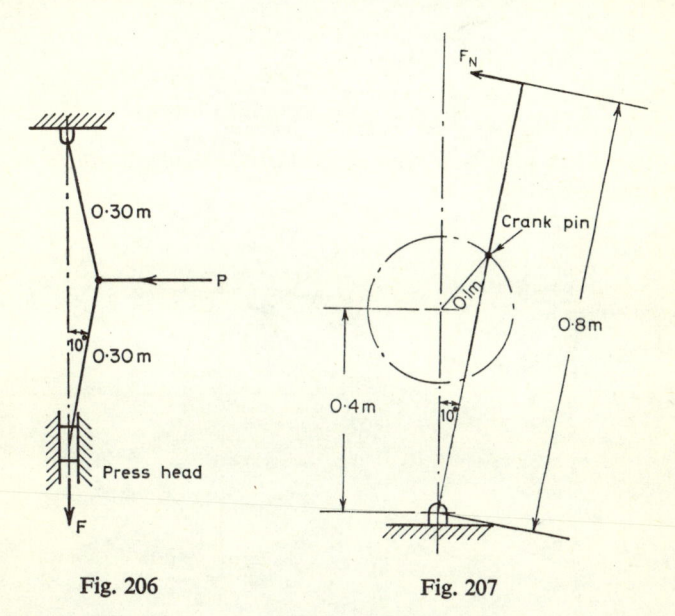

Fig. 206   Fig. 207

7. If the crank pin speed of the shaping machine mechanism shown in Fig. 207 is 4 rev/s, calculate the speed of the ram force F_N, in m/sec.

Note: Tackle this problem in a similar manner as that used when finding the magnitude of F_N, using velocities instead of forces.

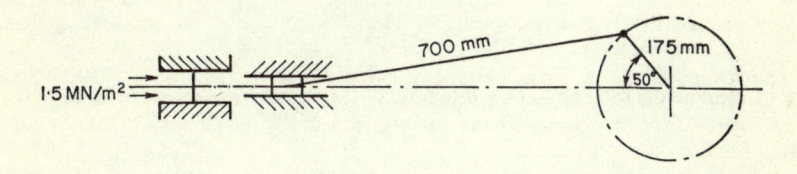

Fig. 208

8. A mechanism for kicking scrap billet ends away from the shear blades is shown in Fig. 208. Compressed air at a pressure of 1.5 MN/m² is supplied to the 200 mm dia. cylinder which, through the connecting rod and crank, operates the shaft to which the kick-off arm is fixed. Calculate the torque on the shaft for the position shown. Neglect friction forces.

238 MATERIALS AND MACHINES IN THE WORKSHOP

9. The mechanism for operating the cutter on a machine tool is shown in Fig. 209. The crank disc is driven at 15 rev/s by a 5 kW motor. Calculate for a crank radius of 125 mm (i) the force, F_N, (ii) the speed at which the tool-head moves in m/s for the position shown.

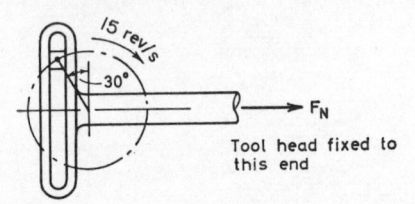

Fig. 209

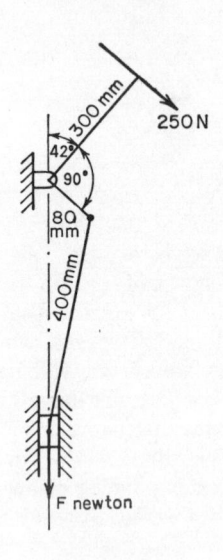

Fig. 210

10. Calculate the force, F, on the tool-head of the hand press shown in Fig. 210.

Chapter 16. Temperature and Heat

There are many industrial processes for which we require *heat*. Some of these are,

1. Heat treatment of metals,
2. Casting,
3. Forging,
4. Riveting,
5. Welding,
6. Soldering,
7. Brazing,
8. Shrinking components together,
9. Moulding Plastic components,
10. Moulding Rubber components, e.g. tyres.

If we are to be competent technicians, then we must understand what *heat* is; what *temperature* is.

Temperature

Temperature is the measure of hotness, or coldness, of a metal, or of any substance. It is *not* a measure of the quantity of heat in the substance, though it has something to do with this. To understand more fully the relationship between temperature and heat we will consider the grinding of a lathe tool on the emery wheel. We know that, as we grind the tool, the particles of metal removed look like sparks, which are at a temperature of about 1200°C. The tool itself, however, will have a mean temperature of little more than 30°C. Those of us who have ground a piece of metal and had one of these sparks settle on our hand know that the burn it makes is of no sigificance. In fact, the sensation is very little more than that caused by a pin prick. This is because the quantity of heat contained by the spark is very small, since its mass is very small. We may understand this a little more fully if we consider what the result would be if the lathe tool were at a temperature of 1200°C. We would burn our hands far more seriously, because there would be a greater quantity of heat in the tool at 1200°C, since its mass is greater, than there is in the spark.

Temperature measurement

We are all familiar with the instrument for measuring temperature. It is the *thermometer*. However, the ordinary glass thermometer is not often used in the workshop, except for measuring bearing temperatures and the temperature of the 'shop' atmosphere. It is fragile, and it cannot measure the high temperatures that are met with, some of which are well above 400°C which is the limit of most ordinary thermometers. The instrument used for measuring higher temperatures is known as a *pyrometer*. There are various types of pyrometers and these are described in the notes on Workshop Processes.

The temperature scale used in the SI system is the Celsius (centigrade) scale. This is shown in Fig. 211. As can be seen, the ice point of water is 0°C and the evaporation (boiling) point is 100° at normal atmospheric pressure of 760 mm of mercury (101·6 kN/m² = 101·6 × 10³ N/m² 1·016 × 10⁵ N/m² ≃ 1 bar).

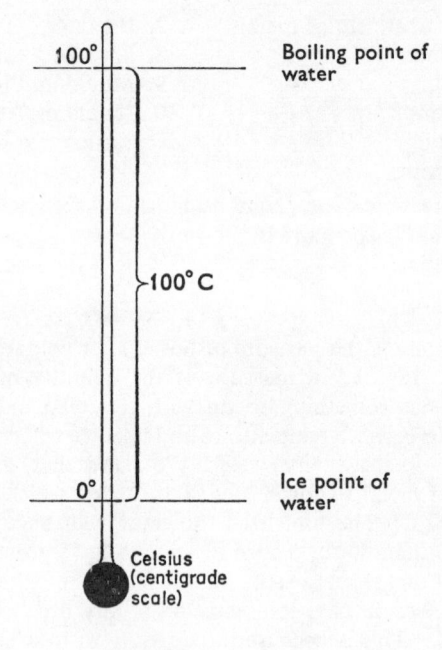

Fig. 211

Heat

Heat is a form of energy and, when we want some of this energy for an industrial process, we obtain it by burning fuel, e.g., coke, coal, gas or oil, or we obtain it from an electricity supply, e.g. heating coil. Some of the heat is lost, usually to the atmosphere, during the combustion, or heating, process.

It has been found useful to divide the effects of heat into two parts, according to the actual result it produces when it is applied to a substance. These parts are

1. The part that causes a *temperature rise,*
2. The part that causes a *change of state.*

To understand this more fully, we will study the following.

TEMPERATURE AND HEAT

When a solid substance, e.g., a piece of metal, is supplied with heat its temperature increases, and continues to increase, until the melting point of the metal is reached. Then the temperature stops rising even though heat is still being supplied. The heat energy is now *changing the state* of the substance, i.e., changing it from a *solid* into a *liquid*. When every particle of the solid metal has changed to liquid, its temperature will again begin to rise and continue to do so until the point at which the liquid changes to a vapour is reached, e.g. the processing of zinc metal from the ore, see page 6. Then again there will be a period of stationary temperature, though energy is still being supplied, until every particle of liquid has changed into vapour. Then the temperature of the substance will again rise and continue to do so whilst heat is supplied.

Sensible heat

The heat that causes the hotness of a substance to increase is known as *sensible* heat. The sense of touch detects a *rise in temperature* and indicates that sensible heat is being supplied. So do thermometers and pyrometers.

Latent heat

The heat that causes a change in state of a substance is known as *latent heat*. There is no increase in the hotness of the substance, i.e., no temperature rise, during a change of state.

From this we deduce that energy, in heat form, is required to change a substance from solid to liquid and liquid to vapour. When the cycle is reversed, e.g. a vapour is cooled and becomes a liquid, then cooled further and becomes a solid, it gives up the *latent heat* that it contains at both stages. The quantity of latent heat required to change the state of a substance differs in value for different substances.

The heat used to dry wet pavements, clothes, etc. is latent heat.

The graph shown in Fig. 211(*a*) illustrates the relationship between *temperature* and *time* as a quantity of tin cools.

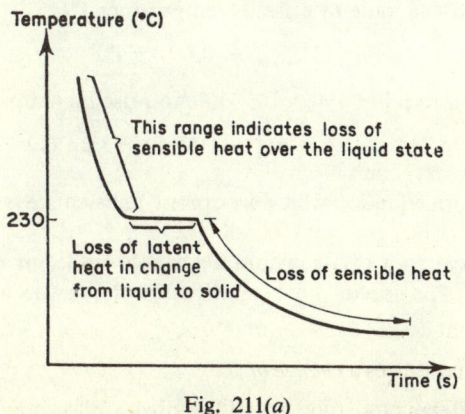

Fig. 211(*a*)

Measurement of heat, heat quantity

The measurement of a *quantity of heat* involves the measuring of two other quantities first. These are, the *mass* of the substance and the *temperature change* caused by the quantity of heat being measured. Having obtained them, they are multiplied and the result multiplied by a quantity known as the *specific heat* capacity.

Unit of Heat Energy

In the SI system of units energy in all its forms is measured in *joules*, or some suitable multiple.

$4 \cdot 187 \times 10^3$ joule is the quantity of energy (heat) required to raise *one kilogramme* of water through *one degree centigrade*.

Specific heat capacity

The *specific heat* capacity of a substance is defined as the amount of heat that is required to raise the temperature of unit mass of a substance by 1°C. Its value differs for different substances. Some *specific heat* capacity values are shown in the following table; the unit of mass being the kilogramme. Specific heat capacity values are measured in joule/kg°C or kilojoule/kg°C.

$$\text{Specific heat capacity (joule/kg°C)} = \frac{\text{Heat quantity (joule)}}{\text{Mass (kg)} \times \text{Temperature change (°C)}}$$

∴ Specific heat capacity, $c_{water} = 4 \cdot 187 \times 10^3$ J/kg°C

1 kg copper requires 398 joule to raise its temperature 1°C,

$$\therefore c_{copper} = 398 \text{ J/kg°C}$$

1 kg steel requires 486 joule to raise its temperature 1°C,

$$\therefore c_{steel} = 486 \text{ J/kg°C}$$

1 kg quenching oil requires $1 \cdot 46 \times 10^3$ joule to raise its temperature 1°C,

$$\therefore c_{oil} = 1 \cdot 46 \times 10^3 \text{ J/kg°C}$$

Let us build up an equation which we can use to calculate *heat quantity*.

Suppose we have 1 kg of water and we supply sufficient *heat* to raise its temperature 1°C. The specific heat capacity of water is $4 \cdot 187 \times 10^3$ joule per kg°C.

Then,

Heat gain by water = 1 (kg) × 1 (°C) × $4 \cdot 187 \times 10^3$ (joule/kg°C).

TEMPERATURE AND HEAT

if we have 6 kg of water heated through 1°C,

Heat gained by water = $6 \times 1 \times 4 \cdot 187 \times 10^3$ joule,

or if we have 21 kg of water (any value could be used) heated through 1°C,

Heat gained by water = $21 \times 1 \times 4 \cdot 187 \times 10^3$ joule.

Heat gained by water = mass of water $\times 1 \times 4 \cdot 187 \times 10^3$ joule.

But, suppose the increase in temperature is 7°C, then

Heat gained by water = mass of water $\times 7 \times 4 \cdot 187 \times 10^3$ joule,

or if temperature rise is 91° (any value could be used) we would have

Heat gained by water = mass of water $\times 91 \times 4 \cdot 187 \times 10^3$ joule.

Heat gained by water = MASS of water × TEMPERATURE CHANGE
$\times 4 \cdot 187 \times 10^3$ J

We can use this formula for any substance, solid or liquid, as long as we remember to use the correct specific heat capacity value, i.e., the specific heat capacity of the substance being dealt with. We can, therefore, write out the formula as follows:

HEAT gained (by any substance)

= MASS (of substance) × TEMPERATURE CHANGE × SPECIFIC HEAT (of substance)

joule = kg × °C × $\dfrac{\text{joule}}{\text{kg°C}}$

Food has a heat energy potential which in future will be measured in joules.

Note: If a substance loses heat, the formula above can be used to determine the loss. Heat contained by a substance is measured from the datum of 0°C. Substances at this temperature have zero heat content. Below this temperature their heat content is negative.

Examples

1. A forging of mass 7 kg is to be annealed by heating it from 20°C to 920°C and allowing it to cool slowly. The specific heat capacity for forging is 460 joule/kg°C. Calculate the quantity of heat that has to be supplied to the forging.

    Begin by writing the formula that we have constructed.

    HEAT gained by forging
    = MASS × TEMPERATURE CHANGE × SPECIFIC HEAT CAPACITY

    Next, we substitute in the formula the values given, noting that the temperature change is (920−20),

    Heat gained by forging = 7(920−20)460
    = 7 × 900 × 460

244   MATERIALS AND MACHINES IN THE WORKSHOP

The *heat gained* will be measured in joule since the mass is measured in kilogramme.

$$\text{Heat gained by forging} = 2898 \times 10^3 \text{J} \ (2\cdot898 \text{ megajoule})$$

2. 10 shafts, of mass 500 kg each and at a temperature of 30°C, are to be charged into a furnace and heated to 930°C, for the purpose of hardening them. If the thermal efficiency of the furnace is 25%, calculate the quantity of heat to be supplied to it, in order to heat the shafts to the required temperature. Specific heat capacity of the shafts is 490 joule/kg°C.

First we must determine what quantity of heat has to be supplied to the shafts.

HEAT SUPPLIED to shafts
= MASS of shafts × TEMPERATURE CHANGE × SPECIFIC HEAT CAPACITY
= 10 × 500(930−30) × 490
= 5 × 10^6 × 9 × 49 = 2205 × 10^6 joule

More heat than this will have to be supplied to the furnace because it is only 25% efficient.

We have already seen that when dealing with *work* and *energy* we used a formula:

$$\text{EFFICIENCY} = \frac{\text{WORK output}}{\text{ENERGY input}} \ (\text{page 144})$$

We can modify this to read,

$$\text{THERMAL EFFICIENCY of furnace} = \frac{\text{HEAT output}}{\text{HEAT input}}$$

In our example the *heat output* is the *heat supplied* to the shafts and the *thermal efficiency* of the furnace is 25%. Therefore we have:

$$\frac{25}{100} = \frac{2205 \times 10^6}{\text{Heat input}}$$

$$\frac{25}{100} \times \text{Heat put into furnace} = 2205 \times 10^6$$

$$\text{Heat put into furnace} = \frac{2205 \times 10^6}{1} \times \frac{100}{25}$$

$$= 8820 \times 10^6 \text{ J } (8820 \text{ megajoule})$$

3. A dozen steel spindles, each of mass 1 kg, are quenched from a temperature of 870°C during a case-hardening process. The temperature of the water into which the spindles are plunged is 20°C and its mass 65 kg. Calculate the final temperature of the water and spindles. The specific heat capacity of the spindle steel is 460 J/kg°C and water 4187 J/kg°C.

Begin by gathering the information we are given.

MASS of water = 65 kg

TOTAL MASS of spindles quenched = 12 × 1 = 12 kg

Initial TEMPERATURE of spindles = 870°C

Initial TEMPERATURE of water = 20°C

TEMPERATURE AND HEAT 245

Let FINAL TEMPERATURE of
spindles and water = $t°C$
SPECIFIC HEAT CAPACITY of spindles = 460 joule/kg°C
SPECIFIC HEAT CAPACITY of water = 4187 joule/kg°C

In a question of this type we must obey the following heat balance equation:

HEAT GAINED = HEAT LOST

In this case the equation becomes:

Heat gained by water = Heat lost by spindles,

Mass of water × Temperature change × Specific heat capacity of water = Mass of spindles × Temperature change × Specific heat capacity of spindles

Now the *temperature change* of the water is $(t-20)$. It is heated up from 20°C to $t°C$. The *temperature change* of the spindles is $(870-t)$, i.e., they are cooled from 870°C to $t°C$.

Therefore we have,

$$65(t-20) \times 4187 = 12(870-t) \times 460$$
$$272 \times 10^3 (t-20) = 5 \cdot 52 \times 10^3 (870-t)$$
$$10^3 (272t - 5440) = 10^3 (4802 - 5 \cdot 52t)$$
$$272t + 5 \cdot 52t = 4802 + 5440$$
$$277 \cdot 5t = 10\,242$$
$$\therefore t = \frac{10\,242}{277 \cdot 5}$$
$$= \underline{37°C}$$

4. A crank shaft of mass 3 megagramme is to be heated to 800°C from a temperature of 30°C in a furnace that is 40% efficient. It is a gas fired furnace and the gas used has a heating value of 12 megajoule per m³. Find the number of cubic metre of gas that must be supplied to the furnace in order to heat the crank shaft to the required temperature. Specific heat capacity of the crank shaft steel is 460 J/kg°C.

First we must calculate the heat supplied to the crank shaft,

HEAT gained by crank shaft = $3 \times 10^3 (800-30) \times 460$
$$= 3 \times 10^3 \times 770 \times 460$$
$$= 1063 \times 10^6 \text{ joule}$$

The quantity of heat supplied to the furnace is greater than this because the furnace wastes some, since its thermal efficiency is only 40%.

We have already learned that:

THERMAL EFFICIENCY of a furnace = $\dfrac{\text{HEAT got out of furnace}}{\text{HEAT put into furnace}}$

In this case, $\dfrac{40}{100} = \dfrac{1063 \times 10^6}{\text{Heat put into furnace}}$

$\dfrac{40}{100} \times$ Heat put into furnace = 1063×10^6

$\therefore$ Heat put into furnace = $\dfrac{1063 \times 10^6}{1} \times \dfrac{100}{40}$

$$= 2660 \times 10^6 \text{ joule}$$

When 1 cubic metre of gas is burnt, 12×10^6 joule are given out. Therefore, if we find how many 12×10^6's there are in 2660×10^6, we shall have obtained the number of cubic metre of gas required.

$$\text{Gas required by furnace} = \frac{2660 \times 10^6}{12 \times 10^6}$$

$$= \underline{222 \text{ m}^3}$$

5. A steel spindle, mass 52 gramme, is heated from 20°C to 750°C. How many joules of heat does it absorb in the process if the specific heat capacity of the steel is 460 joule per kg°C?

$$\text{Heat gained by spindle} = 52 \times 10^{-3}(750-20) \times 460$$
$$= 52 \times 730 \times 0.46$$
$$= \underline{17\,462 \text{ J } (17.5 \text{ kJ})}$$

6. The heating system in a factory comprises a huge 'honeycomb' container, having 2·25 m³ of hot water in it, around which cool air is blown. The temperature of the water falls 15°C. How many megajoule of heat are supplied to the air? One cubic metre of water has a mass 10^3 kg, specific heat capacity of water may be taken as 4200 J/kg°C.

$$\text{Heat supplied to air} = 2 \cdot 25 \times 10^3 \overset{(\text{kg})}{} \times 15 \overset{(°C)}{} \times 4200 \overset{(\text{J/kg°C})}{}$$
$$\text{Heat supplied to air} = 141 \cdot 8 \times 10^6 \text{ J}$$
$$= \underline{142 \text{ MJ}}$$

7. A number of steel components whose total mass is 30 kg are to be quenched at the same time in oil whose temperature must not exceed 45°C for safety reasons. The initial temperature of the oil is 25°C and the components are at a temperature of 750°C, when they are immersed in the oil. The specific heat capacity of the oil is 1·6 kJ/kg°C and that of the steel 0·48 kJ/kg°C. Calculate the least number of litres of oil that must be in the quenching tank if 1 litre of oil has a mass of 0·92 kg.

The final temperature of the oil and the components = 45°C.

First we must calculate the mass of oil required, using the formula:

HEAT GAINED by oil = HEAT LOST by components

Mass oil × temp. change × S.H. = Mass steel × temp. change × S.H.

$$M_{oil}(45-25)1 \cdot 6 \times 10^3 = 30(750-45)0 \cdot 48 \times 10^3$$
$$M_{oil} \times 20 \times 1 \cdot 6 = 30 \times 705 \times 0 \cdot 48$$
$$\text{Mass of oil} = \frac{30 \times 705 \times 0 \cdot 48}{20 \times 1 \cdot 6} = 317 \cdot 3 \text{ kg}$$

$$\text{Volume of 1 kg of oil} = \frac{1}{0 \cdot 92} \text{ litre}$$

$$\text{Volume of 317·3 kg of oil} = \frac{1}{0 \cdot 92} \times 317 \cdot 3$$

$$\text{Quantity of oil required} = \underline{345 \text{ litre}}$$

TEMPERATURE AND HEAT

8. A copper ingot, specific heat capacity 393 J/kg°C and mass 300 kg is heated from 30°C to 720°C for the purpose of forging. Find the heat contained by the ingot before it was put into the furnace and the heat it gained from the furnace.

Heat in ingot before heating = M × temperature above 0°C × S.H.

$$= 300 \times 30 \times 390 \text{ J}$$
$$= \underline{3\cdot 510 \text{ MJ (megajoule)}}$$

Heat gained by ingot from furnace = M × temperature change × S.H.

$$= 300(720-30)393$$
$$= 300 \times 690 \times 393$$
$$= \underline{80\cdot 73 \text{ MJ}}$$

Experiment. *To find a value for the latent heat of fusion (latent heat of melting) of ice.*

This test requires the use of a calorimeter, insulating jacket, centigrade thermometer, tap water, ice and a stirrer. The following readings are to be taken,

$$\text{Mass of empty calorimeter} = M \text{ kg}$$

Mass of calorimeter two thirds filled with cold 'tap water' = M_1 kg

$$\text{Mass of water} = (M_1 - M) \text{ kg}$$

$$\text{Initial temperature of calorimeter and water} = \theta°$$

This initial temperature is to be taken when the calorimeter and water have become adjusted to room conditions. A small piece of melting ice (the temperature of melting ice is 0°C) is to be added to the calorimeter which is in an insulating jacket, and the contents stirred. A close watch on the temperature of the whole is to be maintained. When the ice has completely vanished the temperature of the calorimeter and contents is to be recorded.

$$\text{Final temperature of calorimeter and contents} = \theta_1°$$

$$\text{Mass of calorimeter, water and melted ice} = M_2 \text{ kg}$$

$$\text{Mass of melted ice} = M_3 \text{ kg}$$

Substituting our recorded values in the following equation, we can determine the *latent heat of fusion* of ice.

Latent heat of fusion of ice +
Sensible heat gained by melted ice = Heat lost by calorimeter and water.

$$\begin{array}{l}\text{Latent heat of fusion of ice} + \\ M_3(\theta_1 - 0)\end{array} = \begin{array}{l}M(\theta_1 - \theta) \times \text{S.H. CALORIM.} + \\ (M_1 - M)(\theta_1 - \theta) \times 4187\end{array}$$

$$\text{Latent heat of fusion of ice} = \begin{array}{l}M(\theta_1 + \theta) \times \text{S.H. CALORIM.} + \\ (M_1 - M)(\theta_1 - \theta)\,4187 - M_3(\theta_1 - 0)\end{array}$$

Note: The Specific heat capacity of melted ice and water is 4187 joule per kg°C and of copper 390 joule per kg°C.

$$\frac{\text{(Latent heat of fusion of ice/kg)} \times M_3}{} = \frac{M(\theta_1-\theta) \times \text{S.H. CALORIM.} + (M_1-M)(\theta_1-\theta) - M_3\theta_1}{}$$

Latent heat of fusion of ice/kg

$$= \frac{M(\theta_1-\theta) \times \text{S.H. CALORIM.} + (M_1-M)(\theta_1-\theta)4187 - M_3\theta_1}{M_3}$$

You should obtain a value of 335×10^3 joule per kg°C (335 kJ/kg°C). If your value is on the low side, to what might this be due? What are the possible sources of heat loss, or gain, by the calorimeter and its contents? Give three examples of factory processes in which latent heat of fusion is involved.

Experiment. *To find a value for the latent heat of vaporisation of water at atmospheric pressure.*

The equipment required is as shown in Fig. 212. The following recordings are to be made.

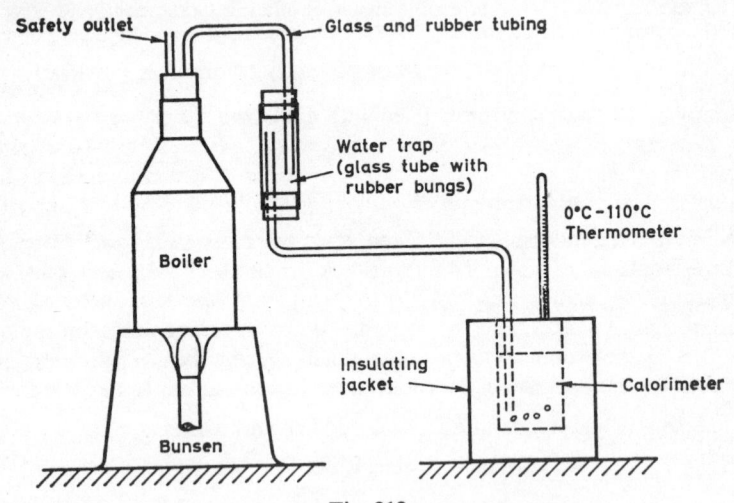

Fig. 212

Mass of empty calorimeter = M kg

Mass of calorimeter two thirds filled with cold 'tap water' = M_1 kg

Mass of water = $(M_1 - M)$ kg

The calorimeter and contents are now placed in an insulating jacket and their temperature taken when they have become adjusted to room conditions.

Initial temperature of calorimeter and contents = θ°C

Steam is now blown into the water until its temperature has risen about 8°C, the final temperature being recorded. The mass of calorimeter and contents is then determined.

Final temperature of calorimeter and contents = $\theta_1 \,°C$

Mass of calorimeter, water and condensate (condensed steam) = M_2 kg

Mass of condensate = M_3 kg

Latent heat of vaporisation of water + Sensible heat lost by condensate
= Heat gained by calorimeter and water

(Latent heat of vaporisation of water/kg) $\times M_3 + M_3(100-\theta_1)4187$

$= M(\theta_1-\theta) \times$ S.H. CALORIM. $+ (M_1-M)(\theta_1-\theta)4187$

Note: The temperature of the steam which is at atmospheric pressure is taken as 100°C, and the specific heat capacities of water and copper are respectively 4187 and 390 J/kg°C.

Latent heat of vaporisation of water/kg, at atmospheric pressure

$$= \frac{M(\theta_1-\theta) \times \text{S.H. CALORIM.} + (M_1-M)(\theta_1-\theta)4187 - M_3(100-\theta_1)4187}{M_3}$$

A suitable value would be 2261×10^3 joule per kg°C (2·261 megajoule/kg°C).

It is important that the test should be carried out in a draught-free atmosphere. Why is this so? Give three examples of industrial processes in which the latent heat of vaporisation of a liquid is involved. Why is it important that the temperature of the water and the calorimeter is not raised to too high a level?

Experiment. *To find the specific heat of a metal.*

Use a number of cubes in turn, e.g. copper, mild steel, etc., of known mass.

Fill a beaker to two thirds of its depth with water and place it on a tripod over a bunsen burner; boil the water. Attach a piece of fine thread to one of the cubes and suspend it in the water being heated.

Whilst this is going on add some cold 'tap water' to a calorimeter, after having determined its mass when empty. Determine the mass of the calorimeter and water, then place the whole in an insulating jacket. After allowing time for adjustment to room conditions, measure the temperature of the calorimeter and contents. The temperature of the cube in the boiling water may be assumed to be 100°C and it should now be removed from the beaker and quickly placed in the calorimeter. Stir the calorimeter contents and at the same time observe the temperature rise. When the temperature reaches its highest level, record this value and the values of masses and temperatures already taken.

Use the values in the equation:

Heat lost by cube = Heat gained by calorimeter and water,

and calculate the specific heat capacity of the cube of metal. How do the values you have obtained compare with accepted standard values? Give three industrial processes in which the specific heat of a metal might be used. Explain how the cooling system of a motor vehicle is dependent on the specific heat capacity of water.

Exercise 16

Section A

1. Calculate the heat content of a block of copper; mass 15 kg, temperature 18°C and specific heat capacity of 390 J/kg°C.

2. How many joules are given up when 35 gramme of steam at 100°C and atmospheric pressure are condensed and then cooled to 18°C? Latent heat of steam at atmospheric pressure is 2·261 MJ/kg°C and specific heat capacity of water 4187 J/kg°C.

3. A steel punch, mass 0·5 kg, is to be annealed. How much heat is required to raise its temperature from 15°C to 850°C? Specific heat capacity of the steel is 486 J/kg°C.

4. A brass forging of mass 7 kg is to be heated from 20°C to 600°C in order to anneal it. If the specific heat capacity of the brass is 392 J/kg°C, calculate how much heat is supplied to the forging.

5. During a case-hardening operation a steel spindle of mass 21 gramme is quenched in water from a temperature of 870°C to 25°C. How many kilojoules of heat are absorbed by the water? Specific heat capacity of steel is 486 J/kg°C.

6. An ingot of zinc of mass 25 kg, temperature 18°C, is to be melted down. Calculate the quantity of heat required, to the melting temperature, if the melting point of the zinc is 420°C and its specific heat capacity is 394 J/kg°C.

7. During a milling operation the cutter, of mass 2 kg, has its temperature increased by 8°C. How many joules of heat does it gain if its specific heat capacity is 482 J/kg°C?

8. To harden the tip of a flat chisel it is heated and then plunged into 2 kg of water at 20°C. The water has to absorb 35·6 kJ from the chisel. Determine the final temperature of the chisel and water. Specific heat capacity of water 4187 J/kg°C.

9. Find the heat content of a spark, from a grinding wheel, that has a mass of 40×10^{-9} kg, temperature 1300°C, and that of a piece of steel bar, 1 kg mass, temperature 40°C. The specific heat capacity of each is 486 J/kg°C.

10. A length, mass 0·2 kg, is cut from a bar of copper, specific heat capacity 390 J/kg°C, by sawing. In the process the temperature of the piece cut off is increased from 15°C to 28°C. Calculate the heat gained by the piece.

Section B

1. A steel billet, mass 1·5 Mg, is heated from 25°C to 1150°C. Calculate the heat contained by the billet before heating, and after heating. Specific heat capacity of steel, 486 J/kg°C.

2. Six steel shafts each of mass 1 Mg are charged into a furnace and heated from 20°C to 800°C. If the thermal efficiency of the furnace is 32%, find the heat supplied in the furnace. Specific heat capacity of the shaft steel is 486 J/kg°C.

3. A component of mass 2 kg, specific heat capacity 486 J/kg°C, is to be hardened by heating it to 890°C and then quenching it in 40 litre of water at 26°C. Calculate the final temperature of the water and the component. One litre of water has a mass of 1 kg and its specific heat capacity is 4187 J/kg°C.

4. During a heat treatment process, a press tool, mass 1 kg, specific heat capacity 486 J/kg°C, is heated from 15°C to 800°C. To do this 0·075 m³ of gas, calorific value 16·8 MJ/m³, is supplied to the furnace in which it is heated. Calculate the thermal efficiency of the furnace.

TEMPERATURE AND HEAT

5. Two litre of coolant, specific heat capacity 3·9 kJ/kg°C, flows every minute during a particular drilling operation and its temperature is increased by 2°C. Find the heat energy carried away by the coolant per s if 1 litre of it has a mass of 1 kg.

6. In an operation that involves the heat treatment of 10 Mg of iron castings, specific heat capacity 486 J/kg°C, the components are heated from 15°C to 845°C. Determine the quantity of heat supplied to the furnace in which the castings are heated if its thermal efficiency is 30%.

7. A steel component, mass 3 kg, is to be quenched in oil. It is essential that the temperature of the oil does not exceed 40°C. The initial temperature of the oil is 15°C, and that of the component 570°C. Determine the quantity of oil, in litre, that must be used. Specific heat capacity of the oil is 1·465 kJ/kg and of the steel 486 J/kg°C, 1 litre of the oil has a mass of 0·94 kg. Give answer in cubic metre.

8. In a cyanide case-hardening process, 3 baskets of components, specific heat capacity 500 J/kg°C, are successively quenched in water. The mass of a basket and its contents is 7 kg. If their temperature before quenching is 910°C and must not exceed 40°C after quenching, how many litre of water are required at an initial temperature of 19°C. One litre of water has a mass of 1 kg and its specific heat capacity is 4187 J/kg°C.

9. An oil fired furnace, 35% thermal efficiency, is used for heating steel billets, each of mass 300 kg, specific heat capacity 486 J/kg°C. Four billets are heated at a time, from 18°C to 900°C. Determine the quantity of oil required if 1 m^3 of it produces 40×10^9 J.

10. What quantity of heat is required to change 1 gramme of ice at -10°C, to steam at 100°C and atmospheric pressure? Plot a graph showing how the heat quantity varies over the temperature rise at 5°-intervals. Specific latent heat of vaporisation of water at atmospheric pressure is 2·261 MJ/kg°C, and of fusion of ice, 335 kJ/kg°C. Specific heat capacity of ice is 2·093 kJ/kg°C and of water 4·187 kJ/kg°C.

11. A locomotive tyre of mass 500 kg is to be shrunk onto its wheel by heating it from 18°C to 750°C. Determine the quantity of heat that must be supplied to the tyre if its specific heat capacity is 486 J/kg°C.

12. An ingot of aluminium, mass 200 kg, is heated from 20°C to 500°C in order to extrude it. Calculate how many joules of heat must be supplied to the ingot if its specific heat capacity is 916 J/kg°C.

Chapter 17. Energy Conversion

ENERGY is in itself a single entity though it may exist in three different forms. Figure 122 on page 141 gives some indication of this.

It may exist in the form of *heat* as in steam supplied to a turbine and fuel burnt in the combustion chamber of a gas turbine, or internal combustion engine.

Energy manifests itself in *mechanical* form at the crankshaft of an engine and a turbine rotor.

A generator (dynamo) produces energy in *electrical* form. An electric motor converts energy from electrical to mechanical form.

All energy is measured in *joules*, or some multiple, e.g. kilojoule (10^3 joule), megajoule (10^6 joule).

The following examples give some indication of the calculations involved in the conversion of energy from one form to another.

The meaning of a Kilowatt-hour

When one kilowatt operates for one hour, one kilowatt-hour of energy is expended. This in terms of joules is as follows:

$$1 \text{ kWh} = 60 \times 60 \times 1000 \text{ joule}$$
$$= 3600 \times 1000 \text{ joule}$$
$$= 3 \cdot 6 \times 10^6 \text{ joule}$$

A kWh is known as a Board of Trade (B.O.T.) unit, and this is the unit used by the people who sell electricity. It is the unit recorded by the meters, and for which we pay at various rates.

Experiment. *To determine the specific heat capacity of water by method of energy conversion.*

Determine the mass of a copper calorimeter, first empty, then half full of water; note the masses. Arrange the calorimeter inside an insulating jacket as shown in Fig. 213, and immerse a heating coil in the water and connect it in series with a switch, 2-volt battery and an ammeter. Connect a voltmeter across the coil. Measure the initial temperature of the water after sufficient time has been allowed for it to adjust itself to the room conditions. Measure it again some little time later as a check. Record the stable value. Close the switch and adjust the rheostat so as to give an ammeter reading of, say, 3 amperes. Allow the switch to remain closed for 5 minutes, ensuring that the ammeter reading remains constant by adjusting the rheostat if necessary. Stir the water during the closure of the switch and take its temperature the instant the switch is opened. Note the voltmeter reading: this gives the p.d. across the coil.

The values recorded should be used in the following equations to determine the value required.

Electrical energy supplied to coil = Volt × Ampere × Time, joule; the time being measured in seconds.

Heat gained by water and calorimeter

$$\begin{array}{cccc} \text{(kg)} & \text{(°C)} & \text{(kg)} & \text{(°C)} \end{array}$$
= (Mass of water × temp. rise × S.H.) + (Mass of calorimeter × temp. rise × S.H.), joule

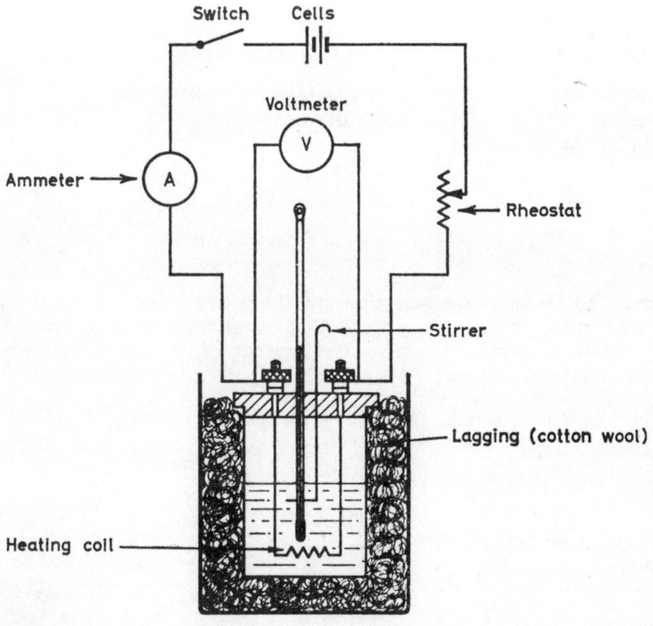

Fig. 213

Assume the specific heat capacity of the copper calorimeter is 390 joule/kg°C.

Combining the two equations we have:

Energy gained by water = Energy supplied by coil

$$M_{water}(\theta_2 - \theta_1)c_{water} + M_{cal}(\theta_2 - \theta_1) \times 390 = E \times I \times t$$

Substitute the recorded values in this equation and determine of value for, c_{water}.

Repeat the experiment a number of times and take an average value for the final result. How does your result compare with the generally accepted value? If it does not agree, what are the possible sources of error?

Examples

1. A steam turbine uses 15 000 kg of steam per hour, the heat energy contained in 1 kg of the steam being 2·8 MJ. Assuming the turbine is 93% efficient, determine the power it develops.

Energy in heat form supplied to turbine per second

$$= \frac{15 \times 10^3 \times 2·8 \times 10^6}{60 \times 60} \text{ joule}$$

93% of this energy is converted into mechanical form.

Energy in mechanical form developed by the turbine per second

$$= 0·011\ 67 \times 10^9 \times 0·93 \text{ joule}$$

Power developed by turbine

$$= 10·8 \times 10^6 \text{ watt (joule/s)}$$

$$= \underline{10·8 \text{ MW (megawatt)}}$$

2. A diesel engine driving a generator uses 15 kg of fuel per hour, calorific value 45 MJ/kg. Of the energy supplied, 63% is lost in heat transfer to the cooling and exhaust systems. Assuming the generator to be 92% efficient, find its output in kilowatt hours.

Energy supplied to engine per hour $= 45 \times 10^6 \times 15$ joule

Of this amount 37% is converted to mechanical form and 92% of the mechanical form is available in electrical form.

$$1 \text{ kWh} = 3·6 \times 10^6 \text{ joule}$$

$$\text{Output of generator} = \frac{45 \times 10^6 \times 15 \times 0·37 \times 0·92}{3·6 \times 10^6}$$

$$= \underline{63·83 \text{ kWh}}$$

3. The power input to an electric motor driving a planing machine is 2 kW and the motor speed is 20 rev/s. If the drive mechanism between the motor and the table is 82% efficient and the speed reduction 20 to 1, calculate the torque available at gear wheel which drives the table.

$$2 \text{ kW} = 2 \times 10^3 \text{ watt}$$

Of this power, 82% is available at the table.

$$\text{Power at table} = 2 \times 10^3 \times 0·82 \text{ watt}$$

$$= 2\pi NT$$

ENERGY CONVERSION 255

where N is speed of table gear wheel and T is the torque at this wheel. Hence,

$$2\pi NT = 2 \times 0\cdot 82 \times 10^3$$

$$T = \frac{1\cdot 64 \times 10^3}{2\pi \times 1}$$

$$T = 0\cdot 2611 \times 10^3 \text{ Nm}$$

$$= \underline{261 \text{ Nm}}$$

4. An electric furnace has a rating of 2·5 kW and a thermal efficiency of 85%. The furnace load is 2·5 kg of steel spindles. Calculate the time taken to heat the spindles from 20°C to 760°C. Specific heat capacity steel 460 J/kg°C.

$$2\cdot 5 \text{ kW} = 2\cdot 5 \times 1\ 000 \text{ watt}$$

$$= 2\ 500 \text{ joule/s}$$

Energy supplied to furnace = 2 500 joule/s

Since thermal efficiency of furnace is 85%,

$$\text{Energy available} = 2\ 500 \times 0\cdot 85$$

$$= 25 \times 85 \text{ joule/s}$$

Energy gained by spindles = 2·5 (760 − 20) 460 joule

If we now divide the value of energy supplied per second into the energy required by the spindles, the result will be the time in seconds to do the operation.

$$\text{Time for operation} = \frac{2\cdot 5 \times 740 \times 460}{25 \times 85}$$

$$= 400 \text{ s}$$

$$= \underline{6 \text{ min } 40 \text{ s}}$$

5. An electric furnace has a rating of 45 kW and its loading capacity is 800 kg. If the furnace charge is steel components of specific heat capacity 460 J/kg°C, determine the temperature rise per minute in °C. Assume a thermal efficiency of 78%.

Electrical power supplied to furnace = 45×10^3 watt

$$= 45 \times 10^3 \text{ joule/s}$$

78% of this is usefully converted to heat.

Energy available at furnace in heat form

$$= 45 \times 10^3 \times 0\cdot 78 \text{ joule/s}$$

Energy required by components/s = 800 × (temp. rise/s) × 460

If we now equate these two expressions we shall obtain an equation for the temperature rise per second.

$$800 \times (\text{temp. rise/s}) \times 460 = 45 \times 10^3 \times 0.78$$

$$\text{Temperature rise per s} = \frac{45 \times 10^3 \times 0.78}{800 \times 460}$$

$$\text{Temperature rise per min} = \frac{45 \times 0.78}{368} \times 60 = \underline{5.7°C}$$

6. In a milling operation 7·6 kilojoule of mechanical energy are converted to heat at the cutter every minute. Assuming that 90% of the heat is carried away by the coolant, specific heat capacity 3·7 kJ/kg°C, calculate the quantity to be supplied in litre per min, if its temperature rise is not to exceed 4°C. One litre of water has a mass of 1 kg. Specific gravity of coolant 0·95.

Energy produced per min in heat form $= 7.6 \times 10^3$ joule

Energy carried away by coolant per min $= 7.6 \times 10^3 \times 0.9$ joule

This may be expressed as follows:

Heat carried away by coolant per min

$$= \text{Mass of coolant/min} \times 4 \times 3.7 \times 10^3 \text{ joule}$$

Equating these two quantities we have,

Mass of coolant/min $\times 4 \times 3.7 \times 10^3$

$$= 7.6 \times 10^3 \times 0.9$$

$$\text{Mass of coolant per min} = \frac{7.6 \times 0.9 \times 10^3}{4 \times 3.7 \times 10^3} \text{ kg}$$

But 1 litre of coolant has a mass of 0·95 kg (specific gravity × 1 kg)

$$\text{Quantity of coolant flowing} = \frac{7.6 \times 0.9}{4 \times 3.7 \times 0.95}$$

$$= \underline{0.49 \text{ litre/min}}$$

To supply 0·5 litre every minute would be satisfactory.

7. On a profiling lathe the equivalent of 0·4 kW is converted to heat at the cutter. Of this amount 92% is carried away by cutting compound which has a specific heat capacity, 3·5 kJ/kg°C. The temperature rise of the coolant is not to exceed 5°C. How many litre per min of the compound must flow if 1 litre has a mass of 0·88 kg?

$$\text{Power converted to heat} = 0.4 \times 10^3 \text{ watt}$$

$$\text{Energy converted to heat} = 0.4 \times 10^3 \text{ joule/s}$$

ENERGY CONVERSION 257

Heat energy produced at cutter per min

$$= 0.4 \times 10^3 \times 60 \text{ joule}$$

Heat carried away by coolant per min $= 0.4 \times 10^3 \times 60 \times 0.92$

This may be expressed as $=$ Mass of coolant/min $\times 5 \times 3.5 \times 10^3$

Equating these expressions,

Mass of coolant per min $\times 5 \times 3.5 \times 10^3$

$$= 24 \times 0.92 \times 10^3$$

$$\text{Mass of coolant flowing per min} = \frac{24 \times 0.92 \times 10^3}{5 \times 3.5 \times 10^3}$$

1 litre of coolant has mass 0·88 kg

$$\text{Quantity of coolant flowing} = \frac{24 \times 0.92}{17.5 \times 0.88}$$

$$= \underline{1.4 \text{ litre/min}}$$

8. A salt bath has to be capable of dealing with three batches of work, mass per batch 8 kg, per hour. The components enter the bath at 20°C and leave it at 760°C. Calculate the power supply to the furnace, and the cost per batch if the power is supplied at 4 pence per unit. Specific heat capacity of components is 460 J/kg°C.

Energy absorbed by 1 batch $= 8(760-20) \times 460$

$$= 8 \times 740 \times 460 \text{ joule}$$

Energy per hour $= 8 \times 740 \times 460 \times 3$

1 joule/s $=$ 1 watt,

1 joule per h $=$ 3 600 watt

$$\text{Power supply to furnace} = \frac{8 \times 740 \times 460 \times 3}{3\,600}$$

$$= \underline{2\,270 \text{ W } (2.27 \text{ kW})}$$

In 1 hour 2.27×1 kWh of energy are used and 3 batches are heated.

$$\text{Units of power used per batch} = \frac{2.27}{3} \text{ kWh}$$

$$\text{Cost per batch} = \frac{2.27}{3} \times 4$$

$$= \underline{3 \text{ pence}}$$

258   MATERIALS AND MACHINES IN THE WORKSHOP

9. Suppose 0·2 kW is absorbed in overcoming friction at the slides of a planing machine table and is converted into energy in heat form. How many units of heat are produced per min? If the drive to the table is 87% efficient, what power allowance must be made at the motor to cover the loss?

$$\text{Energy absorbed by friction} = 0{\cdot}2 \times 10^3 \text{ joule/s}$$

$$= 0{\cdot}2 \times 10^3 \times 60 \text{ J/min}$$

$$= \underline{12 \text{ kJ/min}}$$

$$\text{Power lost at slides} = 0{\cdot}2 \times 10^3 \text{ watt}$$

Since efficiency of drive is 87% more power then $0{\cdot}2 \times 10^3$ watt will be required at the motor.

$$\text{Power at motor to cover slide losses} = \frac{0{\cdot}2 \times 10^3}{0{\cdot}87}$$

$$= \underline{230 \text{ watt}}$$

10. Suppose 1 kW is absorbed in bearing friction on a turbine and it is all converted into heat, 85% of it being carried away by the oil supply. The specific heat capacity of the oil is 1·47 kJ/kg°C and its temperature rise must not exceed 6°C. If the specific gravity of the oil is 0·82, calculate the oil supply required in litre per min. 1 litre of water has a mass of 1kg.

$$\text{Energy lost to friction} = 1 \times 10^3 \text{ joule/s}$$

$$= 1 \times 10^3 \times 60 \text{ J/min}$$

Of this energy, 85% is carried away by the lubricating oil.

$$\text{Heat to oil} = 10^3 \times 60 \times 0{\cdot}85 \text{ joule/min}$$

$$\text{Heat absorbed by oil} = \text{Mass of oil flowing/min} \times 6 \times 1{\cdot}47 \times 10^3$$

Equating these two expressions,

$$\text{Mass of oil flowing per min} \times 6 \times 1{\cdot}47 \times 10^3$$

$$= 10^3 \times 60 \times 0{\cdot}85$$

$$\text{Mass of oil flowing per min} = \frac{10^3 \times 51}{8{\cdot}82 \times 10^3} \text{ kg}$$

1 litre of oil has a mass $1 \times 0{\cdot}82$ kg

$$\text{Quantity of oil flowing} = \frac{51}{8{\cdot}82 \times 0{\cdot}82} \text{ litre/min}$$

$$= \underline{7{\cdot}05 \text{ litre/min}}$$

ENERGY CONVERSION

11. Suppose 120 kg of steel, specific heat capacity 460 J/kg°C, melting point 1500°C, are to be melted down in an electric furnace. Calculate the energy supply, in kWh, to the furnace if the steel enters the furnace at 20°C. Thermal efficiency of furnace is 76%. Assume that the latent heat of fusion for steel is 226 kJ/kg.

Heat to melt steel = Sensible heat from 20° to 1 500°C + Latent heat of fusion

$$\text{Heat required to melt steel} = 120(1\,500 - 20) \times 460 + 120 \times 226 \times 10^3 \text{ joule}$$
$$= (120 \times 1\,480 \times 460) + (120 \times 226 \times 10^3)$$
$$= 120 \times 10^3(680 \cdot 8 + 226) \text{ joule}$$
$$= 120 \times 10^3 \times 906 \cdot 8 \text{ J}$$

$1 \text{ kWh} = 3 \cdot 6 \times 10^6 \text{ joule}$

$$\text{Energy in electrical form} = \frac{120 \times 906 \cdot 8 \times 10^3}{3 \cdot 6 \times 10^6} \text{ kWh}$$

Since the furnace efficiency is 76%, more energy than this will be required

$$\text{Energy supply} = \frac{108 \cdot 8}{3 \cdot 6} \times \frac{100}{76}$$
$$= \underline{40 \text{ kWh.}}$$

Exercise 17

Section A

1. 1·5 Mg of brass, specific heat capacity 393 J/kg°C, melting point 930°C, latent heat capacity 160 kJ/kg, are to be melted down in an electric furnace, thermal efficiency 81%. If the brass is charged into the furnace at 25°C, determine the energy supply to the furnace, in kWh.
2. A gas turbine uses 10 litre of fuel, calorific value 31 MJ/litre, per hour. Of the heat generated 62% is lost to the cooling system and exhaust gases. Calculate the power developed by the turbine.
3. A component, specific heat capacity 460 J/kg°C, temperature 22°C and mass 7 kg, is heated in an electric furnace to 790°C. Calculate the energy supply to the furnace if its thermal efficiency is 75%.
4. 200 kg of aluminium, specific heat capacity 920 J/kg°C, is to be melted down in a furnace of 20 kW rating 85% efficient. Calculate the temperature rise of the metal °C per minute.
5. A diesel engine driven generator is required to supply 15 kW. What power engine is required if the generator is 94% efficient? Determine the fuel supply to the engine in litre per minute if its thermal efficiency is 32% and the calorific value of the fuel, 45 MJ per litre.
6. The tool load on a lathe is 2·7 kN and the cutting speed 0·6 m/s. Gear transmission efficiency is 95%. Calculate the power supply to the motor driving the lathe.

7. The heating coil in a salt bath furnace is rated at 20 kW. Components of specific heat capacity 460 J/kg°C, are loaded in batches of mass 48 kg at a temperature 27°C, and withdrawn at 780°C. Calculate the time cycle for the operation. Assume a thermal efficiency of 75%.
8. On a large machine tool 1 kW is allowed on the motor power supply to cover frictional losses. If 93% of this power is converted to heat, how many joule are produced per minute?
9. In a factory heating system 1 200 m^3 of air, specific heat capacity 962 J/kg°C, at a temperature of 28°C are supplied per hour. The air is heated electrically from 21°C. If electricity costs 3 pence per unit and 1 m^3 of air has a mass of 1·13 kg, what is the cost of heating the air?
10. The rating of a press motor is 18 kW. The press head speed is to be 0·1 m/s. Assuming an efficiency of 87% between the motor and the head, find the average force available at the head.

Section B
1. 22 500 kg/h of steam, heat content 3·85 MJ/kg, are supplied to a turbine whose thermal efficiency is 67%. The turbine drives a generator having an efficiency of 93%. What is the output of the generator in kW?
2. A punching machine punches 25 mm dia. holes in mild steel plate 10 mm thick at the rate of 27 per min. The maximum shear strength of the plate is 350 N/mm^2. Calculate the kW rating of a suitable motor assuming an efficiency of 91% for the drive. The mean shear force is 90% of the maximum value.
3. A lathe motor is taking 1½ kW during a cutting operation. The efficiency of the drive is 92%. Of the tool point power 90% is converted to heat and carried away by the coolant of specific heat capacity 3·5 kJ/kg°C and relative density 0·9. If the coolant is supplied at the rate of 3 litre per minute, by how much does its temperature rise?
4. An electric heat-treatment furnace has to cope with 700 kg per hour of brass forgings, specific heat capacity 393 J/kg°C, and heat them from 25°C to 640°C. Calculate the furnace rating if its thermal efficiency is 76%.
5. A planing machine is driven by a 20 kW motor. Of this power, 92% is available at the table for doing work. If 10% of the table power is absorbed by slide friction, what is the maximum energy available for metal cutting?
6. A pump, 92% efficient, is to deliver 45 m^3 of water per minute to a height of 50 m. Calculate the power rating of a motor suitable to drive the pump. The gravitational pull on 1 m^3 of water is 10^3 × 9·81 newton.
7. A lift 'weight' 4 500 N is hoisted 40 m in 22 s by an electric motor through ropes 'weighing' 5 N/m length. If the drive efficiency 90% find the kW rating of suitable motor to drive the lift with uniform velocity.
8. 27 × 10^3 m^3 of water flow, from a vertical height of 60 m, every hour through a water turbine 76% efficient. The turbine drives a generator 91% efficient. Find the kW output of the generator. One cubic metre of water has a mass of 10^3 kg and the gravitational pull on 1 kg is 9·81 newton.
9. Calculate the quantity of steam, heat content 3·5 MJ/kg, required per hour by a turbine, 67% efficiency, which drives a generator, out power 3 000 kW and 93% efficient.
10. (a) What energy is expended when a 50 kg mass is lifted vertically, at uniform speed, through 25 m?
    (b) How much work is done when a 50 kg metal block is drawn 25 m along a slide, inclined at $\pi/6$ radian, with uniform velocity against a frictional resistance of 2 N per kg?

Chapter 18. Expansion of Solids, Liquids and Gases

SOLIDS, liquids and gases expand when they are heated and contract when they are cooled. This fact can be clearly shown in the laboratory.

Experiment. *To show that a metal expands when heated.*

Use one of the standard pieces of equipment, preferably the one which involves the shearing of a cast iron pin as the bar expands and contracts.

Experiment. *To show that a liquid expands when heated.*

Fill a bulb flask with coloured water and insert a rubber bung, through which a glass tube passes, to the neck of the flask. Heat the filled flask over a bunsen flame and watch the water rise up the bore of the tube.

What happens when the water is cooled?

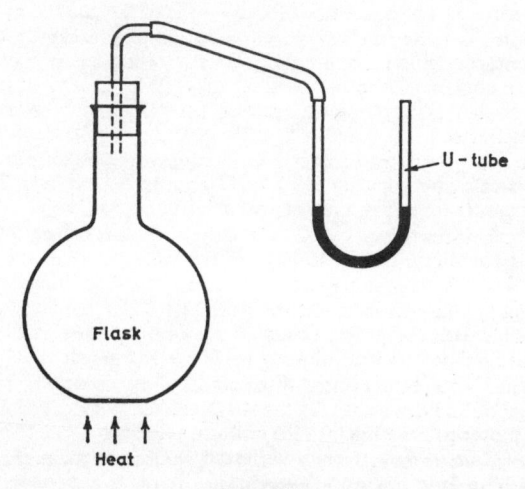

Fig. 214

Experiment. *To show that a gas expands when heated.*

Insert a rubber bung, which has a glass tube passing through it, into the mouth of a bulb flask containing air. By means of rubber tubing connect the glass tube to a U-tube containing coloured water, see Fig. 214. Heat the flask over a bunsen flame and observe the levels of the liquid in the U-tube.

What deductions do you make after observations of U-tube?
What happens to the air in the flask when flask is cooled?

The expansion of air provides the driving force that pushes the piston down the cylinder in internal combustion engines. In fact the engines are air engines. The petrol, or diesel oil, is simply a fuel which on combustion provides the heat to expand the air.

Important as the foregoing is, it is with the expansion of solids, e.g. metals, with which we are chiefly concerned at this stage, for it is with metals in one form or another, which we use or work on.

Different metals expand, or contract, different amounts on a specific length for a given temperature change; e.g. aluminium expands at a greater rate than cast iron. Hence the clearance between the aluminium piston in a motor vehicle engine, and the cast iron cylinder in which it works, will be greater when the engine is cold than it is at its working temperature. We must pay careful attention to the *running clearance* between a steel shaft and its brass, or bronze, bearings on a hot-water pump.

Though expansion takes place on all dimensions of a component, i.e. it is three-dimensional, we are chiefly concerned with the expansion in the *linear* sense, rather than cubical, e.g. the expansion that takes place on a length, or diameter, rather than volumetric expansion.

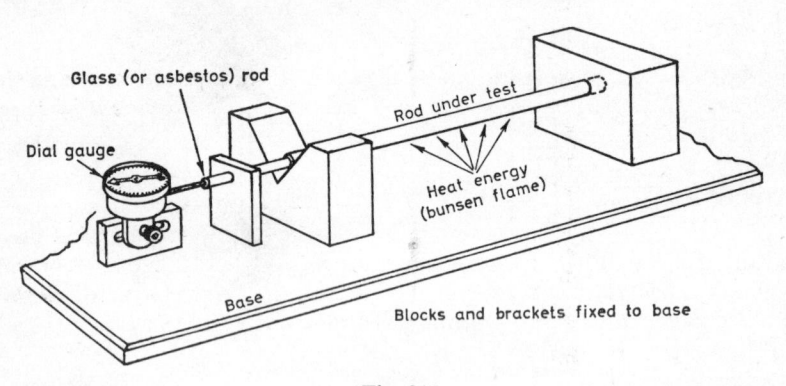

Fig. 215

Experiment. *To show that rods of different metals, having the same length, expand different amounts for a given temperature rise.*

Place in turn, rods of cast iron, brass and aluminium, etc. in the apparatus shown in Fig. 215; apply heat for a specific length of time and observe the dial gauge.

What does the dial gauge indicate as the rods cool?

The expansion of a metal is put to use in shrinking operations, e.g. shrinking a locomotive tyre on to its wheel, or a crank pin into the crank web. The tyre is heated until the bore becomes large enough to go on to the wheel. It is then contracted on to the wheel by allowing it to cool. In the case of the crank pin, if it is small, it may be contracted to fit the hole by refrigeration.

EXPANSION OF SOLIDS, LIQUIDS AND GASES

For large pins, the holes are usually expanded by heating the metal around the holes.

If we could calculate the amount of expansion on a specific length for a given temperature rise, it would assist us greatly in the practical applications of expansion and contraction.

Suppose we have a bar of metal 1 unit long, and its temperature is increased by 1°C (see Fig. 216). It does not matter for the time being what type of metal it is, or what the unit of length is.

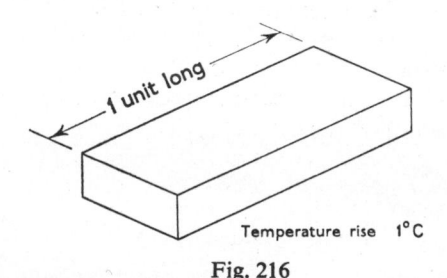

Fig. 216

For this 1°C rise in temperature, there will be a definite increase in the *length* of the bar, and this increase is known as the *coefficient of linear expansion* of the metal.

Coefficient of linear expansion

This is the amount of expansion, or increase in length, that takes place when a bar 1 unit long has its temperature raised by 1°C. We must bear in mind that different metals expand at different rates, i.e. the increase in length per degree rise in temperature is different for the various metals.

For steel the *coefficient of linear expansion* is 0·000 011 9/°C

For copper the value is 0·000 016 7/°C

The unit in which the expansion of a component is measured may be the millimeter, or metre, according to whether the length on which the expansion is taking place is in millimetre or metre.

$$\text{Coefficient of linear expansion} = \frac{\text{Expansion (mm)}}{\text{Original length (mm)}} \text{ (a ratio) per °C}$$

The units cancel.

Returning to the bar shown in Fig. 216. Let us suppose it is 2 units long, but still has its temperature raised 1°C, then the amount of expansion is double what it was before. If its length is 27 units, and the temperature rise remains at 1°C, the amount of expansion is 27 times as much as it was at first, i.e., 27 times the coefficient of linear expansion. What we are really saying is,

AMOUNT of EXPANSION
$= 27 \times 1 \times$ COEFFICIENT OF LINEAR EXPANSION of the bar.

Suppose that the length of the bar is 1 unit but, instead of the temperature rise being 1°C, it is 2°C. Then the amount of expansion is twice as much as it was for 1°C temperature rise. If the temperature increase is 57°C, and the original length remains at 1 unit, the amount of expansion is 57 times what it would be for a 1°C rise, i.e., 57 times the coefficient of linear expansion. If we state this in formula fashion, what we are saying can be written thus:

AMOUNT OF EXPANSION
$= 1 \times 57 \times$ COEFFICIENT OF LINEAR EXPANSION

If the length of the bar is 7 units and the temperature rise 9°C (any numerical values could be used),

AMOUNT OF EXPANSION on bar
$= 7 \times 9 \times$ COEFFICIENT OF LINEAR EXPANSION

We can now state that the equation for giving the *amount of linear expansion* is:

EXPANSION
$=$ ORIGINAL LENGTH $\times$ TEMP. CHANGE $\times$ COEFF. LINEAR EXPANSION

Or by similar reasoning:

CONTRACTION
$=$ ORIGINAL LENGTH $\times$ TEMP. CHANGE $\times$ COEFF. LINEAR EXPANSION

Experiment. Use one of the standard pieces of apparatus fitted with a micrometer, or sphereometer, attachment to determine the coefficient of linear expansion of steel, brass, aluminium, etc. Alternatively the apparatus given in Fig. 215 could be used.

Examples

1. A 300 mm steel rule is guaranteed correct at a temperature of 20°C. What would be the error on a 300 mm length if the rule were used at a temperature of 40°C? Give the answer correct to two significant figures. Coefficient of linear expansion for the rule is $11 \cdot 9 \times 10^{-6}$/°C.

    The error will be caused by the expansion of the metal. Thus,

    Expansion on rule $=$ Original length $\times$ temperature change
    $\times$ coefficient of linear expansion

    $= 300(40-20) \times 11 \cdot 9 \times 10^{-6}$

    $= 300 \times 20 \times 11 \cdot 9 \times 10^{-6}$

    $= 0 \cdot 071\ 40$

    $= 0 \cdot 071$ mm correct to two significant figures.

EXPANSION OF SOLIDS, LIQUIDS AND GASES 265

2. An aluminium piston is turned exactly 100 mm dia. at a temperature of 22°C. When it is working in its cylinder, its average temperature is 322°C. Calculate the piston dia. at this higher temperature, giving the value correct to five significant figures. Coefficient of linear expansion for aluminium is 0·000 025 5/°C.

In problems where we are given a diameter, we take it as the original length for the purpose of calculating the amount of expansion.

We first find how much the piston diameter expands. Then, by adding this to the piston diameter, at 22°C we have the new piston diameter, i.e., its diameter at 322°C.

$$\text{EXPANSION on piston dia.} = \text{Original dia.} \times \text{temperature change} \\ \times \text{coefficient of linear expansion}$$
$$= 100(322-22)25\cdot5 \times 10^{-6}$$
$$= 100 \times 300 \times 25\cdot5 \times 10^{-6}$$
$$= 0\cdot765 \text{ mm}$$

Hence, the piston dia. at 322°C is $100 + 0\cdot765$
$$= 100\cdot77 \text{ to five sig. fig.}$$

3. A collar which has a 75·00 mm dia. bore at 20°C is to be shrunk on to a shaft which is 75·15 mm at the same temperature. The collar is to be expanded by heating, until it will go on to the shaft leaving a 0·05 mm clearance all round between the shaft and the collar bore. Calculate the temperature to which the collar must be heated. Coefficient of linear expansion for the collar material is 0·000 011 9/°C.

Begin by making a note of the information given.

Original diameter of collar = 75·00 mm
Original diameter of shaft = 75·15 mm

Difference between shaft and collar diameter = 0·15 mm interference.

Now, the collar bore has to be expanded until there is a 0·05 mm clearance all round between it and the shaft, i.e., 0·10 mm on the diameter.

Amount collar bore must be expanded = $0\cdot15 + 0\cdot10$
$$= 0\cdot25 \text{ mm.}$$

Remember, in this type of problem we use the *original diameter* as the original length. Therefore our formula becomes,

$$\text{EXPANSION on collar diameter} = \text{Original dia.} \times \text{temperature change} \times \\ \text{coefficient of linear expansion}$$
$$0\cdot25 = 75\cdot00 \times \text{temperature change} \times 0\cdot000\,011\,9$$

Let the temperature to which the collar has to be heated be θ°C.

Temperature change = $(\theta - 20)$
$$0\cdot25 = 75\cdot00(\theta - 20) \times 0\cdot000\,011\,9$$
$$\frac{0\cdot25}{75 \times 11\cdot9 \times 10^{-6}} = \theta - 20$$
$$280\cdot1 = \theta - 20$$
$$280\cdot1 + 20 = \theta$$
$$300\cdot1 = \theta$$

We should have to heat the collar until its temperature was at least 300°C in order to obtain the required conditions.

266   MATERIALS AND MACHINES IN THE WORKSHOP

4. An aluminium piston is turned to 87·09 mm dia., so that it makes a running fit with a cast iron cylinder of 87·50 mm dia. Both dimensions are taken at 20°C. The working temperature of piston and cylinder is 320°C. Calculate the clearance, to the nearest 0·01 mm, between the piston and the cylinder at 20°C and the clearance between the two at the working temperature. Coefficient of linear expansion for aluminium is 0·000 025 5/°C and for cast iron 0·000 010 2/°C.

$$\text{CLEARANCE at } 20°C = 87·50 - 87·09 = 0·41 \text{ mm}$$
$$\text{EXPANSION on piston} = 87·09(320-20) \times 25·5 \times 10^{-6}$$
$$= 87·09 \times 300 \times 25·5 \times 10^{-6}$$
$$= 0·666\,2 \text{ mm}$$
$$\text{Piston dia. at } 320°C = 87·09 + 0·666 = 87·756 \text{ mm}$$
$$\text{EXPANSION on cylinder} = 87·5(320-20) \times 10·2 \times 10^{-6}$$
$$= 87·5 \times 300 \times 10·2 \times 10^{-6}$$
$$= 0·267\,7 \text{ mm}$$
$$\text{Cylinder dia. at } 320°C = 87·5 + 0·268 = 87·768 \text{ mm}$$
$$\text{CLEARANCE at } 320°C = 87·768 - 87·756$$
$$= 0·012 \text{ mm}$$
$$= \underline{0·01 \text{ mm}} \text{ to the nearest } \tfrac{1}{100} \text{ mm.}$$

We have to study our workshop technology notes to check whether or not this clearance would be great enough to provide a satisfactory running fit. We can see from this type of problem the usefulness of the 'Expansion Formula' that we have constructed.

5. A collar of bore diameter 100·00 mm is to be fitted to a shaft 100·15 dia. by heating the collar to 200°C and then cooling the shaft by refrigeration until its diameter is 0·100 mm smaller than that of the collar bore when its temperature is 200°C. If the initial dimensions are taken when the collar and shaft are at a temperature of 22°C, find to what temperature the shaft must be cooled. Coefficient of linear expansion for collar and shaft is 0·000 012/°C.

First we determine how much the bore diameter of the collar expands when it is heated from 22°C to 200°C.

$$\text{EXPANSION of collar bore} = 100(200-22) \times 12 \times 10^{-6}$$
$$= 100 \times 178 \times 12 \times 10^{-6}$$
$$= 0·213\,6 \text{ mm}$$
$$\text{Diameter of collar bore at } 200°C = 100 + 0·213\,6$$
$$= 100·213\,6 \text{ mm}$$

The shaft has to be cooled until its diameter is 0·100 mm smaller than 100·214 mm, i.e., the shaft has to be cooled until its diameter is,

$$100·214 - 0·100 = 100·114$$

As the initial diameter of the shaft is 100·150 mm, it would have to be contracted from 100·150 to 100·114 mm, i.e., it will have to be contracted 0·036 mm. Hence,

$$\text{CONTRACTION of shaft} = \text{Original dia.} \times \text{temperature change} \times$$
$$\text{coefficient of linear expansion}$$
$$= 100·15(22-\theta) \times 12 \times 10^{-6}$$

EXPANSION OF SOLIDS, LIQUIDS AND GASES

where $\theta°C$ is the temperature to which the shaft has to be cooled and has some value less than 22°C.

Then,
$$0.036 = 100 \cdot 15(22-\theta) \times 12 \times 10^{-6}$$

$$\frac{0.036}{100 \cdot 15 \times 12 \times 10^{-6}} = 22-\theta$$

$$30 = 22-\theta$$

$$30-22 = -\theta$$

Multiplying both sides of this equation by -1, we get,

$$-8 = \theta$$

$$\theta = \underline{-8°C}$$

Therefore, the shaft has to be cooled until its temperature is 8°C below zero.

6. A steel bar 250 mm long and a copper bar 249·5 mm long at 15°C, are to be expanded by heating. If the coefficient of linear expansion of the steel is $11·9 \times 10^{-6}/°C$ and that of copper is $16·7 \times 10^{-6}/°C$, at what temperature will both be the same length, and what will this length be?

Let us think what will happen when these bars are heated. For every 1°C rise in temperature the copper bar expands more than the steel bar by the amount,

$$(249·5 \times 16·7 \times 10^{-6}) - (250 \times 11·9 \times 10^{-6})$$
$$= 10^{-6}(4\,167 - 2\,975)$$
$$= 1\,192 \times 10^{-6} \text{ mm}$$

Now, the difference between the lengths of the bars at a temperature of 15°C is:

$$250 - 249·5 = 0·5 \text{ mm}$$

The gain in expansion of the copper bar over the steel bar will have to continue until this 0·5 mm difference has been made up.

We know that $1\,192 \times 10^{-6}$ mm is gained for 1°C rise in temperature. Therefore, if we divide $1\,192 \times 10^{-6}$ into 0·5, it will give us the number of degrees through which the bars will have to be heated in order to bring them to the same length.

Increase in temperature of the two bars

$$\frac{0·5}{1\,192 \times 10^{-6}}$$
$$= 419·6°C$$

Temperature when the two bars are at the same length

$$= 15 + 420$$
$$= \underline{435°C}$$

We can find the lengths of the bars at this temperature as follows:

EXPANSION of steel bar $= 250 \times 419·6 \times 11·9 \times 10^{-6}$
$$= 1·248 \text{ mm}$$

EXPANSION of copper bar $= 249·5 \times 419·6 \times 16·7 \times 10^{-6}$
$$= 1·748$$

Length of steel bar at 419·6°C $= 250 + 1·248 = 251·248$ mm
and Length of copper bar at 419·6°C $= 249·5 + 1·748 = 251·248$ mm

This is as it should be.

Exercise 18

Section A

1. A pin gauge 820·015 mm long at its normal temperature of 20°C is found to be at a temperature of 15°C when it is brought out of storage. Calculate how much the gauge is in error, correct to two significant figures. Coefficient of linear expansion for the gauge steel is $11·9 \times 10^{-6}$/°C.

2. The coefficient of linear expansion for cast iron is 0·000 010 2/°C and for wrought iron 0·000 011 9/°C. Bars of wrought iron and cast iron, each 250 mm long, are heated from 20°C to 100°C. Find which will expand most and by how much in excess of the other.

3. The piston of a steam engine is 148·75 mm dia., when it is finished turned at a temperature of 18°C. When it is working in its cylinder its temperature is 200°C. Calculate the working diameter of the piston. Coefficient of linear expansion is 0·000 010 2/°C. Give answer to three decimal places.

4. The finished diameter of a brass spindle being turned in a lathe must be 100·015 mm at 20°C. By mistake the spindle is made 100·015 mm at 36°C, due to heat generated during the cutting of the metal. What error will there be on the diameter? Give answer correct to three decimal places. Coefficient of linear expansion is 0·000 018 9/°C.

5. The bore diameter of a locomotive tyre at 20°C is 1 575 mm and it has to be expanded 4·850 mm in order to shrink it on to its wheel. To what temperature must the tyre be heated? Coefficient of linear expansion for the tyre is $11·9 \times 10^{-6}$/°C.

6. Because a 12 mm dia. twist drill is used at too fast a speed and without coolant, its temperature is increased from 20°C to 320°C. If the coefficient of linear expansion for the drill is 0·000 011 9/°C, determine how much it expands on its diameter.

7. A 20 mm dia. rivet is heated to 900°C from 20°C before it is put into its 22 mm dia. hole. What is the clearance between the rivet and the hole? Give your answer correct to two significant figures. Coefficient of linear expansion for the material is 0·000 012/°C.

8. The two main joists of a railway bridge are 40 metre long at 15°C. On a summer day the temperature of the joists may reach 35°C. How much do the joists expand? Give your answer to the nearest millimetre. Coefficient of linear expansion for the joists is $11·9 \times 10^{-6}$/°C.

9. By using slip gauges, a length of 60·937 mm is made up, the slip sizes being correct at 20°C. When the slips are used their temperature is 13°C. Calculate the error in the dimension if the coefficient of linear expansion is $11·9 \times 10^{-6}$/°C. Give answer correct to three decimal places.

10. A copper rod 500 mm long at 19°C is placed in vee-blocks with one of its ends up against a stop and the other touching the plunger of a dial test indicator that is set at zero. What will be the temperature of the rod when the dial gauge reading is 0·088 mm? Coefficient of linear expansion for the rod is 0·000 016 7/°C.

Section B

1. In order to shrink a spindle that measures 77·650 mm dia. at 20°C to 77·625 mm dia., it is placed in a refrigerator. To what temperature must the spindle be cooled if its coefficient of linear expansion is $11·9 \times 10^{-6}$/°C?

EXPANSION OF SOLIDS, LIQUIDS AND GASES

2. A collar, bore diameter 262·500 mm, is to be shrunk on to a shaft 263·0 mm dia., the dimensions being taken at 20°C. The collar is to be heated until its bore is 0·450 mm bigger than the shaft diameter. Calculate the temperature to which the collar has to be heated. If its coefficient of linear expansion is 0·000 011 9/°C.

3. If the coefficient of linear expansion for brass is 0·000 018 9/°C and for cast iron 0·000 010 2/°C, explain why a greater allowance for contraction has to be made on brass castings than has to be made on iron castings.

   A 75 mm dia. bar of phosphor bronze for making bushes is 500 mm long when it solidifies at 1 080°C after casting. What will its length be, to the nearest millimetre when it has cooled to 18°C? Coefficient of linear expansion for the bronze is 0·000 018 1/°C.

4. When at its average working temperature of 285°C, the cylinder of a motor cycle engine is 76·550 mm dia. The clearance between the cylinder and the aluminium piston at this temperature must be at least 0·075 mm on the diameter. Calculate the diameter to which the piston must be turned at a temperature of 15°C. Coefficient of linear expansion for the piston is 22×10^{-6}/°C.

5. During the rough turning of a steel shaft 5 metre long between centres, its temperature is increased by 22°C. How much does the length of the shaft increase? Give your answer to the nearest 0·5 mm. Coefficient of linear expansion for the shaft is 0·000 011 9/°C.

   What bad effect could this increase in length have, and how could it be remedied?

6. The portion of a crank pin that is to be shrunk into a locomotive wheel is 87·650 mm dia. and the hole into which it is to be shrunk is 87·500 dia., at 15°C. The metal round the hole is to be heated until the hole diameter is increased to a value 0·375 mm bigger than the diameter of the pin. Calculate the temperature to which the metal must be heated. Coefficient of linear expansion is 10×10^{-6}/°C.

7. A stainless steel valve seat is turned to 75·125 mm dia at 20°C and pressed into a recess in a bronze valve body which is 75·000 mm dia., at 20°C. The working temperature of the valve is 170°C. What will be the size of the interference, or clearance, between the two diameters at 170°C? Coefficient of linear expansion for bronze is 0·000 018 2/°C and for stainless steel 0·000 011 4/°C.

8. A number of pins 15·68 mm diameter and 254 mm long at 20°C are to be produced by die-casting them in zinc alloy. The working temperature of the die is 420°C. Calculate to the nearest 0·01 mm the length and diameter of the impression in the die when it is at a temperature of 20°C. Coefficient of linear expansion for the die material is 0·000 013/°C and for the alloy 0·000 026/°C.

9. A die for extruding tube 56 mm outside diameter and 53 mm bore diameter, at 15°C, is at an average temperature of 510°C when in the extrusion machine. If the tube dimensions are guaranteed to within 0·01 mm, calculate the diameters to which the die and plug must be turned, at a temperature of 20°C. Coefficient of linear expansion of the die steel is 12×10^{-6}/°C and for tube 17×10^{-6}/°C.

10. A spindle of 25·075 mm dia. is to be shrunk into a flywheel which has 25·00 mm bore diameter, both diameters being taken at 22°C. The process is carried out by cooling the spindle to 0°C, then heating the flywheel until it is 0·080 mm bigger in diameter than the spindle is at 0°C. Find the temperature to which the flywheel must be heated. Coefficient of linear expansion for the flywheel is 0·000 010 2/°C and for the spindle 0·000 011 9/°C.

REVISION EXERCISES

Grade 1
1. Explain, giving workshop examples, the difference between (i) an element and a compound, (ii) a mixture and an alloy.
2. A standards room contains very accurate and sensitive gauges and measuring instruments. Describe in detail what efforts are made to ensure a satisfactory atmosphere in the room.
3. A lathe operator on measuring the diameter of the work he is turning finds it to be 90·22 mm. Its temperature is 96°C. What will be the diameter of the work at 20°C. Give answer correct to four significant figures. Coefficient of linear expansion 0·000 011 25/°C.
4. In making a welded joint the quantity of heat involved can be divided into two distinct parts. State what these are and explain the difference between them.
5. A spindle is hardened by quenching it in water at 18°C. The final temperature of the water is 20°C. If there are 30 kg of water, how much heat has it gained? Specific heat capacity of water is 4187 J/kg°C.
6. A 3 kW motor drives a lathe and the transmission to the spindle is 92% efficient. Of the energy available at the spindle, 13% is converted to heat through frictional resistances. Calculate the heat produced per minute when maximum power is used.
7. Two resistors of 6 ohms and 4 ohms are connected (i) in series, (ii) in parallel, with a 6 volt battery. Find the current flowing through each resistor in the two instances.
8. A 2 kW heating coil is connected to a 250 volt supply. Find the current flowing through the coil. If the coil was immersed in 80 kg of water at 15°C, to what temperature would the water be heated in one hour? Specific heat capacity of water 4187 J/kg°C.
9. A billet 'weighing' 7·5 kN is slung horizontally from a two leg chain sling. Determine the force in each chain when the angle between them is (i) 25°, (ii) 47°.
10. Figure 217 shows a beam which 'weighs' 200 N/m length. A set of chain blocks is attached to the right-hand end to lift a load of 5·5 kN. Find the reactions in the supports A and B.

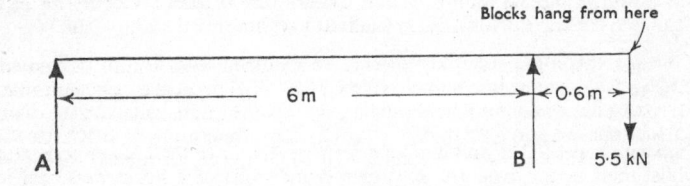

Fig. 217

11. Find the centroid of an I-shaped template. The cross limbs are 100 mm × 20 mm and 110 mm × 12 mm respectively and the vertical limb 200 mm × 12 mm.
12. Find the centre of gravity of a shaft 1·5 m long. It is 80 mm diameter for 0·7 m, 75 mm dia. for 0·5 m, and 50 mm dia. for the remaining length.

REVISION EXERCISES 271

13. In a turning operation the cutting force is 800 N and cutting speed 0·5 m/s. Calculate the power required at the lathe motor in kW if the transmission efficiency is 92%.

14. A steel casting, specific heat capacity 460 J/kg°C, mass 25 Mg, is heated to 850°C from 20°C in an electric furnace, thermal efficiency 80%, in 10 hour. How many kW of power are required?

15. A machine table is raised by applying a force of 50 N to a handle 250 mm long. How much work is done in 1 rev of the handle? If the lifting were done by an electric motor, in 2 s what power would be used?

16. A 90° bell-crank lever has arms 300 mm and 240 mm long respectively. What is the normal force on the shorter arm when a normal force of 200 N is applied to the other arm? What is the reaction through the hinge? *Note:* A bell-crank lever is shown in Fig. 186.

17. The power supply to a machine tool motor is 3 kW, the tool load 8 kN and cutting speed 0·3 m/s. What is the efficiency of the drive?

18. What is corrosion? Explain fully how a coat of zinc on a mild steel component prevents, or deters atmospheric corrosion.

19. Explain what is meant by a bonderized surface. Describe the process of bonderizing.

20. Describe briefly, two workshop processes in each instance which involve (i) a physical change, (ii) a chemical change.

21. A solid metal cylinder and two sealed hollow ones, one containing gas, the other a liquid, are heated. What effect would this have on the solid metal, liquid and gas. State how the effects could be demonstrated in the laboratory.

22. A locomotive tyre 1362·55 mm in dia. bore is heated from 20°C to 550°C. How much does the bore diameter increase? Coefficient of linear expansion $11·9 \times 10^{-6}$/°C.

23. When tightening the jaws of a chuck, the operator applies a force of 70 N to each end of a key handle 280 mm long. Calculate the moment of the couple on the key assuming the handle is central.

24. What is an electrical insulator? Explain why these are necessary on switches and electrical machines.

25. A 2 ohm resistor is connected in series with three resistors, 3, 4 and 7 ohm, which are in parallel, and a 12 volt battery whose resistance may be neglected. Calculate the current flowing through each resistor.

26. A bronze forging, mass 8 kg, specific heat capacity 393 J/kg°C, is heated from 20°C to 740°C in a gas furnace. How many cubic metre of gas, calorific value 17·5 MJ, are required if the thermal efficiency of the furnace is 57%?

27. Two wire ropes 2·3 m and 2 m long respectively, have their lower ends attached, whilst their upper ends are fixed to the underside of a horizontal roof joist at points 2·7 m apart. A load of 15 kN is slung from the ropes. Find the forces in the two ropes.

28. In a slotting machine operation ½ kW is being used by the motor which drives the tool head through a 97% efficient transmission. What is the average cutting force at the head when its speed is (i) 0·3 m/s, (ii) 0·1 m/s?

29. Figure 218 shows a joist which is fixed to a roof frame at points A and B. The 'weight' of the joist is 600 N/m. Calculate the reactions at the fixing points.

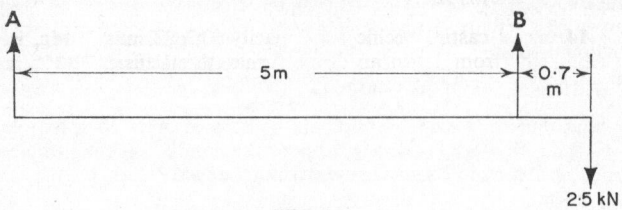

Fig. 218

30. Explain why open-jaw spanners are made in different lengths. A nut is tightened by applying a 150 N force to a spanner at a point 290 mm from the centre of the nut. Find the turning moment on the nut.
31. A steel plate 300 mm × 250 mm has two holes 90 mm and 45 mm dia. respectively cut in it on the centre line parallel to the longest edges. The hole centres are centrally situated 150 mm apart. Calculate the centroid of the plate.
32. The cutting pressure on a planing machine tool is 2·6 kN/mm². A 12 mm depth of cut in used with a 6 mm feed rate per stroke. Find the cutting force on the tool point and the work done per stroke if the surface being machined is 1·8 m long. What energy in electrical form does this represent?
33. In a gang milling operation three cutters are used, one 80 mm diameter, the others 110 mm dia. The cutting force on each cutter is 480 N, and the arbor speed 4 rev/s. Find the torque on the arbor and the power supply to the motor if it drives the arbor through a 90% efficient transmission.
34. A beam 10 m long and 'weighing' 300 N/m length is simply supported at each end. It carries a load of 50 kN at a point 1·1 m from the left hand support and one of 30 kN, 2·3 m from the right hand support. Calculate the reactions in the supports.
35. A steel bar of uniform diameter and 4·8 m long 'weighs' 1·3 kN/m length. The bar is supported in a horizontal position by a wire rope at the right-hand end which is inclined at 25° to the vertical and one at the left-hand inclined at 35° to the vertical. Find the tensions in the ropes. If the ropes are 10 mm dia. and 70% of the cross-sectional area may be taken as load carrying calculate the stresses in them.
36. A steel collar 75 mm dia. bore is shrunk on to a shaft 75·10 mm dia. by heating it until the bore dia. is 0·38 mm bigger than the shaft diameter. If the initial dimensions were taken at 20°C, find the temperature to which the collar must be heated. Coefficient of linear expansion 0·000 010 9/°C.
37. A punching machine is driven by a 10 kW motor through a transmission which is 91% efficient and 12% of the power available at the punch head is lost in friction. Assuming all the energy absorbed by friction is converted to heat, how many units are produced per minute when maximum power is being used?
38. What are the two main elements in coal gas. Explain why a gas burner gives a hotter flame if air is supplied to the gas just before combustion takes place, rather than being supplied from the atmosphere around the flame. On some furnaces the air supply is pre-heated, what advantage does this have?
39. An ammeter in the circuit of a press motor which is wired to a 240 volt d.c. supply shows a reading of 31·5 amp. If the transmission to the press head is 93% efficient and the head speed is 0·08 m/s, calculate the press force.

REVISION EXERCISES 273

40. In a milling operation the cutting force is 700 N and the cutting speed 0·4 m/s. The energy used in actual cutting is 85% of the energy available at the cutter, the other 15% being converted to heat by frictional resistances. If the transmission is 89% efficient, how many kW are required at the motor to cover the frictional losses? What would be the temperature rise, in °C per min, per kg of coolant supplied, if all the heat generated were carried away by a liquid of specific heat capacity 3·76 kJ/kg°C?

Grade 2

1. Explain with the aid of a sketch, how a magnetic chuck having permanent magnets works.
2. A thermostat is a type of automatic switch which cuts off the power supply to an electric furnace when a certain temperature has been reached. Design such a switch and give a sketch of it and a brief description of how it works.
3. Explain how the flux ' killed spirits ' could be made and state the chemical formula involved.
4. A press head is driven by a 150 mm long crank through a connecting rod 0·65 m long. Calculate the press force when there is an 80 Nm torque on the crank and it is 45° from the inner dead centre on the downward stroke.
5. 6 mm diameter holes are punched at the rate of 16 per second in 6 mm thick mild steel plate, maximum shear strength 320 N/mm². If the gear transmission is 93% efficient, calculate the power supply to the motor assuming the mean shear force is 80% of the maximum value.
6. A 150 mm dia. slitting saw is held by friction on a milling machine arbor, it being gripped between collars, 26 mm dia bore, 40 mm outside diameter. When the force on the cutter is 200 N and the coefficient of friction 0·12, calculate the gripping force required, and the direct stresses, in N/m², in the arbor and collars.
7. A casting 'weighing' 10·63 kN, and supported on packing pieces is clamped to the side of a drilling machine table by two 25 mm dia. bolts exerting a combined force of 27 kN. Coefficient of friction between casting and table is 0·27. Calculate the shear stress in each bolt if the packing pieces were removed. Give the answer in MN/m².
8. What power is required to haul a train of 10 trucks, 'weight' 120 kN each, up an 8° incline, at a speed of 0·2 m/s, if the tractive resistance is 5 N/kN 'weight' of truck. Give answer correct to 2 significant figures.
9. A milling machine table, 'weight' 1·6 kN, coefficient of sliding friction 0·22, is power traversed horizontally at 0·01 m/s by a screw mechanism which is 25% efficient. What electrical power is required at the motor which drives the screw if it is supplied through a 96% efficient transmission?
10. A 450 mm dia. belt pulley, 76 mm bore dia., is keyed to a shaft by a key 12 mm wide, 90 mm long. When the effective pull in the belt is 1·2 kN, find the shear stress in the key.
11. A goods lift, mass 1·2 Mg, is raised 40 m at a speed of 0·7 m/s by an electric motor through a rope of 'weight' 4 N/m length. If the gear transmission is 90% efficient, what electrical power is being used at the motor?
12. A 3 Mg steel ingot, specific heat capacity 460 J/kg°C is heated from 30°C to 930°C in an electric furnace rated at 90 kW and 80% efficient. How long will the operation take?
13. What effort is required to lift a casting 'weighing' 2 kN with a 6-pulley set of rope blocks 90% efficient?

14. A 3 kW motor, speed 24 rev/s, drives a lathe spindle through a 92% efficient gear transmission in which a 32 tooth pinion meshes with a 51 tooth intermediate shaft gear. Also on this shaft is a 45-tooth wheel meshing with a 35-tooth spindle gear. Calculate the maximum power, and speed available at the spindle.

15. A set of chain blocks comprises a single start worm and 40-tooth worm wheel. Effort wheel and load drum are 190 mm and 150 mm dia. respectively. What effort is required to lift a 5 kN load? What effort would be required if the worm were double start? Assume an efficiency of 35% at this load.

16. The tooth load on a gearwheel 350 mm dia. pitch circle, 150 mm wide, is 100 N/mm width. If its speed is 12 rev/s, calculate the torque and power on the wheel.

17. The bronze bush in the little end of a connecting rod is a 0·025 mm interference fit in a hole 49·98 mm dia., at 20°C. What will be the interference at 270°C? Coefficient of linear expansion, rod 0·000 011/°C, bush 0·000 018/°C.

18. A suds-pump motor is rated at 0·01 kW. How many litre/s of coolant will it supply through a pump 65% efficient, over a height of 1·2 m, if 1 litre of coolant has a mass of 0·82 kg? Gravitational pull on 1 kg is 9·81 N.

19. A 300 mm hacksaw blade, 12 mm wide, 1 mm thick, is subjected to a 1 kN load when tightened in its frame on pegs 3 mm dia. Find the maximum tensile stress in the blade, the shear stress in the pegs and how much the blade stretches. Modulus of elasticity 210×10^9 N/m².

20. A 'knock off' gear is actuated by compressed air through a piston and cylinder. The inlet valve to the cylinder is operated electrically, the movement being executed as rapidly as possible. Describe with the aid of sketches a suitable device for operating the valve.

21. State for the following the chemical formula and whether they are elements or compounds: zinc oxide, oxygen, copper oxide, forging scale, hydrogen, rust, tap water, distilled water.

22. In four hours 1500 kg of copper, melting point 1083°C, initially at 23°C are melted in an electric furnace, thermal efficiency 85%, which is coupled to a 550 volt d.c. supply. How much current is the furnace taking? Specific heat capacity of copper 376 J/kg°C, latent heat of fusion 180 kJ/kg.

23. Explain with the aid of a sketch how a circular electro-magnetic chuck works.

24. A component of mass 2·8 Mg is suspended from two sets of blocks whose directions of pull are inclined 30° to the right and 60° to the left of the vertical. Calculate the load supported by each block. Check your answers by graphical means.

25. In a shaping machine mechanism, see Fig. 207, the bull wheel has a torque of 80 Nm on it and a speed of 0·5 rev/s. The crank pin is at 100 mm radius and 60° from its top position, moving in an anti-clockwise direction. The rocker arm is 1·05 m long and pivoted 0·7 m below the bull wheel centre. What are the normal force and velocity values at the top of the rocker arm?

26. A mild steel test piece, 25 mm dia., 200 mm gauge length, tested to destruction gave the following results: Elastic limit load 118 kN, maximum load 221 kN, extension at 92 kN load 0·18 mm, total extension 52·5 mm, fracture dia. 16·5 mm. Calculate the modulus of elasticity, stress at elastic limit, ultimate tensile stress, percentage elongation and percentage reduction in area.

REVISION EXERCISES 275

27. A lathe input pulley is 240 mm dia. and is driven by a 110 mm dia. pulley fixed to a motor rated at 2·5 kW, speed 25 rev/s. The transmission to the spindle has a velocity ratio of 3 to 1 and efficiency 91%. Find the speed, power and torque available at the spindle at maximum conditions.
28. A planing machine table has a mass of 3·4 Mg and the coefficient of friction for its slides is 0·085. The rack bolted to the underside is driven by a pinion having a pitch circle 260 mm dia. and making 1·2 rev/s. Calculate the torque on the pinion, power required at the motor to cover frictional losses if the transmission efficiency is 93%.
29. Holes 25 mm square are punched in fishplates 6 mm thick, maximum shear strength 320 N/mm² with a punch having a 30 mm dia. shank. Calculate the maximum compressive stress in the shank of the punch and the heat equivalent of the energy used punching 1 hole if the average punch force is 88% of the maximum value.
30. Figure 219 shows the drive, 95% efficient, to a planing table which with the job has a mass of 12 Mg. If the coefficient of friction for the table slides is 0·07, find the power required at the motor to cover frictional losses. What is the heat equivalent of this power over a 1 h period?

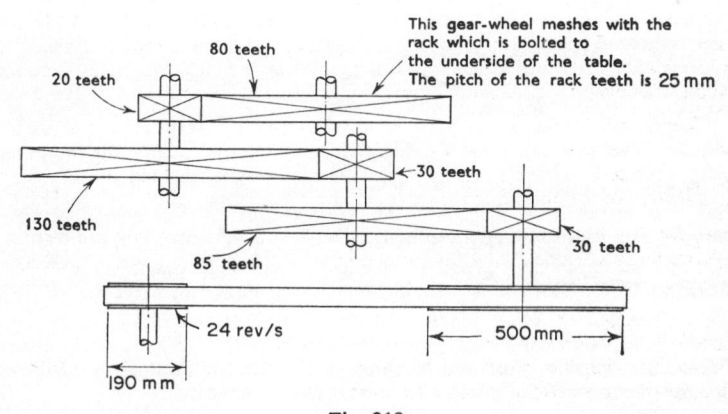

Fig. 219

31. A drilling machine motor, speed 25 rev/s, connected to a 250 volt d.c. supply is taking 5 A of current. The motor drives the spindle through a gear box, 6½ to 1 ratio, which is 89% efficient. What is the couple on the drill at maximum radius if its dia. is 20 mm?
32. Calculate the reactions in the bearings A and B of the lathe spindle shown in Fig. 220. The bearing diameters are 90 mm, spindle speed 7 rev/s, and coefficient of friction 0·04. What power is required to overcome bearing friction?
33. A collar, mass 15 kg, bore diameter 162·5 mm at 18°C is to be heated, until its bore diameter is 163·12 mm, in an electric furnace, which is connected to a 500 volt d.c. supply, thermal efficiency 82%. Find the current required, to the nearest amp., by the furnace if the operation is to take half an hour. Coefficient of linear expansion, $11·9 \times 10^{-6}$/°C, specific heat capacity 465 J/kg°C.
34. Explain with the aid of a suitable sketch the principle of electro-plating.

35. During a milling operation 1 kW is converted to heat at the cutter and 90% of this is taken away by the coolant whose temperature rise is restricted to 3°C. Calculate the quantity of coolant required, in m^3/s, if its specific gravity is 0·94. Specific heat capacity of coolant is 3·76 kJ/kg°C, mass of 1 m^3 of water 10^3 kg.
36. The carriage of a lathe weighs 600 N and during the turning of a roller, 250 mm mean dia., the feed force is 200 N, cutting force 550 N, spindle speed 3·2 rev/s, and feed rate 0·5 mm per rev. Coefficient of friction for carriage slide is 0·10. Assuming an overall efficiency of 86%, calculate the power required at the motor.

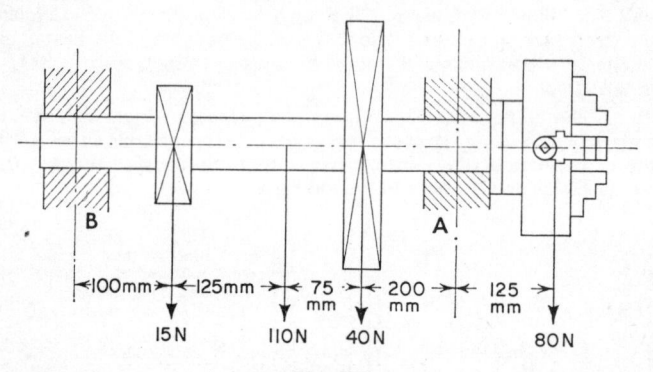

Fig. 220

37. A Weston differential chain block has pulleys which have 12 and 16 chain link flats respectively. Calculate the effort required to lift a load of 1·8 kN assuming the blocks are 60% efficient.
38. Ninety steel components, each of mass 1·5 kg, are to be quenched from 840°C in oil at 19°C. The rise in the oil temperature must not exceed 12°C. Find the quantity of oil required, specific heat capacities are, oil 1·5 kJ/kg°C, steel 0·460 kJ/kg°C, 1 m^3 of the oil has a mass of 900 kg.
39. Figure 221 shows a drive to a machine tool; calculate the highest and lowest spindle speeds when the clutch is (i) disengaged, (ii) engaged.
40. Figure 222 shows the safety coupling fitted to a 25 mm dia. lead screw in which the shear pins have a 3 mm mean dia. and maximum shear strength 320 N/mm^2. What torque at the spindle would cause the coupling to fail?
41. Make a sketch of a device which could be used to lift the tool clear of the work, on the return stroke, on the tool box of a large planing machine. Explain how the mechanism works.
42. Copper sulphate may be used to prevent a surface being carburised. Explain fully the chemistry behind the coating process and the carburising process, giving chemical formulae where applicable.
43. With the aid of a sketch, explain fully how a simple electric motor works. How could the energy available from the motor be increased?
44. A heating coil 96% efficient, is rated at 4 kW and is fitted to a tank of thermal efficiency 92% which contains 190×10^{-3} m^3 of water at atmospheric pressure. How long will it take to heat the water from 14°C to 38°C? How many kWh would be required to boil the water? One cubic metre of water has a mass of 10^3 kg. The specific heat capacity of water is 4187 J/kg°C.

REVISION EXERCISES 277

45. A rod, modulus of elasticity 210×10^9 N/m², 4 m long, 40 mm dia. is subjected to a gradually applied pull of 180 kN. How much energy is stored in the rod assuming that it is not stretched beyond the elastic limit?

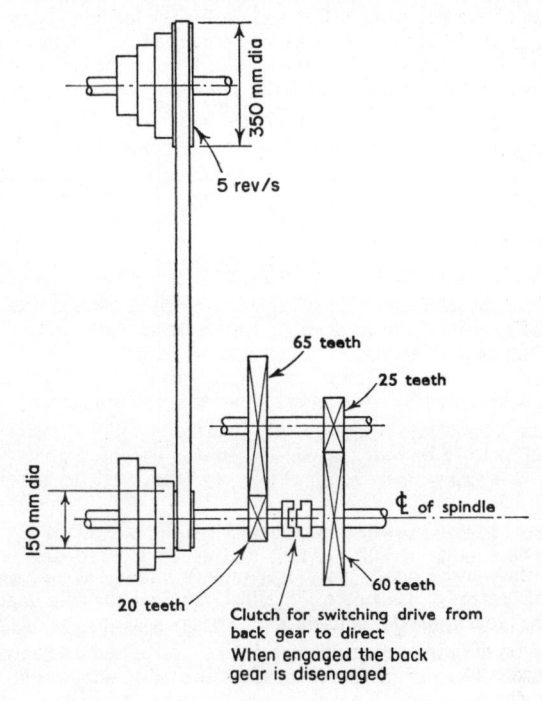

Fig. 221

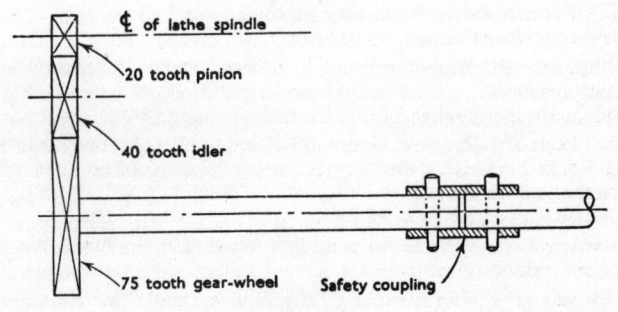

Fig. 222

46. Explain the following terms: ultimate tensile stress, proof stress, limit of proportionality, yield stress, longitudinal strain, factor of safety. Indicate the practical value of each.

47. The following table gives values from a tensile test on an aluminium-alloy specimen 9 mm dia., gauge length 100 mm. Calculate the modulus of elasticity,

Extension mm	0	0·025	0·075	0·100	0·150	0·200	0·250	0·275
Load kilonewton	0	2·07	6·24	8·34	12·6	16·7	20·9	22·1

Extension mm	0·300	0·325	0·350	0·375	0·400	0·425	0·450
Load kilonewton	23·2	23·8	24·3	24·6	25·0	25·2	25·3

and the proof stress of the alloy assuming 0·1 % plastic deformation.

48. A press head moves in vertical guides, being driven by a crank 60 mm long, speed 3 rev/s, through a connecting rod 200 mm long. The torque on the crank is 300 Nm. Determine: the force to overcome guide friction if the coefficient of friction is 0·07, the speed in m/s of the head and the force at the press head, when the crank has turned 25° past top dead centre in a clockwise sense.

49. To hand traverse a lathe carriage, the operator applies a force of 20 N to a hand wheel, bore 18 mm dia., at 70 mm radius. The hand wheel is keyed to its spindle by, a 3 mm wide by 15 mm long, key. Calculate the shear stress in the key.

50. Two tool boxes are in use when machining a casting, 'weight' 30 kN, on a planing machine whose table 'weighs' 10 kN, and which has forward and backward speeds of 0·2 m/s and 0·38 m/s respectively. The pressure on both tool points due to cutting is $1·2 \times 10^9$ N/m^2, depth of cut 20 mm, feed rate 16 mm per stroke and coefficient of friction for slides 0·06. What percentage of the total power being used on the forward stroke is lost to slide friction on (i) forward stroke (ii) backward stroke?

51. Explain the differences between a solution, mixture and alloy, give examples of each. If a piece of pure copper plate were heated in the open atmosphere what chemical reaction would take place and what would it produce? Could the same effect be produced without heat?

52. Describe three chemical reactions which could impair the accuracy and finish of a component during its stages of production.

53. The motor driving a punching machine is rated at 0·25 kW. The machine is to punch 12 mm dia. holes in 6 mm thick brass plate having maximum shear strength of 260 N/mm^2. How many holes can be punched per minute? Assume the average punch force to be 88 % of the maximum value and the power is transmitted with an efficiency of 90 %.

54. What power is required to lift a milling machine table, which has a mass of 350 kg, by means of a single start screw 6 mm pitch, if the mechanism is 30 % efficient and the speed ratio between motor and screw is 60 to 1. Motor speed is 17 rev/s.

55. Figure 223 shows a brake mechanism in which a force F newton is applied by a solenoid. The combined pulley and brake is driven by a motor, speed 17 rev/s through a pulley 106 mm dia. If coefficient of friction for brake is 0·45, find the value of F required to stop the pulley assuming the equivalent of 1·1 kW has to be absorbed.

REVISION EXERCISES 279

56. A number of grinding wheels, 1 Mg mass, specific heat capacity 837 J/kg°C, are heated in a furnace from 25°C to 1250°C. 95 m³ of gas, calorific value 17·5 MJ/m³, are used. What is the thermal efficiency of the furnace?
57. A crane designed to carry a maximum load of 900 kN has 8 falls of wire rope, ultimate tensile strength 620 N/mm², on which to support the load. Assuming a factor of safety of 4, calculate a suitable rope diameter if 0·7 of the cross-section of each rope is effective.
58. Figure 224 shows an automatic feed on a centre lathe. The velocity ratio between the spindle and feed shaft is 32. What is the rate of feed per rev of spindle?

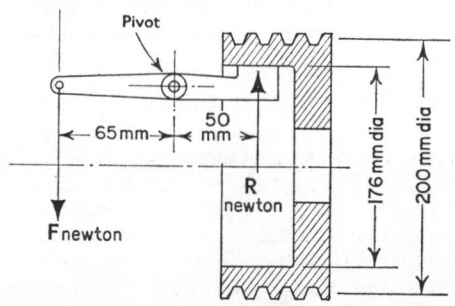

Fig. 223

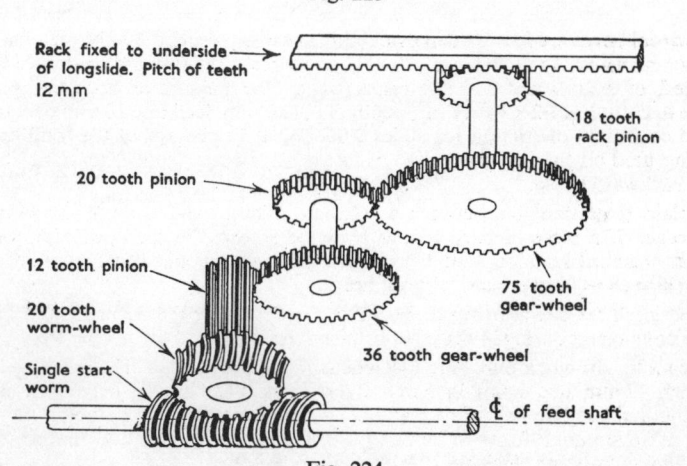

Fig. 224

59. A motor with a pulley 260 mm dia., speed 24 rev/s, drives a 1·2 m dia. pulley on a countershaft by means of a flat belt. The pull on the tight side of the belt is 760 N and on the slack side 215 N. Calculate the countershaft speed and power transmitted by the belt.
60. The mass of a lathe carriage is 330 kg. The normal cutting force is 5·4 kN, the cutting speed 1·2 m/s and the coefficient of friction for the slide 0·16. Find the power required at the motor, for a feed rate of 1·6 mm/s, to overcome slide friction and express this as a percentage of the total power used. Transmission efficiency is 93%.

ANSWERS TO EXERCISES

Exercise 4
Section A
7. 4·31 kW. **9.** 3·26 amps, 8·71 amps. **10.** 0·24 amps, 0·6 amps, 0·8 amps.

Section B
1. 1·3 amps. **2.** 80 amps. **3.** 4·41 ohms. **4.** 60 volts, 20 volts. **5.** 4·28 ohms, 11·7 amps.
6. 0·57 kW. **7.** 3·22 kW. **8.** 26·2 watts, 579 watts. **10.** 22, 13·2, 8·8 amps.

Exercise 5
Section A
1. $12·5 \times 10^6$ N/m², tensile. **2.** 133×10^6 N/m², comp. **3.** 81 N. **4.** 48 kN.
5. 56×10^3 N/m². **6.** 225 N. **7.** 1 kN, 2×10^6 N/m², tensile. **8.** $1·74 \times 10^6$ N/m².
9. 15·7 kN. **10.** $1·62 \times 10^6$ N/m² comp.

Section B
1. 16 mm, shear. **2.** 7·3 mm. **3.** $20·4 \times 10^6$ N/m². **4.** 13 mm. **5.** $3·22 \times 10^6$ N/m².
6. 110 kN. 350 N/mm². **7.** 12 mm. **8.** 19·9 kN, 79·5 kN. **9.** 235×10^6 N/m², Brass.
10. $84·5 \times 10^6$ N/m², 140×10^6 N/m².

Exercise 6
Section B
1. 940×10^6 N/m², 732×10^6 N/m², 726×10^6 N/m², 19%, 57%. **2.** 28 mm.
3. 0·15 mm, **4.** 297×10^6 N/mm². **5.** 25 mm. **6.** 0·18 mm. **7.** 0·049 mm, Cu;
0·046 mm phos. bronze. **8.** 37 mm. **9.** 216 kN. **10.** 79·2 kN, 6·25 mm.

Exercise 7(a)
Section A
1. 37 N. **2.** 125 N, 157 N. **3.** 846 N. **4.** $R_R = 5·28$ kN, $R_L = 4·92$ kN.
5. $R_R = 695$ N, $R_L = 555$ N. **6.** 131 N. **7.** 21×10^6 N/m². **8.** $R_A = 54$ N, $R_B = 456$ N. **9.** 1·24 MN/m². **10.** 63·2 N.

Section B
1. 103·5 mm. **2.** 65 mm. **3.** 70 mm. **4.** 46 mm, 89 mm. **5.** 300 mm. **6.** 130 mm.
7. 115 mm. **8.** 1·39 m. **9.** 0·87 m. **10.** 172 mm, 65·5 mm.

Exercise 7(b)
Section A
1. 400 N, 60 Nm. **2.** 400 N, 50 Nm. **3.** 1·5 kN, 1·386 kN. **4.** 500 N, $2·6 \times 10^6$ N/m².
5. 30 Nm. **6.** 11·38 kN. **7.** 65 N. **8.** 48·7 Nm. **9.** 300 Nm. **10.** 1·87 Nm.

Section B
1. 11 Nm, 55 Nm. **2.** 13 Nm, 10 Nm. **3.** 1·05 kNm, 1·14 kNm. **5.** 6288 Nm, 6132 Nm, 6000 Nm. **7.** 213 N, 160 N, 128 N, 106·8 N. **10.** 282 Nm, 207 Nm.

ANSWERS TO EXERCISES 281

Exercise 8

Section A
1. 128 J. **2.** 2·40 kJ. **3.** 3·53 kJ. **4.** 240 J. **5.** 9·6 J/s. **6.** 45π J. **7.** 57·6 J. **8.** 300·3 J. **9.** 375 W. **10.** 36·7 kW.

Section B
1. 2 W. **2.** 3·24 kJ. **3.** 33 W. **4.** 19·5 kW. **5.** 590 W. **6.** 118 mm. **7.** 7·67 kW. **8.** 80 J. **9.** 2·55 kW. **10.** 11·25 kW. **11.** 8·5 kN, $3·8 \times 10^3$ Nm, 326 kJ. **12.** 1·575 J, 44%, 2·79 J. **13.** 2·5 kN, 5 kN. **14.** 0·72 J. **15.** 5·3 J. **16.** 20 kW. **17.** 5·7 m³/s. **18.** 470 kN. **19.** 29·4 kW. **20.** 596 kW.

Exercise 9

Section A
1. 14 rev/s. **2.** 21·2 m/s. **3.** 57 rev/s, 36 m/s. **4.** 17 m/s. **5.** 9·5 rev/s. **6.** 46 rev/s. **7.** 14 rev/s, 20 rev/s. **8.** 8 rev/s. **9.** 39 rev/s, 36 rev/s. **10.** 1·17 m.

Section B
1. 3·67 rev/s. **2.** 0·5 m. **3.** 4·7, 7·0, 10·5 rev/s. **4.** 557 mm. **5.** 1·9 m. **6.** 10, 16·2 rev/s. **7.** 10·7 rev/s. **8.** 2·9, 2·7 rev/s. **9.** 5·96, 6·15 m/s, 7·3 rev/s. **10.** 1·4, 2·0, 2·6 rev/s.

Exercise 10

Section A
1. 3·6 rev/s. **2.** 2·9 radian. **3.** 1·12 m/s. **4.** 2·55, 2·95 m/s. **5.** 1·33 mm. **6.** 3·24 stroke/s. **7.** 0·3 rev/s. **8.** 1·1 rev/s, 1·2 rev/s. **9.** 16·8 rev/s. **10.** 22 rev/s.

Section B
1. 4·8 rev/s. **2.** 0·35 mm. **3.** 19 rev. **4.** 0·35 m/s. **5.** 3 rev/s. **6.** 1·4 to 1, 90 drives 30 and 50 drives 110. **7.** 29·1 mm. **8.** 26·5 rev/s. **9.** 0·83 m/s. **10.** 31 rev/s.

Exercise 11

Section A
3. 277 N. **4.** 835 N. **5.** 99·2 N. **6.** 134 J. **7.** 306 N. **8.** 2·9 MN/m². **9.** 61·4%. **10.** 1·25 kN, $2·55 \times 10^6$ N/m².

Section B
1. 2·8 kN. **2.** 2×10^6 N/m². **3.** 763 J. **4.** 5·5 N. **5.** 1·08 kN, $1·52 \times 10^6$ N/m². **6.** 203 N. **7.** 271 N. **8.** 204 J, 42·1 J. **9.** 1·0 kW. **10.** $3·7 \times 10^6$ N/m². **11.** 5·5 kJ.

Exercise 12

Section B
1. 0·6 Nm. **2.** 1·02 kN. **3.** 11·4 Nm. **4.** 21 rev/s. **5.** 1·35 kNm, 29·6 kW. **6.** 20 Nm, 640 Nm. **7.** 2·85 kN. **8.** 9·82 kNm. **9.** 3·19 kN. **10.** 129 kN. **11.** 246 kW, 6·8 kNm. **12.** 2·3 rev/s, 832 Nm. **13.** 80 Nm, 2 kW. **14.** 5·19 Nm, 11·6 kJ. **15.** 22·8 Nm, 4·67 kW. **16.** 46%. **17.** 3·3 kW, 500 Nm. **18.** 2·57 kW, 27 Nm. **19.** 1·34 kNm. **20.** 848 W, 14·7 rev/s.

Exercise 13

Section A
1. 3·3, 4. **2.** 12. **3.** 50. **4.** 31. **5.** 200 N, 180 N. **6.** 415 N. **7.** 447 N. **8.** 53%. **9.** 82 N.
10. 37·7 N.

Section B
1. 69·2 N. **2.** 26·3 N. **3.** 78 Nm. **4.** 248 N. **5.** 79·7 N. **6.** 482 N. **7.** 7·9 N.
8. P = 0·1 W+1·5. **9.** 80, 12·8 N. **10.** 20, 96%.

Exercise 14

Section A
1. 427 N, 20·6° a.c.w. to 400 N force. **2.** 86 N. **3.** 131 N, 281 N. **4.** W = 3·64 kN, F = 2·10 kN. **5.** 1·044 kN, 823 N. **6.** 15·5 kN in jib, 11·9 kN in tie. **7.** 204 N. **8.** 1·87 kN, 4·95 kN. **9.** $2·32 \times 10^6$ N/m², 295 N. **10.** F = 1·12 kN, H = 970 N.

Section B
1. 386 N, 373 N. **2.** 6·67 kN, 206° a.c.w. from horizontal member. **3.** 80 N, 81·8 N, 95·4 N. **4.** 54 N, 44 N. **5.** 26 kN horizontal member, 30 kN. **6.** 5·4 kN. **7.** 525 N.
8. 450 N. **9.** 429 N. **10.** 171 N at 148° to the 95 N force.

Exercise 15

Section B
1. 147·4 N, 144·5 N. **2.** 13·77 kN. **3.** 250 kN. **4.** 5·12 kN, 645 N. **5.** 10·21 kN, 10·36 kN.
6. 293 N. **7.** 3 m/s. **8.** $7·347 \times 10^3$ Nm. **9.** 490 N, 13·6 m/s. **10.** 1·11 kN.

Exercise 16

Section A
1. 105 kJ. **2.** 91×10^3 J. **3.** 203 kJ. **4.** 1·59 MJ. **5.** 8·61 kJ. **6.** 3·96 MJ. **7.** 7·71 kJ.
8. 24°C. **9.** 25·2 mJ, 19·4 kJ. **10.** 1·013 kJ.

Section B
1. 18·2 MJ, 820 MJ. **2.** 7·11 GJ. **3.** 32°C. **4.** 31%. **5.** 260 J. **6.** 13·5 GJ. **7.** $22·6 \times 10^{-3}$ m³ **8.** 104 litre. **9.** $36·8 \times 10^{-3}$ m³. **10.** 3·04 kJ. **11.** 177·6 MJ. **12.** 88 MJ.

Exercise 17

Section A
1. 266 kWh. **2.** 33·5 kW. **3.** 3·3 MJ. **4.** 5·5°C/min. **5.** 0·067 l/min. **6.** 1·7 kW.
7. 18·5 min. **8.** $55·8 \times 10^3$ J. **9.** 7 p. **10.** 156 kN.

Section B
1. 15 MW. **2.** 1·2 kW. **3.** 7°C. **4.** 62 kW. **5.** 16·5 kJ/s. **6.** 400 kW. **7.** 9·3 kW. **8.** 3 MW.
9. 4·95 Mg. **10.** 12·3 kJ, 8·63 kJ.

Exercise 18

Section A
1. 0·049 mm. **2.** W.I., 0·034 mm. **3.** 149·026 mm. **4.** 0·030 24 mm. **5.** 279°C.
6. 0·042 8 mm. **7.** 1·8 mm. **8.** 10 mm. **9.** 0·005 mm. **10.** 29·5°C.

ANSWERS TO EXERCISES

Section B
1. $-7°C$. 2. $324°C$. 3. 490 mm. 4. 76·021 mm. 5. 1·5 mm. 6. $615°C$.
7. Interference = 0·048 6 mm. 8. Length = 255·31 mm, 15·76 mm dia.
9. Die = 56·15 mm dia., Plug = 53·13 mm dia. 10. $605°C$.

REVISION EXERCISES
Grade 1
3. 90·14 mm. 5. 251·2 kJ. 6. 21·7 kJ. 7. (i) 0·6 A, (ii) 1·0, 1·5 A. 8. 8 A, $37°C$.
9. (i) 3·84 kN, (ii) 4·09 kN. 10. $R_A = 50$ N, $R_B = 6650$ N. 11. 126 mm.
12. 0·653 m. 13. 0·435 kW. 14. 332 kW. 15. 39·3 W. 16. 250 N, 321 N. 17. 80%.
22. 8·59 mm. 23. 19·6 Nm. 25. 3·55, 1·63, 1·2, 0·70 A. 26. 0·227 m^3. 27. 8·45 kN,
10·7 kN. 28. 1·61 kN, 4·85 kN. 29. $R_A = 1·12$ kN, $R_B = 4·8$ kN. 30. 43·5 Nm.
31. 155 mm. 32. 187 kN, 337 kJ, 337 kJ. 33. 72 Nm, 2·0 kW. 34. $R_A = 52·9$ kN, $R_B = 30·1$ kN. 35. LH support = 3·8 kN and $69·4 \times 10^6$ N/m^2, RH support 2·0 = 3·4 kN and $62·8 \times 10^6$ N/m^2. 36. $606°C$. 37. 65·5 kJ. 39. 88 kN. 40. 0·056 kW, $8°C$/min.

Grade 2
4. 648 N. 5. 3 kW. 6. 3·8 kN, $7·16 \times 10^6$ N/m^2, $5·24 \times 10^6$ N/m^2. 7. 3·4 MN/m^2.
8. 35 kW. 9. 15 W. 10. 6·57 MN/m^2. 11. 9·22 kW. 12. 4·8 h. 13. 371 N. 14. 2·76 kW,
19 rev/s. 15. 283 N, 566 N. 16. $2·63 \times 10^3$ Nm, 198 kW. 17. 0·113 mm. 18. 0·674 l/s.
19. 111×10^6 N/m^2, 142×10^6 N/m^2, 0·12 mm. 22. 129 A. 24. 23·8 kN, 13·8 kN.
25. 343 N, 0·26 m/s. 26. 208×10^9 N/m^2, 242×10^6 N/m^2, 450×10^6 N/m^2, 26%,
56%. 27. 3·8 rev/s, 2·27 kW, 95 Nm. 28. 368 Nm, 3 kW. 29. 272 MN/m^2, 1·01 kJ.
30. 3·22 kW, 12·98 MJ. 31. 2·3 kN. 32. $R_A = 176·5$ N, $R_B = 68·5$ N, 19·4 W. 33. 3 A.
35. $0·084 \times 10^{-3}$ m^3/s. 36. 1·6 kW. 37. 375 N. 38. 3·1 m^3. 39. (i) 1·5 rev/s,
0·28 rev/s, (ii) 11·7 rev/s, 2·14 rev/s. 40. 15·1 Nm. 44. 1·5 h, 20 kWh. 45. 245 J.
47. 109×10^9 N/m^2, 396×10^6 N/m^2. 48. 145 N, 3·7 m/s, 16·09 kN. 49. $3·47 \times 10^6$ N/m^2. 50. (i) 0·62%, (ii) 1·2%. 53. 43. 54. 19·5 W. 55. 378 N. 56. 62%. 57. 36 mm.
58. 0·03 mm/rev. 59. 5·2 rev/s, 10·7 kW. 60. 2·4 W, 0·034%.

INDEX

Acetylene, 23
Action (force), 101
Alloy, 5, 10
Aluminium, 28, 29, 30, 32, 33, 52
  ore, 9
  oxide, 9
Ammeter, 68
Ampere, 34
Anode, 8
Antimony, 52
  ore, 10
  sulphide, 10
Atmosphere, 4. 14
  composition of, 15
Atom, 2, 32
  splitting of, 2

Barium carbonate, 21
Battery, 33
  internal resistance of, 39
Bauxite, 9
Belt drive, 143
  efficiency of, 149
  power transmitted by, 143
  slip, 157
  speed, 152
Bismuth, 52
Board of Trade unit, 252
Bonderising, 29
Brine solution, 28
Bull wheel, 171, 176
Bunsen flame, 23

Calcium carbonate, 21
  chloride, 27
  phosphate, 21
Carbon, 4, 21, 30
  dioxide, 7, 14, 20 to 24
  monoxide, 7, 21, 22
Carburisation, 4, 21
Carburising atmosphere, 30
Cathode, 8
Caustic soda, 28, 29
Cell, electric, 32, 35

Cementite, 21
Centre of gravity, 114
  calculation of position of, 116
Centroid, 109
  calculation of position of, 110
  experimental determination of, 109
Ceramic tips, 9
Chalk, 21
Charcoal, 21
Chemical reaction, 3, 5
Chlorine, 3, 20, 21
Chromium, 29
Circular pitch, 166
Coal gas, 30
Cobalt, 52
Combustion, 22
Compass, 53, 60
Component forces, 231
  calculation of, 232
Compound, 3
  preparation of, 4
Compression test, 88
Contraction of a liquid, 261
  of a solid, 262, 266
Co-planar concurrent system of forces, 222
Copper, 3, 5, 20, 28, 52
  ore, 8
  oxide, 15, 20
  sulphate, 3, 4, 6, 8
Corrosion, 25
Corundum, 9
Coulomb, the, 34
Couple, 108
  moment of, 108
Cryolite, 9
Cutting compound, 29
  speed, 141

Dew, 18
Dial gauge mechanism, 173
Distilled water, 4
Displacement ratio, 213, 214
Dust, 17
Dynamo, 33, 68

INDEX

Efficiency, mechanical, 144
  thermal (of a furnace), 244
Elastic limit, 82
  strain energy, 139
Elasticity, 81
  modulus of, 83
Electric motor, principle of, 67
Electricity, 32
  alternating current, 34
  conductors, 33
  direct current, 34
  generator, 68
  insulators, 33
  resistance to flow of, 34
Electro-chemistry, 8
Electrolysis, 9
Electrolyte, 9
Electro-magnetic brake, 62
  field, 59
Electro-magnets, 63–66
Electro-motive force, 33
Electrons, 2, 32
Electro-plating, 8
Elements, chemical, 2
Elongation, percentage, 87, 94
Emulsion, 4
Energy, 140
  conversion of form, 252
  —experiment, 253
  cycle, 141
  electrical form, 36, 140
  mechanical form, 140
  units of, 37, 140, 252
Equilibrant force, 222
Equilibrium, 106
Evaporation, 18
Expansion, 261, 264
  linear coefficient of, 263
Experiments:
  centroid, 109, 111
  chemistry, 3, 5, 8, 13–16, 24–29
  electrical, 35, 38–43
  expansion, 261, 262
  friction, 181
  heat, 247–249, 252
  lifting machine, 216
  magnetism, 51, 53–64, 67
  materials, 81–83, 88
  torque, 128, 129
  work done, 138

Factor of safety, 89
Ferric (iron) chloride, 20
  oxide, 6

Ferric (iron) sulphate, 21
Ferrous (iron), carbonate, 6
  chloride, 20
  sulphate, 6, 21
Fog, 18
Force, 70
  compressive, 71
  diagram, 223
  measurement of, 70
  shear, 71
  tensile, 70
Friction, 179
  advantages of, 180
  and lubrication, 182
  coefficient of, 182
  disadvantages of, 179
  force, 181
  increasing of, 180
  on a lifting machine, 208
  reducing of, 180
Frost, 18

Gas, a, 1
Gauge length, 84
Gear wheels, compound train, 176
  simple train, 166, 172
  pitch circle, 166
  speed of, 166, 167
Gradient of a graph, 93

Heat, balance rule, 247, 249
  definition of, 240
  latent, 241
  measurement of, 242
  sensible, 241
  units of, 242
  uses of, 239
Heat-treatment of metals, 29–30
Hooke's law, 82
Hounsfield, extensometer, 88
  tensometer, 88
Humidity, 19
Hydro-carbon, 23
Hydrochloric acid, 3, 20, 28
Hydrogen, 3, 21, 23, 30
  atom of, 3
  peroxide, 20
  sulphide, 21

Idler wheel, 172
Inert gas, 30
Ions, 8

INDEX

Iron, 5, 10, 26, 51
   carbide, 10, 21
   ore, 6
   oxide, 25–27, 30
   sulphate, 6
   sulphide, 5, 10

Joule, 36, 135, 252

Kaolin, 9
Kilowatt-hour, 252

Lead, 15, 28, 52
   ore, 7
   oxide, 7, 15
   sulphide, 7
Lifting machine, 208
   definition of, 208
   efficiency of, 210
   law of, 216
   mechanism, screw and nut, 215
Limit of proportionality, 84
Liquid, 1
Litharge, 7
Load-extension graph, 93
Lubricating film, 183
Lubrication, 182–184

Magnesium, 25, 29, 52
   oxide, 25
Magnetic axis, 52
   chuck, 58, 66
   circuit, 65, 66
   field, round bar magnet, 54
   —round coil conductor, 61
   —round ring conductor, 60
   —round straight conductor, 59
   lines of force, 58, 65
   materials, 52
   poles, 54
Magnetism, 51
Magnets, bar type, 52–57
   electro-, 58–69
   for crane, 67
   horse-shoe type, 56, 65
   keeper, 56
   permanent, 52–58
   —care of, 55
Manganese, 52
Marking-out paint, 5
Matter, 1
Mechanical advantage, 208, 209

Metrology laboratory, 18
Mica, 9
Mixture, 4
   perfect, 5
Moisture, 18
Molecule, 2
Moment of an area, 111
   of a force, 97
   types of, 98
   units of, 97

Neutrons, 2, 32
Nickel, 51, 52
Nitric acid, 20
Nitrogen, 4, 15, 22
   peroxide, 22

Ohm, 34
Ohm's law, 35
   application of, 44–48
Oiliness, 182
Oxidation, 20, 26, 29
Oxidising agents, 20
   atmosphere, 30
Oxygen, 3, 14, 20, 26, 30

Parallel circuit, 39, 45–47
   resistance of, 47
Parallelogram of forces, 220
Pitch, circular, 166
   line of rack teeth, 170, 172
   of rack teeth, 169, 171
Plasticity, 81
Potential drop, 34
Power, electrical, 36
   mechanical, 141
   unit of, 37, 141
Pressure-force relationship, 147
Protons, 2, 32
Pulley speeds, 154
Pyrometer, 69

Quenching solution, 29

Reaction (force), 100
Reciprocal, 210
Reducing atmosphere, 29
Reduction, 20, 29
   in area, percentage, 87, 95
Resolution of forces, 231
Resultant force, 220
Rheostat, 42

INDEX

Rope blocks, 211
Rust, 25, 27

Scale (iron oxide), 15, 25
Screw and nut mechanism, 215
  jack, 212
  thread, double start, 216
  —lead, 215
  —pitch, 215
  —single start, 215
Screwcutting, 167
Seger cone, 9
Series circuit, 38, 42–45
  resistance of, 44
Silver, 32
Simultaneous equations, 217
SI units, table of, vii
Sodium cyanide, 4, 21
Solenoid, 62, 63
Solid, a, 1
  solution, 5, 10
Solute, 4
Solution, 4
Solvent, 4
Space diagram, 223
Specific gravity, 258
  —heat capacity, 242
Spring stiffness, 150
Steel, 10, 27, 51
  stainless, 29
Straight-line law, 216
Strain, 80, 81–83
  calculation of, 80
Stress, 72, 80–87
  breaking, 84
  calculation of, 73–78
  compressive, 73
  maximum, 84, 87
  proof, 86, 87, 94
  safe working, 89
  shear, 73
  tensile, 73
  ultimate tensile (U.T.S.) 84, 87
  units of, 73
  yield point, 84
Stress–strain diagram, 84
Sublimation, 1
Substance, a pure, 4
Sulphur, 3, 5
  dioxide, 7
Sulphuric acid, 29

Temperature, 239
Tests on materials, 82
  laboratory for, 83
Thermo-couple, 69
Tin, 52
  ore, 9
  oxide, 9, 10
Tinplate, 28
Tinstone, 9
Torque, 121
  calculation of, 122–127
  constant, 122, 195
  experiments, 128, 129
  variable, 128
Triangle of forces, 222
Turning moment, calculation of, 97, 120, 122

Ultimate tensile strength (stress), 84, 87

Velocity ratio, 167, 202, 209
  of rope blocks, 211
  of screw jack, 212
  of Weston differential pulleys, 213
  of worm and wheel, 214
Verdigris, 28
Viscosity, 182
Vitriol, blue, 6
  green, 6
Volt, 34
Voltmeter, 69

Water, 3
  vapour (steam), 18, 23
Watt, the, 37, 141
White metal, 10
Work done by constant force, 135
  —by variable force, 137
  —representation by area under graph 138–140
Worm and wheel, 214
  reduction gear, 204

Yield point, 84

Zinc, 2, 20
  carbonate, 7
  chloride, 2, 20
  ore, 6
  oxide, 7, 20
  sulphide, 7

008e# TABLES

LOGARITHMS

	0	1	2	3	4	5	6	7	8	9	1	2	3	4	5	6	7	8	9
10	0000	0043	0086	0128	0170	0212	0253	0294	0334	0374	4	8	12	17	21	25	29	33	37
11	0414	0453	0492	0531	0569	0607	0645	0682	0719	0755	4	8	11	15	19	23	26	30	34
12	0792	0828	0864	0899	0934	0969	1004	1038	1072	1106	3	7	10	14	17	21	24	28	31
13	1139	1173	1206	1239	1271	1303	1335	1367	1399	1430	3	6	10	13	16	19	23	26	29
14	1461	1492	1523	1553	1584	1614	1644	1673	1703	1732	3	6	9	12	15	18	21	24	27
15	1761	1790	1818	1847	1875	1903	1931	1959	1987	2014	3	6	8	11	14	17	20	22	25
16	2041	2068	2095	2122	2148	2175	2201	2227	2253	2279	3	5	8	11	13	16	18	21	24
17	2304	2330	2355	2380	2405	2430	2455	2480	2504	2529	2	5	7	10	12	15	17	20	22
18	2553	2577	2601	2625	2648	2672	2695	2718	2742	2765	2	5	7	9	12	14	16	19	21
19	2788	2810	2833	2856	2878	2900	2923	2945	2967	2989	2	4	7	9	11	13	16	18	20
20	3010	3032	3054	3075	3096	3118	3139	3160	3181	3201	2	4	6	8	11	13	15	17	19
21	3222	3243	3263	3284	3304	3324	3345	3365	3385	3404	2	4	6	8	10	12	14	16	18
22	3424	3444	3464	3483	3502	3522	3541	3560	3579	3598	2	4	6	8	10	12	14	15	17
23	3617	3636	3655	3674	3692	3711	3729	3747	3766	3784	2	4	6	7	9	11	13	15	17
24	3802	3820	3838	3856	3874	3892	3909	3927	3945	3962	2	4	5	7	9	11	12	14	16
25	3979	3997	4014	4031	4048	4065	4082	4099	4116	4133	2	3	5	7	9	10	12	14	15
26	4150	4166	4183	4200	4216	4232	4249	4265	4281	4298	2	3	5	7	8	10	11	13	15
27	4314	4330	4346	4362	4378	4393	4409	4425	4440	4456	2	3	5	6	8	9	11	13	14
28	4472	4487	4502	4518	4533	4548	4564	4579	4594	4609	2	3	5	6	8	9	11	12	14
29	4624	4639	4654	4669	4683	4698	4713	4728	4742	4757	1	3	4	6	7	9	10	12	13
30	4771	4786	4800	4814	4829	4843	4857	4871	4886	4900	1	3	4	6	7	9	10	11	13
31	4914	4928	4942	4955	4969	4983	4997	5011	5024	5038	1	3	4	6	7	8	10	11	12
32	5051	5065	5079	5092	5105	5119	5132	5145	5159	5172	1	3	4	5	7	8	9	11	12
33	5185	5198	5211	5224	5237	5250	5263	5276	5289	5302	1	3	4	5	6	8	9	10	12
34	5315	5328	5340	5353	5366	5378	5391	5403	5416	5428	1	3	4	5	6	8	9	10	11
35	5441	5453	5465	5478	5490	5502	5514	5527	5539	5551	1	2	4	5	6	7	9	10	11
36	5563	5575	5587	5599	5611	5623	5635	5647	5658	5670	1	2	4	5	6	7	8	10	11
37	5682	5694	5705	5717	5729	5740	5752	5763	5775	5786	1	2	3	5	6	7	8	9	10
38	5798	5809	5821	5832	5843	5855	5866	5877	5888	5899	1	2	3	5	6	7	8	9	10
39	5911	5922	5933	5944	5955	5966	5977	5988	5999	6010	1	2	3	4	5	7	8	9	10
40	6021	6031	6042	6053	6064	6075	6085	6096	6107	6117	1	2	3	4	5	6	8	9	10
41	6128	6138	6149	6160	6170	6180	6191	6201	6212	6222	1	2	3	4	5	6	7	8	9
42	6232	6243	6253	6263	6274	6284	6294	6304	6314	6325	1	2	3	4	5	6	7	8	9
43	6335	6345	6355	6365	6375	6385	6395	6405	6415	6425	1	2	3	4	5	6	7	8	9
44	6435	6444	6454	6464	6474	6484	6493	6503	6513	6522	1	2	3	4	5	6	7	8	9
45	6532	6542	6551	6561	6571	6580	6590	6599	6609	6618	1	2	3	4	5	6	7	8	9
46	6628	6637	6646	6656	6665	6675	6684	6693	6702	6712	1	2	3	4	5	6	7	7	8
47	6721	6730	6739	6749	6758	6767	6776	6785	6794	6803	1	2	3	4	5	5	6	7	8
48	6812	6821	6830	6839	6848	6857	6866	6875	6884	6893	1	2	3	4	4	5	6	7	8
49	6902	6911	6920	6928	6937	6946	6955	6964	6972	6981	1	2	3	4	4	5	6	7	8
50	6990	6998	7007	7016	7024	7033	7042	7050	7059	7067	1	2	3	3	4	5	6	7	8
51	7076	7084	7093	7101	7110	7118	7126	7135	7143	7152	1	2	3	3	4	5	6	7	8
52	7160	7168	7177	7185	7193	7202	7210	7218	7226	7235	1	2	2	3	4	5	6	7	7
53	7243	7251	7259	7267	7275	7284	7292	7300	7308	7316	1	2	2	3	4	5	6	6	7
54	7324	7332	7340	7348	7356	7364	7372	7380	7388	7396	1	2	2	3	4	5	6	6	7

LOGARITHMS

	0	1	2	3	4	5	6	7	8	9	1 2 3 4	5	6 7 8 9
55	7404	7412	7419	7427	7435	7443	7451	7459	7466	7474	1 2 2 3	4	5 5 6 7
56	7482	7490	7497	7505	7513	7520	7528	7536	7543	7551	1 2 2 3	4	5 5 6 7
57	7559	7566	7574	7582	7589	7597	7604	7612	7619	7627	1 2 2 3	4	5 5 6 7
58	7634	7642	7649	7657	7664	7672	7679	7686	7694	7701	1 1 2 3	4	4 5 6 7
59	7709	7716	7723	7731	7738	7745	7752	7760	7767	7774	1 1 2 3	4	4 5 6 7
60	7782	7789	7796	7803	7810	7818	7825	7832	7839	7846	1 1 2 3	4	4 5 6 6
61	7853	7860	7868	7875	7882	7889	7896	7903	7910	7917	1 1 2 3	4	4 5 6 6
62	7924	7931	7938	7945	7952	7959	7966	7973	7980	7987	1 1 2 3	3	4 5 6 6
63	7993	8000	8007	8014	8021	8028	8035	8041	8048	8055	1 1 2 3	3	4 5 5 6
64	8062	8069	8075	8082	8089	8096	8102	8109	8116	8122	1 1 2 3	3	4 5 5 6
65	8129	8136	8142	8149	8156	8162	8169	8176	8182	8189	1 1 2 3	3	4 5 5 6
66	8195	8202	8209	8215	8222	8228	8235	8241	8248	8254	1 1 2 3	3	4 5 5 6
67	8261	8267	8274	8280	8287	8293	8299	8306	8312	8319	1 1 2 3	3	4 5 5 6
68	8325	8331	8338	8344	8351	8357	8363	8370	8376	8382	1 1 2 3	3	4 4 5 6
69	8388	8395	8401	8407	8414	8420	8426	8432	8439	8445	1 1 2 2	3	4 4 5 6
70	8451	8457	8463	8470	8476	8482	8488	8494	8500	8506	1 1 2 2	3	4 4 5 6
71	8513	8519	8525	8531	8537	8543	8549	8555	8561	8567	1 1 2 2	3	4 4 5 5
72	8573	8579	8585	8591	8597	8603	8609	8615	8621	8627	1 1 2 2	3	4 4 5 5
73	8633	8639	8645	8651	8657	8663	8669	8675	8681	8686	1 1 2 2	3	4 4 5 5
74	8692	8698	8704	8710	8716	8722	8727	8733	8739	8745	1 1 2 2	3	4 4 5 5
75	8751	8756	8762	8768	8774	8779	8785	8791	8797	8802	1 1 2 2	3	3 4 5 5
76	8808	8814	8820	8825	8831	8837	8842	8848	8854	8859	1 1 2 2	3	3 4 5 5
77	8865	8871	8876	8882	8887	8893	8899	8904	8910	8915	1 1 2 2	3	3 4 4 5
78	8921	8927	8932	8938	8943	8949	8954	8960	8965	8971	1 1 2 2	3	3 4 4 5
79	8976	8982	8987	8993	8998	9004	9009	9015	9020	9025	1 1 2 2	3	3 4 4 5
80	9031	9036	9042	9047	9053	9058	9063	9069	9074	9079	1 1 2 2	3	3 4 4 5
81	9085	9090	9096	9101	9106	9112	9117	9122	9128	9133	1 1 2 2	3	3 4 4 5
82	9138	9143	9149	9154	9159	9165	9170	9175	9180	9186	1 1 2 2	3	3 4 4 5
83	9191	9196	9201	9206	9212	9217	9222	9227	9232	9238	1 1 2 2	3	3 4 4 5
84	9243	9248	9253	9258	9263	9269	9274	9279	9284	9289	1 1 2 2	3	3 4 4 5
85	9294	9299	9304	9309	9315	9320	9325	9330	9335	9340	1 1 2 2	3	3 4 4 5
86	9345	9350	9355	9360	9365	9370	9375	9380	9385	9390	1 1 2 2	3	3 4 4 5
87	9395	9400	9405	9410	9415	9420	9425	9430	9435	9440	0 1 1 2	2	3 3 4 4
88	9445	9450	9455	9460	9465	9469	9474	9479	9484	9489	0 1 1 2	2	3 3 4 4
89	9494	9499	9504	9509	9513	9518	9523	9528	9533	9538	0 1 1 2	2	3 3 4 4
90	9542	9547	9552	9557	9562	9566	9571	9576	9581	9586	0 1 1 2	2	3 3 4 4
91	9590	9595	9600	9605	9609	9614	9619	9624	9628	9633	0 1 1 2	2	3 3 4 4
92	9638	9643	9647	9652	9657	9661	9666	9671	9675	9680	0 1 1 2	2	3 3 4 4
93	9685	9689	9694	9699	9703	9708	9713	9717	9722	9727	0 1 1 2	2	3 3 4 4
94	9731	9736	9741	9745	9750	9754	9759	9763	9768	9773	0 1 1 2	2	3 3 4 4
95	9777	9782	9786	9791	9795	9800	9805	9809	9814	9818	0 1 1 2	2	3 3 4 4
96	9823	9827	9832	9836	9841	9845	9850	9854	9859	9863	0 1 1 2	2	3 3 4 4
97	9868	9872	9877	9881	9886	9890	9894	9899	9903	9908	0 1 1 2	2	3 3 4 4
98	9912	9917	9921	9926	9930	9934	9939	9943	9948	9952	0 1 1 2	2	3 3 4 4
99	9956	9961	9965	9969	9974	9978	9983	9987	9991	9996	0 1 1 2	2	3 3 3 4

ANTILOGARITHMS

	0	1	2	3	4	5	6	7	8	9	1 2 3 4	5	6 7 8 9
·00	1000	1002	1005	1007	1009	1012	1014	1016	1019	1021	0 0 1 1	1	1 2 2 2
·01	1023	1026	1028	1030	1033	1035	1038	1040	1042	1045	0 0 1 1	1	1 2 2 2
·02	1047	1050	1052	1054	1057	1059	1062	1064	1067	1069	0 0 1 1	1	1 2 2 2
·03	1072	1074	1076	1079	1081	1084	1086	1089	1091	1094	0 0 1 1	1	1 2 2 2
·04	1096	1099	1102	1104	1107	1109	1112	1114	1117	1119	0 1 1 1	1	2 2 2 2
·05	1122	1125	1127	1130	1132	1135	1138	1140	1143	1146	0 1 1 1	1	2 2 2 2
·06	1148	1151	1153	1156	1159	1161	1164	1167	1169	1172	0 1 1 1	1	2 2 2 2
·07	1175	1178	1180	1183	1186	1189	1191	1194	1197	1199	0 1 1 1	1	2 2 2 2
·08	1202	1205	1208	1211	1213	1216	1219	1222	1225	1227	0 1 1 1	1	2 2 2 3
·09	1230	1233	1236	1239	1242	1245	1247	1250	1253	1256	0 1 1 1	1	2 2 2 3
·10	1259	1262	1265	1268	1271	1274	1276	1279	1282	1285	0 1 1 1	1	2 2 2 3
·11	1288	1291	1294	1297	1300	1303	1306	1309	1312	1315	0 1 1 1	2	2 2 2 3
·12	1318	1321	1324	1327	1330	1334	1337	1340	1343	1346	0 1 1 1	2	2 2 2 3
·13	1349	1352	1355	1358	1361	1365	1368	1371	1374	1377	0 1 1 1	2	2 2 3 3
·14	1380	1384	1387	1390	1393	1396	1400	1403	1406	1409	0 1 1 1	2	2 2 3 3
·15	1413	1416	1419	1422	1426	1429	1432	1435	1439	1442	0 1 1 1	2	2 2 3 3
·16	1445	1449	1452	1455	1459	1462	1466	1469	1472	1476	0 1 1 1	2	2 2 3 3
·17	1479	1483	1486	1489	1493	1496	1500	1503	1507	1510	0 1 1 1	2	2 2 3 3
·18	1514	1517	1521	1524	1528	1531	1535	1538	1542	1545	0 1 1 1	2	2 2 3 3
·19	1549	1552	1556	1560	1563	1567	1570	1574	1578	1581	0 1 1 1	2	2 3 3 3
·20	1585	1589	1592	1596	1600	1603	1607	1611	1614	1618	0 1 1 1	2	2 3 3 3
·21	1622	1626	1629	1633	1637	1641	1644	1648	1652	1656	0 1 1 2	2	2 3 3 3
·22	1660	1663	1667	1671	1675	1679	1683	1687	1690	1694	0 1 1 2	2	2 3 3 3
·23	1698	1702	1706	1710	1714	1718	1722	1726	1730	1734	0 1 1 2	2	2 3 3 4
·24	1738	1742	1746	1750	1754	1758	1762	1766	1770	1774	0 1 1 2	2	2 3 3 4
·25	1778	1782	1786	1791	1795	1799	1803	1807	1811	1816	0 1 1 2	2	2 3 3 4
·26	1820	1824	1828	1832	1837	1841	1845	1849	1854	1858	0 1 1 2	2	3 3 3 4
·27	1862	1866	1871	1875	1879	1884	1888	1892	1897	1901	0 1 1 2	2	3 3 3 4
·28	1905	1910	1914	1919	1923	1928	1932	1936	1941	1945	0 1 1 2	2	3 3 4 4
·29	1950	1954	1959	1963	1968	1972	1977	1982	1986	1991	0 1 1 2	2	3 3 4 4
·30	1995	2000	2004	2009	2014	2018	2023	2028	2032	2037	0 1 1 2	2	3 3 4 4
·31	2042	2046	2051	2056	2061	2065	2070	2075	2080	2084	0 1 1 2	2	3 3 4 4
·32	2089	2094	2099	2104	2109	2113	2118	2123	2128	2133	0 1 1 2	2	3 3 4 4
·33	2138	2143	2148	2153	2158	2163	2168	2173	2178	2183	0 1 1 2	2	3 3 4 4
·34	2188	2193	2198	2203	2208	2213	2218	2223	2228	2234	1 1 2 2	3	3 4 4 5
·35	2239	2244	2249	2254	2259	2265	2270	2275	2280	2286	1 1 2 2	3	3 4 4 5
·36	2291	2296	2301	2307	2312	2317	2323	2328	2333	2339	1 1 2 2	3	3 4 4 5
·37	2344	2350	2355	2360	2366	2371	2377	2382	2388	2393	1 1 2 2	3	3 4 4 5
·38	2399	2404	2410	2415	2421	2427	2432	2438	2443	2449	1 1 2 2	3	3 4 4 5
·39	2455	2460	2466	2472	2477	2483	2489	2495	2500	2506	1 1 2 2	3	3 4 5 5
·40	2512	2518	2523	2529	2535	2541	2547	2553	2559	2564	1 1 2 2	3	4 4 5 5
·41	2570	2576	2582	2588	2594	2600	2606	2612	2618	2624	1 1 2 2	3	4 4 5 5
·42	2630	2636	2642	2649	2655	2661	2667	2673	2679	2685	1 1 2 2	3	4 4 5 6
·43	2692	2698	2704	2710	2716	2723	2729	2735	2742	2748	1 1 2 3	3	4 4 5 6
·44	2754	2761	2767	2773	2780	2786	2793	2799	2805	2812	1 1 2 3	3	4 4 5 6
·45	2818	2825	2831	2838	2844	2851	2858	2864	2871	2877	1 1 2 3	3	4 5 5 6
·46	2884	2891	2897	2904	2911	2917	2924	2931	2938	2944	1 1 2 3	3	4 5 5 6
·47	2951	2958	2965	2972	2979	2985	2992	2999	3006	3013	1 1 2 3	3	4 5 5 6
·48	3020	3027	3034	3041	3048	3055	3062	3069	3076	3083	1 1 2 3	4	4 5 6 6
·49	3090	3097	3105	3112	3119	3126	3133	3141	3148	3155	1 1 2 3	4	4 5 6 6

ANTILOGARITHMS

	0	1	2	3	4	5	6	7	8	9	1 2 3 4	5	6 7 8 9
·50	3162	3170	3177	3184	3192	3199	3206	3214	3221	3228	1 1 2 3	4	4 5 6 7
·51	3236	3243	3251	3258	3266	3273	3281	3289	3296	3304	1 2 2 3	4	5 5 6 7
·52	3311	3319	3327	3334	3342	3350	3357	3365	3373	3381	1 2 2 3	4	5 5 6 7
·53	3388	3396	3404	3412	3420	3428	3436	3443	3451	3459	1 2 2 3	4	5 6 6 7
·54	3467	3475	3483	3491	3499	3508	3516	3524	3532	3540	1 2 2 3	4	5 6 6 7
·55	3548	3556	3565	3573	3581	3589	3597	3606	3614	3622	1 2 2 3	4	5 6 7 7
·56	3631	3639	3648	3656	3664	3673	3681	3690	3698	3707	1 2 3 3	4	5 6 7 8
·57	3715	3724	3733	3741	3750	3758	3767	3776	3784	3793	1 2 3 3	4	5 6 7 8
·58	3802	3811	3819	3828	3837	3846	3855	3864	3873	3882	1 2 3 4	4	5 6 7 8
·59	3890	3899	3908	3917	3926	3936	3945	3954	3963	3972	1 2 3 4	5	5 6 7 8
·60	3981	3990	3999	4009	4018	4027	4036	4046	4055	4064	1 2 3 4	5	6 6 7 8
·61	4074	4083	4093	4102	4111	4121	4130	4140	4150	4159	1 2 3 4	5	6 7 8 9
·62	4169	4178	4188	4198	4207	4217	4227	4236	4246	4256	1 2 3 4	5	6 7 8 9
·63	4266	4276	4285	4295	4305	4315	4325	4335	4345	4355	1 2 3 4	5	6 7 8 9
·64	4365	4375	4385	4395	4406	4416	4426	4436	4446	4457	1 2 3 4	5	6 7 8 9
·65	4467	4477	4487	4498	4508	4519	4529	4539	4550	4560	1 2 3 4	5	6 7 8 9
·66	4571	4581	4592	4603	4613	4624	4634	4645	4656	4667	1 2 3 4	5	6 7 9 10
·67	4677	4688	4699	4710	4721	4732	4742	4753	4764	4775	1 2 3 4	5	7 8 9 10
·68	4786	4797	4808	4819	4831	4842	4853	4864	4875	4887	1 2 3 4	6	7 8 9 10
·69	4898	4909	4920	4932	4943	4955	4966	4977	4989	5000	1 2 3 5	6	7 8 9 10
·70	5012	5023	5035	5047	5058	5070	5082	5093	5105	5117	1 2 4 5	6	7 8 9 11
·71	5129	5140	5152	5164	5176	5188	5200	5212	5224	5236	1 2 4 5	6	7 8 10 11
·72	5248	5260	5272	5284	5297	5309	5321	5333	5346	5358	1 2 4 5	6	7 9 10 11
·73	5370	5383	5395	5408	5420	5433	5445	5458	5470	5483	1 3 4 5	6	8 9 10 11
·74	5495	5508	5521	5534	5546	5559	5572	5585	5598	5610	1 3 4 5	6	8 9 10 12
·75	5623	5636	5649	5662	5675	5689	5702	5715	5728	5741	1 3 4 5	7	8 9 10 12
·76	5754	5768	5781	5794	5808	5821	5834	5848	5861	5875	1 3 4 5	7	8 9 11 12
·77	5888	5902	5916	5929	5943	5957	5970	5984	5998	6012	1 3 4 5	7	8 10 11 12
·78	6026	6039	6053	6067	6081	6095	6109	6124	6138	6152	1 3 4 6	7	8 10 11 13
·79	6166	6180	6194	6209	6223	6237	6252	6266	6281	6295	1 3 4 6	7	9 10 11 13
·80	6310	6324	6339	6353	6368	6383	6397	6412	6427	6442	1 3 4 6	7	9 10 12 13
·81	6457	6471	6486	6501	6516	6531	6546	6561	6577	6592	2 3 5 6	8	9 11 12 14
·82	6607	6622	6637	6653	6668	6683	6699	6714	6730	6745	2 3 5 6	8	9 11 12 14
·83	6761	6776	6792	6808	6823	6839	6855	6871	6887	6902	2 3 5 6	8	9 11 13 14
·84	6918	6934	6950	6966	6982	6998	7015	7031	7047	7063	2 3 5 6	8	10 11 13 15
·85	7079	7096	7112	7129	7145	7161	7178	7194	7211	7228	2 3 5 7	8	10 12 13 15
·86	7244	7261	7278	7295	7311	7328	7345	7362	7379	7396	2 3 5 7	8	10 12 13 15
·87	7413	7430	7447	7464	7482	7499	7516	7534	7551	7568	2 3 5 7	9	10 12 14 16
·88	7586	7603	7621	7638	7656	7674	7691	7709	7727	7745	2 4 5 7	9	11 12 14 16
·89	7762	7780	7798	7816	7834	7852	7870	7889	7907	7925	2 4 5 7	9	11 13 14 16
·90	7943	7962	7980	7998	8017	8035	8054	8072	8091	8110	2 4 6 7	9	11 13 15 17
·91	8128	8147	8166	8185	8204	8222	8241	8260	8279	8299	2 4 6 8	9	11 13 15 17
·92	8318	8337	8356	8375	8395	8414	8433	8453	8472	8492	2 4 6 8	10	12 14 15 17
·93	8511	8531	8551	8570	8590	8610	8630	8650	8670	8690	2 4 6 8	10	12 14 16 18
·94	8710	8730	8750	8770	8790	8810	8831	8851	8872	8892	2 4 6 8	10	12 14 16 18
·95	8913	8933	8954	8974	8995	9016	9036	9057	9078	9099	2 4 6 8	10	12 15 17 19
·96	9120	9141	9162	9183	9204	9226	9247	9268	9290	9311	2 4 6 8	11	13 15 17 19
·97	9333	9354	9376	9397	9419	9441	9462	9484	9506	9528	2 4 7 9	11	13 15 17 20
·98	9550	9572	9594	9616	9638	9661	9683	9705	9727	9750	2 4 7 9	11	13 16 18 20
·99	9772	9795	9817	9840	9863	9886	9908	9931	9954	9977	2 5 7 9	11	14 16 18 20